MW00998812

Oracle9*i* Developer:
PL/SQL Programming

Joan Casteel

THOMSON

COURSE TECHNOLOGY

Australia • Canada • Mexico • Singapore • Spain • United Kingdom • United States

THOMSON

COURSE TECHNOLOGY

Oracle9i Developer: PL/SQL Programming

by Joan Casteel

Senior Vice President, Publisher:
Kristen Duerr

Executive Editor:
Jennifer Locke

Product Manager:
Barrie Tysko

Developmental Editor:
Jill Batistick

Production Editor:
Elena Montillo

Acquisitions Editor:
Bill Larkin

Associate Product Manager:
Janet Aras

Editorial Assistant:
Christy Urban

Marketing Manager:
Jason Sakos

Cover Designer:
Betsy Young

Cover Art:
Rakefet Kenaan

Manufacturing Coordinator:
Laura Burns

Compositor:
GEX Publishing Services

BRIEF

Contents

PREFACE   xv

CHAPTER ONE
Introduction to PL/SQL   1

CHAPTER TWO
Handling Data in PL/SQL Blocks   27

CHAPTER THREE
PL/SQL Processing   83

CHAPTER FOUR
Procedures   139

CHAPTER FIVE
Functions   179

CHAPTER SIX
PL/SQL Packages   213

CHAPTER SEVEN
Program Unit Dependencies   255

CHAPTER EIGHT
Database Triggers   293

CHAPTER NINE
Oracle-Supplied Packages   335

CHAPTER TEN
Introduction to Dynamic SQL and Object Technology   381

CHAPTER ELEVEN
Performance Tuning   425

APPENDIX A
Tables for the Brewbean's Database   465

APPENDIX B
Procedure Builder   477

APPENDIX C
TOAD (Tool for Oracle Application Developers)   499

APPENDIX D
PL/SQL and Oracle9i Developer Forms (available at *www.course.com* via
the "Student Downloads" link on the Web page for this textbook)

GLOSSARY   521

INDEX   525

TABLE OF
Contents

PREFACE   **XV**

CHAPTER ONE
Introduction to PL/SQL   **1**
Programming and Procedural Languages   2
Application Programming and PL/SQL   2
    Application Programming   2
    History of PL/SQL   4
Application Models   5
    Two-Tier or Client/Server Application Model   6
    Three-Tier Application Model   7
Documentation on the Web and on CD-ROM   8
SQL and PL/SQL Tools   9
    Software Tools Used in this Book   9
    Third-Party PL/SQL Development Tools   12
Databases Used in this Book   13
Chapter Summary   17
Review Questions   18
Hands-on Assignments   20
    Assignment 1-1: Review Data in the Brewbean's Database   20
    Assignment 1-2: Review Third-Party Software Tools   21
    Assignment 1-3: Identify Processing Steps   22
    Assignment 1-4: Use OTN Documentation   23
Case Projects   23
    Case 1-1: Review Procedure Builder Documentation   23
    Case 1-2: The More Movies Database   24

CHAPTER TWO
Handling Data in PL/SQL Blocks   **27**
The Current Challenge in the Brewbean's Application   28
PL/SQL Block Structure   29
Working with Variables   30
Working with Scalar Variables   31
    Variable Declarations in Code   32
    Variable Initialization   32
    NOT NULL and CONSTANT   33
    The Role of Scalar Variables in the BEGIN Section   34
    Using DBMS_OUTPUT to Check Values   34
Including SQL Within a PL/SQL Block   35
Executing a PL/SQL Block with Errors   38
Working with Host or Bind Variables   42
Using the %TYPE Attribute   43

Working with Composite Data Types   44
  Record Data Type   44
  The Role of the %ROWTYPE Attribute   46
  Table of Records   47
Processing with IF Statements and Looping Actions   51
  Simple IF Statements   51
  Loops: Basic and FOR   52
Working with Collections   53
  Index-by Tables   53
  Other Collections: VARRAYS and Nested Tables   56
Working with Cursors   56
  Implicit Cursors   57
  Explicit Cursors   58
  Cursor Variables   66
Working with Variable Scope   67
Chapter Summary   71
Review Questions   72
Advanced Review Questions   74
Hands-on Assignments   75
  Assignment 2-1: Use Scalar Variables   75
  Assignment 2-2: Use a Record Variable   77
  Assignment 2-3: Use an Explicit Cursor   78
  Assignment 2-4: Use a CURSOR FOR loop   80
  Assignment 2-5: Use Implicit Cursors   81
  Assignment 2-6: Use Variable Scope   81
  Assignment 2-7: Use Scalar Variables for Data Retrieval   81
  Assignment 2-8: Use a Record Variable for Data Retrieval   81
Case Projects   82
  Case 2-1: Variable Types   82
  Case 2-2: More Movie Rentals   82

CHAPTER THREE
PL/SQL Processing   **83**

The Current Challenge in the Brewbean's Application   84
Rebuilding Your Database   84
Control Structures   85
  IF Statement Logic   85
  CASE Statements   93
Looping Constructs   98
  Basic Loop   98
  WHILE Loop   100
  FOR Loop   101
  Common Errors While Using Looping Statements   103
GOTO Statement   104
Exception Handlers   105
  Predefined Oracle Errors   105
  Non-Predefined Oracle Errors   112
  User-Defined Exception   113
Additional Exception Concepts   115
  WHEN OTHERS, SQLCODE, and SQLERRM   115
  RAISE_APPLICATION_ERROR   119
  Exception Propagation   120
Commenting Code   124

Chapter Summary	125
Review Questions	126
Advanced Review Questions	128
Hands-on Assignments	129
Assignment 3-1: Use IF Statements	129
Assignment 3-2: Use Searched CASE Statements	130
Assignment 3-3: Use a WHILE Loop	132
Assignment 3-4: Use Exception Handling	133
Assignment 3-5: Work with IF Statements	135
Assignment 3-6: Perform Exception Handling of Predefined Errors	135
Assignment 3-7: Perform Exception Handling of Non-Predefined Errors	136
Assignment 3-8: Perform Exception Handling with User-Defined Errors	136
Case Projects	136
Case 3-1: Brewbean's Application Exception Handlers	136
Case 3-2: More Movie Rentals	137

CHAPTER FOUR
Procedures   139

The Current Challenge in the Brewbean's Application	140
Rebuilding Your Database	141
Introduction to Named Program Units	142
Client and Server Considerations	143
Types of Named Program Units	143
Making Procedures Reusable: Parameters	144
Create Procedure Statement	145
The Name	146
The Mode	146
The Data Type	146
Create a Procedure in SQL*Plus	147
When a CREATE PROCEDURE Statement Produces Errors	148
Testing a Procedure	149
Using the IN OUT Parameter Mode	152
Calling a Procedure from Another PL/SQL Block	153
Creating a Procedure That Calls Another Procedure	153
Testing the Procedure	154
DESCRIBE Command	156
Debugging in SQL*Plus by Displaying Messages to the Screen	157
Software Utilities Available to Assist in Program Unit Development	162
Subprograms	162
Exception Handling and Transaction Scope	163
Error Handling Using RAISE_APPLICATION_ERROR	167
Removing Procedures	169
Chapter Summary	169
Review Questions	170
Advanced Review Questions	172
Hands-on Assignments	173
Assignment 4-1: Create a Procedure	173
Assignment 4-2: Use a Procedure with IN Parameters	174
Assignment 4-3: Calculate the Tax on an Order	174
Assignment 4-4: Update Columns in a Table	175
Assignment 4-5: Update the Status of an Order	175
Assignment 4-6: Return Order Status Information	176
Assignment 4-7: Identify Customers	176

Assignment 4-8: Add Items to a Cart   177
Assignment 4-9: Create a Logon Procedure   177
Case Projects   178
Case 4-1: Reporting and Analysis Summary Tables   178
Case 4-2: The More Movie Rentals Company Rental Process   178

CHAPTER FIVE
Functions   **179**

The Current Challenge in the Brewbean's Application   180
Rebuilding Your Database   181
An Introduction to Functions   182
Creating a Stored Function in SQL*Plus   184
Invoking and Testing a Created Function   185
Using a Function in an SQL Statement   187
Building and Testing a Function for the Brewbean's Member Name Display   188
Using the OUT Parameter Mode in a Function   190
Multiple RETURN Statements   192
Using a RETURN Statement in a Procedure   193
Actual and Formal Parameter Constraints   194
Techniques of Passing Parameter Values   196
Controlling Which Value Passing Technique Is Used   196
Function Purity Levels   198
Data Dictionary Information on Program Units   201
Deleting Program Units   202
Chapter Summary   203
Review Questions   204
Advanced Review Questions   206
Hands-on Assignments   208
Assignment 5-1: Format Numbers as Currency   208
Assignment 5-2: Calculate Total Shopper Spending   209
Assignment 5-3: Calculate the Count of Orders by a Shopper   209
Assignment 5-4: Identify the Day of the Week for the Order Date   210
Assignment 5-5: Calculate Days Between Ordering and Shipping   210
Assignment 5-6: Identify the Description of an Order Status Code   210
Assignment 5-7: Calculate the Tax Amount for an Order   211
Assignment 5-8: Identify Products That Are on Sale   211
Case Projects   211
Case 5-1: Update Basket Data Upon Order Completion   211
Case 5-2: More Movies Rentals   212

CHAPTER SIX
PL/SQL Packages   **213**

The Current Challenge in the Brewbean's Application   214
Rebuilding Your Database   214
Package Specification   215
Declarations in a Package Specification   215
Ordering of Items Within a Specification   216
Package Body   217
Invoking Package Constructs   220
Package Construct Scope   222

Package Global Constructs   224
    Testing the Persistence of Packaged Variables   225
    Package Specifications with No Body   227
    Improving Processing Efficiency   228
Forward Declaration in Packages   231
One Time Only Procedure   234
Overloading Program Units in Packages   237
Managing Packaged Function SQL Restrictions   240
    Why Developers Indicate Purity Levels   241
    PRAGMA RESTRICT_REFERENCES in Action   242
    Default Purity Level for Packaged Functions   243
    Functions Written in External Languages   243
Program Unit and Package Execute Privileges   244
Data Dictionary Information for Packages   244
Deleting Packages   246
Chapter Summary   246
Review Questions   247
Advanced Review Questions   250
Hands-on Assignments   251
    Assignment 6-1: Create a Package   251
    Assignment 6-2: Use Packaged Program Units   251
    Assignment 6-3: Create a Package with Private Program Units   252
    Assignment 6-4: Use Packaged Variables   252
    Assignment 6-5: Package Overloading   253
    Assignment 6-6: Create a Package with a Specification Only   253
    Assignment 6-7: Use a Cursor in a Package   253
    Assignment 6-8: Use a One Time Only Procedure in a Package   254
Case Projects   254
    Case 6-1: Brewbean's Order Checkout Package   254
    Case 6-2: More Movies Program Unit Packaging   254

CHAPTER SEVEN
Program Unit Dependencies   **255**

The Current Challenge in the Brewbean's Application   256
Rebuilding Your Database   256
Local Dependency Activity   257
Identifying Direct and Indirect Dependencies   261
Data Dictionary Views for Dependencies   262
The Dependency Tree Utility   263
Package Dependencies   268
Remote Object Dependencies   272
Remote Dependency Invalidation Methods   276
Tips to Avoid Recompilation Errors   278
Granting Program Unit Privileges   279
Chapter Summary   281
Review Questions   283
Advanced Review Questions   286
Hands-on Assignments   288
    Assignment 7-1: Review Dependency Information in the
        Data Dictionary   288
    Assignment 7-2: Test Dependencies on Stand-alone Program Units   289
    Assignment 7-3: Test Dependencies on Packaged Program Units   290
    Assignment 7-4: Test Remote Object Dependencies   291

Assignment 7-5: Identify All Dependencies Using the Dependency Tree Utility   291
Assignment 7-6: Review the utldtree.sql Script   292
Assignment 7-7: Avoid Recompilation Errors   292
Assignment 7-8: Identify the Types of Dependencies   292
Case Projects   292
Case 7-1: The Brewbean's Application Maintenance   292
Case 7-2: The More Movies Rental Application   292

CHAPTER EIGHT
Database Triggers   **293**
The Current Challenge in the Brewbean's Application   294
Rebuilding Your Database   296
Introduction to Database Triggers   296
Database Trigger Syntax and Options   297
Database Trigger Code Example   298
Trigger Timing and Correlation Identifiers   298
Trigger Events   301
Trigger Body   301
Conditional Predicates   303
Creating and Testing a DML Trigger in SQL*Plus   304
Creating and Testing an Instead-Of Trigger   307
System Triggers   311
Applying Triggers to Address Processing Needs   313
Restrictions of Trigger Use Including Mutating Tables   316
The ALTER TRIGGER Statement   320
Delete a Trigger   321
Data Dictionary Information for Triggers   321
Chapter Summary   322
Review Questions   323
Advanced Review Questions   326
Hands-on Assignments   327
Assignment 8-1: Create a Trigger to Address Product Restocking   327
Assignment 8-2: Update Stock Information When a Product Request Is Filled   328
Assignment 8-3: Update the Stock Level If a Product Fulfillment Is Cancelled   330
Assignment 8-4: Update Stock Levels When an Order Is Cancelled   331
Assignment 8-5: Process Discounts   331
Assignment 8-6: Use Triggers to Maintain Referential Integrity   332
Assignment 8-7: Update Summary Data Tables   332
Assignment 8-8: Maintain an Audit Trail of Product Table Changes   333
Case Projects   333
Case 8-1: Map the Flow of Database Triggers   333
Case 8-2: More Movies Inventory Processing   334

CHAPTER NINE
Oracle-Supplied Packages   **335**
The Current Challenge in the Brewbean's Application   336
Rebuilding Your Database   337
Communications   337
DBMS_PIPE Package   338
DBMS_ALERT Package   340
UTL_SMTP Package   341
UTL_HTTP Package   344
UTL_TCP Package   345

Generating Output   346
    DBMS_OUTPUT Package   346
    UTL_FILE Package   351
Large Objects   354
    DBMS_LOB Package   355
    Using DBMS_LOB to Manipulate Images   355
Dynamic SQL and PL/SQL   357
Miscellaneous Packages   359
    DBMS_JOB Package   359
    DBMS_DDL Package   366
    Exploring Additional Oracle-Supplied Packages   368
Chapter Summary   369
Review Questions   370
Advanced Review Questions   372
Hands-on Assignments   373
    Assignment 9-1: Use the DBMS_PIPE Package   373
    Assignment 9-2: Use the DBMS_ALERT Package   374
    Assignment 9-3: Use the DBMS_DDL Package   375
    Assignment 9-4: Use the UTL_FILE Package to Read and Insert Data   377
    Assignment 9-5: Use the UTL_FILE Package to Insert Data Columns   377
    Assignment 9-6: Send E-mail Using UTL_SMTP   378
    Assignment 9-7: Use the DBMS_JOB Package   378
    Assignment 9-8: Use DBMS_OUTPUT   379
Case Projects   379
    Case 9-1: Search Oracle Built-In Packages   379
    Case 9-2: The More Movies Company   379

CHAPTER TEN
Introduction to Dynamic SQL and Object Technology 381

The Current Challenge in the Brewbean's Application   382
Rebuilding Your Database   382
Dynamic SQL   383
    The DBMS_SQL Package   385
    Native Dynamic SQL   394
    DBMS_SQL Versus Native Dynamic SQL   398
Object Technology   399
    Creating Object Types   399
    Using an Object Type   400
    Object Methods   403
    Object Relations   406
    REF Pointers Versus Foreign Keys   408
    Object Views   410
    Sorting and Comparing Object Type Columns   412
Chapter Summary   414
Review Questions   415
Hands-on Assignments   419
    Assignment 10-1: Use the DBMS_SQL Package   419
    Assignment 10-2: Use Native Dynamic SQL   420
    Assignment 10-3: Create an Object Type   420
    Assignment 10-4: Create Object Views   421
    Assignment 10-5: Create a Product Object Type with Sort Capability   422
    Assignment 10-6: Use Native Dynamic SQL   423
    Assignment 10-7: Object-oriented Programming   423
    Assignment 10-8: Business Intelligence   423

Case Projects   423
    Case 10-1: The Brewbean's Ad Hoc Query System   423
    Case 10-2: The More Movies Database   423

CHAPTER ELEVEN
Performance Tuning   **425**
The Current Challenge in the Brewbean's Application   426
Rebuilding Your Database   426
Tuning Concepts and Issues   426
    Identifying Problem Areas in Coding   427
    Processing and the Optimizer   432
    The Cost-Based Optimizer   434
    The Explain Plan and AUTOTRACE   434
    Timing Feature   438
SQL Statement Tuning   439
    Avoiding Unnecessary Column Selection   439
    Cost Versus Rule Basis   440
    Index Suppression   442
    Concatenated Indexes   445
    Subqueries   446
    Joins   448
    Optimizer Hints   448
PL/SQL Statement Tuning   452
    Program Unit Iterations   452
    Using ROWID When Updating   454
    Variable Comparisons with the Same Data Type   454
    Ordering Conditions by Frequency   455
    Using the PLS_INTEGER Data Type   455
    Pinning Stored Program Units   456
Chapter Summary   456
Review Questions   457
Hands-on Assignments   459
    Assignment 11-1: Review Statement Execution Plans 459
    Assignment 11-2: Use the Timing Feature in SQL*Plus 460
    Assignment 11-3: Compare Explain Plans   461
    Assignment 11-4: Use ROWID to Improve Updates   461
    Assignment 11-5: Index Suppression   462
    Assignment 11-6: Optimizer Hints   462
    Assignment 11-7: Execution Plan   462
    Assignment 11-8: Focus Tuning Efforts   463
Case Projects   463
    Case 11-1: Brewbean's Professional Development   463
    Case 11-2: More Movies Performance Tuning   463

APPENDIX A
Tables for the Brewbean's Database   **465**
    BB_SHOPPER   466
    BB_BASKET   467
    BB_BASKETITEM   468
    BB_PRODUCT   470
    BB_PRODUCTOPTION   471

BB_PRODUCTOPTIONDETAIL   472
BB_ PRODUCTOPTIONCATEGORY   473
BB_DEPARTMENT   473
BB_BASKETSTATUS   474
BB_TAX   474
BB_SHIPPING   475

APPENDIX B
Procedure Builder   **477**
   Rebuilding Your Database   478
   Creating a Procedure Using Procedure Builder   479
      Starting Procedure Builder   479
      Using the Program Unit Editor   480
      Testing the Procedure in the Interpreter Panel 483
   Debugging with Procedure Builder   485
      Working with Breakpoints   485
      Displaying Values to the Screen   493
   Appendix Summary   497

APPENDIX C
TOAD (Tool for Oracle Application Developers)   **499**
   Rebuilding Your Database   500
   Create a Procedure Using TOAD   500
      Starting TOAD   500
      Using the Procedure Editor   502
      Testing a Procedure with TOAD   506
      Debugging with TOAD   508
      Creating Breakpoints   509
      Displaying Variables   517
   Appendix Summary   519

APPENDIX D
PL/SQL and Oracle9*i* Developer Forms (available at
***www.course.com* via the "Student Downloads" link on**
the Web page for this textbook)

GLOSSARY   **521**

INDEX   **525**

Preface

Almost every organization depends on a relational database to meet their information system needs. One significant challenge faced by these organizations is providing productive and user-friendly application interfaces to allow users to work efficiently. To facilitate logical processing and user interaction with the Oracle9i database, Oracle provides a robust procedural language extension, PL/SQL, to complement the industry standard SQL. PL/SQL is an implicit part of the Oracle9i database, and the PL/SQL compiler and interpreter are also embedded in the Oracle9i Developer Suite of tools. This provides a consistent development environment on both the client and server side. PL/SQL knowledge leads to opportunities not only in the development of applications with Oracle Developer tools, but also in supporting existing Oracle applications that are developed and marketed by Oracle in a number of industries. In addition to database software and application development tools, Oracle is one of the world's leading suppliers of application software.

The purpose of this textbook is to introduce the student to using the PL/SQL language to interact with an Oracle9i database and to support applications in a business environment. In addition, concepts relating specifically to the objectives of the Oracle9i PL/SQL certification exams have been incorporated in the text for those individuals wishing to pursue certification.

The Intended Audience

This textbook has been designed for students in technical two-year or four-year programs who need to learn how to develop application code with Oracle9i databases. This textbook assumes that the student already has an understanding of relational database design and SQL commands.

Oracle Certification Program (OCP)

This textbook covers objectives of *Exam 1Z0-101, Develop PL/SQL Program Units* and *Exam 1Z0-147, Oracle9i: Program with PL/SQL*. Successful completion of Exam 1Z0-101 can be applied toward certification as an Oracle Certified Professional Internet Application Developer, Oracle Forms Developer Release 6/6i. Those pursuing certification as either an Oracle9i Forms Developer Certified Professional or an Oracle9i Application Developer Certified Associate will need to take Exam 1Z0-147 rather than Exam 1Z0-101. At the time of publication, Exam 1Z0-147 is available only in beta form, as Exam 1Z1-147 (beta means the exam is in the last stage of testing). Information about these exams, including registration and other reference material, can be found at *www.oracle.com/education/certification*. In addition, grids presenting the Oracle exam objectives and the chapters of this book that address them, will be available for download at *www.course.com* via the "Student Downloads" link on the Web page for this book.

The Approach

The concepts introduced in this textbook are presented in the context of a hypothetical "real world" business—an online coffee goods retailer named Brewbean's. First, the business operation and the database structure are introduced and analyzed. Each chapter begins with a description of the current application challenge Brewbean's needs to address. Then, PL/SQL statements are introduced and used throughout the chapter to solve the Brewbean's application challenge. This allows students to not only learn the syntax of PL/SQL statements, but also to apply them in a real-world environment. A script file that generates the needed database objects is provided for each chapter to allow students hands-on practice in re-creating the examples and practicing variations of PL/SQL statements to enhance their understanding.

The core PL/SQL language elements are presented in Chapters 4 through 9, including procedures, functions, packages, and database triggers. However, before jumping into this section, Chapters 2 and 3 introduce the basic PL/SQL language elements and processing techniques. The last two chapters introduce more advanced topics including dynamic SQL, object-oriented features, and performance tuning. These topics go beyond certification content; however, they are important in gaining an appreciation of the many development features available to the Oracle developer.

To reinforce the material presented, there is a topic summary, review questions, assignments, and case projects at the end of each chapter. They test the students' knowledge and challenge them to apply that knowledge to solving business problems. In addition, to further expand their appreciation and knowledge of the PL/SQL arena, the appendices provide tutorials in using software utilities available to assist in PL/SQL code development. They also provide an overview of Oracle application development using Oracle9i Forms software.

Overview of This Book

The examples, assignments, and cases in this book will help students to achieve the following objectives:

- Create PL/SQL blocks
- Use a variety of variable types to handle data in a block
- Conditionally process statements using control structures

- Reuse lines of code with looping structures
- Manage errors with exception handlers
- Create and use procedures and functions
- Bundle program units with packages
- Develop database triggers
- Leverage the features of Oracle-supplied packages
- Identify program unit dependencies
- Use dynamic SQL and object technology features
- Perform code tuning

The contents of the chapters build in complexity while reinforcing previous ideas. **Chapter 1** introduces what PL/SQL is and how it fits into application programming. It also introduces the Brewbean's database that is used throughout this textbook. **Chapter 2** shows how to retrieve and handle data in a block using scalar variables, composite variables, and cursors. **Chapter 3** demonstrates processing with control and loop structures and handling exceptions. **Chapter 4** presents how to create a procedure and pass values using parameters. **Chapter 5** shows how to create functions and return values with functions. **Chapter 6** covers package creation, including the package specification and body. **Chapter 7** covers methods to identify program unit dependencies. **Chapter 8** reveals how to develop DML, Instead-Of, and system triggers. **Chapter 9** presents a sampling of the Oracle-supplied packages available. **Chapter 10** introduces the use of dynamic SQL and object technology. **Chapter 11** introduces performance-tuning concepts. The Appendices provide further support and expansion of chapter materials. **Appendix A** provides a printed version of the tables and data in the Brewbean's database. This database serves as a sustained example from chapter to chapter. **Appendix B** provides a tutorial on using the Procedure Builder utility, available in Oracle Forms Developer 6*i,* to create and debug program units. **Appendix C** provides a tutorial on using a third-party utility, TOAD, to create and debug program units; note that Procedure Builder is not available in the Developer 9*i* Suite. **Appendix D** provides an overview of how PL/SQL is used in application development by providing a demonstration of Oracle9*i* Forms development. This appendix is available for download at *www.course.com* via the "Student Downloads" link on the Web page for this textbook.

Features

To enhance students' learning experience, each chapter in this book includes the following elements:

- **Chapter Objectives:** Each chapter begins with a list of the concepts to be mastered by the chapter's conclusion. This list provides a quick overview of chapter contents as well as a useful study aid.

- **Running Case:** An application development challenge for the Brewbean's Internet coffee retail company is presented at the beginning of each chapter.

- **Methodology:** New concepts are introduced in the context of resolving the application challenge encountered by Brewbean's in that chapter. The chapter explains the role of the concepts and provides working step-by-step exercises to illustrate the concepts. Many screen snapshots and code examples are provided to assist in demonstrating the concepts. As you proceed through the tutorials, less detailed instructions are provided for familiar tasks as more detailed instructions are provided for the new tasks being introduced.

- **Tip:** This feature, designated by the *Tip* icon, provides students with practical advice or information. In some instances, Tips explain how a concept applies in the workplace.

- **Note:** These explanations, designated by the *Note* icon, provide further information about files and operations being presented.

- **Chapter Summaries:** Each chapter's text is followed by a summary of chapter concepts. These summaries are a helpful recap of chapter contents.

- **Review Questions:** End-of-chapter assessment begins with a set of 15 review questions that reinforce the main ideas introduced in each chapter. These questions ensure that students have mastered the concepts and understand the information presented. This includes 10 multiple choice and 5 short answer questions.

- **Advanced Review Questions:** Chapters 1 through 9 contain five multiple-choice questions that cover the material presented within the chapter and that are more similar to Oracle certification exam-type questions. These are included to prepare students for the type of questions that can be expected on a certification exam. They are also included to measure the students' level of understanding. This section is not available in the last two chapters as these materials go beyond the certification objectives.

- **Hands-on Assignments:** Along with conceptual explanations and examples, each chapter provides eight to nine hands-on assignments related to the chapter's contents. The purpose of these assignments is to provide students with practical experience. In most cases, the assignments are based on the Brewbean's database application and serve as a continuation of the examples given within the chapter.

- **Case Projects:** Two major cases are presented at the end of each chapter. These cases are designed to help students apply what they have learned to real-world situations. The cases give students the opportunity to independently synthesize and evaluate information, examine potential solutions, and make recommendations, much as students will do in an actual business situation. One case expands on the on-going Brewbean's company application development, while a second case presents the challenges of another company that is developing database application code to support their business processes.

In addition to book-based features, *Oracle9i Developer: PL/SQL Programming* offers this software component:

- The *Course Technology Kit for Oracle9i Software* is available when purchased as a bundle with this book. It provides the Oracle9i Enterprise Edition, Standard Edition, and Personal Edition database software, Release 2 (Version 9.2.0.1.0), on three CDs. You can use the software included in the kit with Microsoft Windows NT, Windows 2000 Professional or Server, and Windows XP Professional operating systems. The installation instructions for Oracle9i and the log on procedures are available at *www.course.com/cdkit*. Look for this book's title and front cover, and click the link to access the information specific to this book.

SQL*Plus Versus iSQL*Plus:

Oracle9i introduced a browser-based version of SQL*Plus, which operates much like the client SQL*Plus software product and is named iSQL*Plus. However, there are a few differences in commands and settings. A brief user's guide is available to introduce students to this tool. This user's guide is available for download at *www.course.com* via the "Student Downloads" link on the Web page for this textbook. Please note: the displays in this text show SQL*Plus, not iSQL*Plus screens.

Teaching Tools

The following supplemental materials are available when this book is used in a classroom setting. All teaching tools available with this book are provided to the instructor on a single CD-ROM.

- **Electronic Instructor's Manual:** The Instructor's Manual that accompanies this textbook includes the following elements:

  - Additional instructional material to assist in class preparation, including suggestions for lecture topics.

  - Solutions to all the in-chapter tutorials and end-of-chapter materials, including the Review Questions, Hands-on Assignments, and Case Projects.

- **ExamView®:** This objective-based test generator lets the instructor create paper, LAN, or Web-based tests from testbanks designed specifically for this Course Technology text. Instructors can use the QuickTest Wizard to create tests in fewer than five minutes by taking advantage of Course Technology's question banks—or they can create customized exams.

- **PowerPoint Presentations:** Microsoft PowerPoint slides are included for each chapter. Instructors might use the slides in three ways: as teaching aids during classroom presentations, as printed handouts for classroom distribution, or as network-accessible resources for chapter review. Instructors can add their own slides for additional topics introduced to the class.

- **Data Files:** The script files necessary to create all needed database objects are provided through the Course Technology Web site at *www.course.com* and are also available on the Teaching Tools CD-ROM.

- **Solution Files:** Solutions to the chapter examples, end-of-chapter review questions, multiple-choice questions, Hands-On Assignments, and the Case Projects are provided on the Teaching Tools CD-ROM. Solutions may also be found on the Course Technology Web site at *www.course.com*. The solutions are password protected.

ACKNOWLEDGMENTS

First and foremost, I need to thank my best friend Scott for the endless support and encouragement I needed to accomplish this goal. Scott has the strongest shoulders and most patient ears known to mankind. I must be the luckiest person in the world. I am also grateful to my family, friends, and colleagues who continue to put up with me and to Tidewater Community College for providing me with many opportunities to grow and learn.

And, as many of my students know, I am entirely too excited about PL/SQL to be completely human. However, my experience in producing this textbook has proven just how human I am. The usefulness of this book is due to the many efforts from Course Technology employees to entertain my ideas, mold my writing into a comprehendible learning style, and test the many snippets of code used throughout the book. Many individuals have made hands-on contributions in the creation of this book and I will never be able to repay their incredible efforts.

I would like to express my appreciation to Barrie Tysko, the Product Manager who always respected and entertained my opinions and ideas, and Jill Batistick, Developmental Editor, who had the grand task of managing my writing. Also, hats off to Serge Palladino and Vitaly Davidovich for having to work through the examples and material presented in each chapter during the two quality assurance stages of the process. Many other individuals were involved that I did not get the opportunity to work with directly. I would like to thank— Bill Larkin, Acquisitions Editor; Jennifer Locke, Executive Editor; Elena Montillo, Production Editor; Nicole Ashton, Manuscript Quality Assurance Lead, and Janet Aras, Associate Product Manager.

In addition to a number of my students who have provided valuable feedback in my classes, the following reviewers also provided helpful suggestions and insight into the development of this textbook: Jason C.H. Chen, Gonzaga University; Judith A. Dunn, Laramie County Community College; Dean Jefferson, Madison Area Technical College; Debbie Meyer, St. Louis Community College–Forest Park; Gayle Moody, Valencia Community College–East Campus; Angela Norville, J. Sargeant Reynolds Community College; Eli Weissman, DeVry Institute of Technology; and David Welch, Nashville State Technical Community College.

Read This Before You Begin

TO THE USER

Data Disks

To work through the examples and complete the projects in this book, you will need to load the data files created for this book. Your instructor will provide you with those data files, or you can obtain them electronically from the Course Technology Web site by accessing *www.course.com* and then searching for this book's title. Each chapter in the book has its own set of data files. These include files used in the tutorials and any files that may be needed for the end of chapter Hands-on Assignments or Case Projects.

The database objects used throughout the book are created with a script file provided for each chapter. The files are named in the format of c#Dbcreate.sql, where the (#) indicates the chapter number. These files are located in the data folder for each chapter on your Data Disk. Steps are provided at the beginning of each chapter providing instructions on executing the script files. The data folder for each chapter also contains any other files referenced throughout the chapter, including assignments at the end of the chapter. If the computer in your school lab—or your own computer—has Oracle9*i* database software installed, you can work through the chapter examples and complete the Hands-on Assignments and Case Projects. At a minimum, you will need the Oracle9*i* Release 2 Personal Edition of the software to complete the examples and assignments in this textbook.

Using Your Own Computer

To use your own computer to work through the chapter step-by-step tutorials and to complete the Hands-on Assignments and Case Projects, you will need the following:

- **Hardware:** A computer capable of using the Microsoft 2000 Professional, 2000 Server, or XP Professional operating system. Microsoft 2000 installations should have service pack 1 or higher installed. You should have at least 256MB of RAM, 140MB of system drive space, and 4.75GB of drive space available for the Oracle Home before installing the software.

- **Software:** Oracle9*i* Database, Release 2 (Version 9.2.0.1.0) Personal, Standard, or Enterprise Edition.

When you install the Oracle9*i* software, you will be prompted to change the password for certain default administrative user accounts. Make certain that you record the names and passwords of the accounts because you may need to log on to the database with one of these administrative accounts in later chapters. After you install Oracle9*i*, you will be required to enter a user name and password to access the software. One default user name created during the installation process is "scott". The default password for the user name is "tiger". If you have installed the Personal Edition of Oracle9*i*, you will not need to enter a Connect String during the log on process.

The *Course Technology Kit for Oracle9i Software*, which is available as a bundle with this book, contains the database software necessary to perform all the tasks shown in this textbook. Detailed installation, configuration, and logon information for the software in this kit are provided at *www.course.com/cdkit*. Look for this book's title and front cover, and click the link to access the information specific to this book.

■ **Data Files:** You will not be able to use your own computer to work through the chapter examples and complete the projects in this book unless you have the data files. You can get the data files from your instructor, or you can obtain the data files electronically by accessing the Course Technology Web site at *www.course.com* and then searching for the Student Downloads link under this book's title.

When you download the data files, they should be stored in a directory separate from any other files on your hard drive or diskette. For example, create a directory named plsql class to contain all the data files for this book. You will need to remember the path or folder containing the files as this will be referenced to execute or retrieve files in SQL*Plus. (SQL*Plus is the interface tool you will use to interact with the database.)

Naming Conventions Used

Every programming shop should follow a variable naming convention for easier readability of code. The following tables outline the naming conventions used throughout this book. They include variable type prefixes, scalar variable data type suffixes, and program unit type suffixes.

Variable Type Prefixes

Variable Type	Prefix
PL/SQL block local scalar	lv_
Package	pv_
Packaged	pvg_
Parameter (program unit)	p_
Cursor	cur_
User defined data type	type_
Record	rec_

Scalar Variable Data Type Suffixes

Data Type	Suffix
Character	_txt
Number	_num
Date	_date
Boolean	_bln

Program Unit Type Suffixes

Program Unit Type	Suffix
Stored procedure	_sp
Stored function	_sf
Stored package	_pkg
Packaged procedure	_pp
Packaged function	_pf
Database trigger	_trg

Visit Our World Wide Web Site

Additional materials designed especially for you might be available on the World Wide Web. Go to *www.course.com* periodically and search this site for more details. An example of the information made available is the "iSQL*Plus User's Guide," which is discussed in the Preface of this book on page xix.

TO THE INSTRUCTOR

To complete the chapters in this book, your users must have access to a set of data files. These files are included in the Instructor's Resource Kit. They may also be obtained electronically through the Course Technology Web site at *www.course.com*.

The database objects used throughout the book are created with a script file provided for each chapter. The files are named in the format of c#Dbcreate.sql, where the (#) will indicate the chapter number. These files are located in the data folder for each chapter on the Data Disk. Steps are provided at the beginning of each chapter providing instructions on executing the script files. The data folder for each chapter also contains any other files referenced throughout the chapter, including assignments at the end of the chapter. If your users are connecting to a server running the Standard or Enterprise version of the Oracle database (rather than Personal Oracle), a separate account or schema should be created for each student. An example of a user creation command is included in the Instructor's Manual. Also, to complete all exercises in Chapter 11, "Performance Tuning," the student will need to have DBA privileges.

The book uses a fading strategy in regards to SQL*Plus usage. In other words, the step-by-step sections in the first few chapters are explicit regarding the opening, closing, and usage of SQL*Plus. Beyond this point, it is assumed that the user is comfortable with SQL*Plus and does not need explicit directions of this type.

The book demonstrates all the chapter content using a client SQL*Plus installation. To assist in exposing users to other interface tools, appendices demonstrating Procedure Builder and TOAD are provided. In addition, an "iSQL*Plus User's Guide" is available at *www.course.com* via the "Student Downloads" link on the Web page for this book.

I have discovered that one of the largest hurdles students have in regards to PL/SQL is visualizing how PL/SQL fits into application development. In order to assist users in this regard, I demonstrate the development of a simple application screen using Oracle9*i* Developer Forms software. This demonstration is included as Appendix D, which is available at *www.course.com,* via the "Student Downloads" link, on the Web page for this book.

THE ORACLE SERVER AND CLIENT SOFTWARE

This book was written and tested using the following software:

- Oracle9*i* Enterprise Edition Server, Release 2, Version 9.2.0.1.0, installed on a Windows 2000 Server workstation, and Personal Oracle9*i*, Release 2, Version 9.2.0.1.0, installed on a Windows XP Professional workstation.

- Oracle9*i* SQL*Plus, Release 2, Version 9.2.0.1.0, installed on a Windows 2000 Professional workstation, and connecting to an Oracel9*i* Enterprise Edition database, running on a Windows 2000 server and installed on Windows XP Professional connecting to Personal Oracle 9*i* database on the same machine.

- SQL*Plus is used throughout the textbook to provide a common mechanism or interface to the Oracle database regardless of the client tools available or version of software. Procedure Builder and TOAD are presented in appendices to provide an introduction to these tools.

SQL*PLUS VERSUS ISQL*PLUS

Oracle9*i* introduced a browser-based version of SQL*Plus, which operates much like the client SQL*Plus software product and is named iSQL*Plus. However, there are a few differences in commands and settings. iSQL*Plus can be used in conjunction with this textbook, and a brief user's guide is available to introduce students to this tool. This user's guide is available for download at *www.course.com* via the "Student Downloads" link on the Web page for this textbook.

COURSE TECHNOLOGY DATA FILES

You are granted a license to copy the data files to any computer or computer network used by individuals who have purchased a copy of this book.

1

INTRODUCTION TO PL/SQL

> **In this chapter, you will:**
> ♦ Understand what programming and procedural languages provide
> ♦ Understand PL/SQL and application programming
> ♦ Learn about application models
> ♦ Understand how documentation can be used
> ♦ Learn about the SQL and PL/SQL tools
> ♦ Understand the databases used in this book

In this chapter, we explore the definition of programming and procedural languages, what PL/SQL is and why we need it, basic application models, the role of documentation, SQL and PL/SQL tools, and the example databases used throughout this text. When you have finished this chapter, you will be ready to work your way through the examples and concepts in the rest of the book.

PROGRAMMING AND PROCEDURAL LANGUAGES

In its simplest definition, a **programming language** allows the actions of an end user to be converted into instructions that a computer can understand. All programming languages share some basic capabilities, such as manipulating data, using variables to hold data for processing, and making code reusable.

Structured Query Language (SQL) is a programming language. It allows us to add, delete, or change data in a database. However, it isn't a **procedural language** that allows a programmer to code a logical sequence of steps to make decisions and to instruct the computer as to the tasks that need to be accomplished. For that kind of functionality, you need to use Oracle's procedural language—PL/SQL.

APPLICATION PROGRAMMING AND PL/SQL

So what is PL/SQL? What role does it play in application programming? Why do I need to learn it? These are probably some of the questions you have right now and they need to be answered before we dive into the depths of Oracle programming. We begin with a discussion of application programming.

Application Programming

First, let's clarify what we mean by application programming. Let's say you are an employee of a company named Brewbean's that retails coffee products on the Internet and the company needs to develop a software application to support the business. One part of the application consists of the user interface or screens that allow the customers to access the product catalog and place an order. One of the ordering screens is shown in Figure 1-1.

This screen displays the shopping cart containing all the items a shopper has selected thus far. Notice that the shopper has a number of choices at this point, such as to continue shopping, to remove items from the cart, to change item quantities, or to check out to complete the order.

Consider the potential processing that could be required in this application. If, for example, the Check Out link is clicked, what should happen? Processing activities may include the following:

- Verify quantity ordered is greater than zero for each item
- Calculate taxes
- Calculate shipping charges
- Check and/or update product inventory
- Determine if the shopper already has credit card information stored

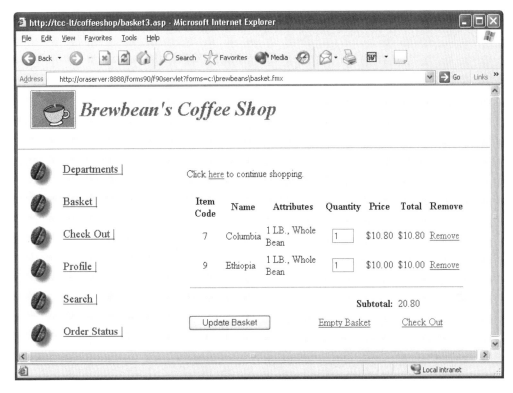

Figure 1-1   Brewbean's Coffee Shop shopping cart screen

Behind the user interface in Figure 1-1 is where a programming language instructs the computer system what to do in response to an application user action, such as clicking an item. In this case, if the user clicks the Check Out link, several things might need to happen. First, we need to check the completeness of (or validate) the user information. For example, we might need to verify that every line item has a quantity entered on the screen. We might also want to check the database to ensure all items are in stock. This is where PL/SQL enters the scene.

Note that Oracle considers PL/SQL to be a procedural language extension of SQL, hence the name. Recall that a procedural language allows a programmer to code a logical sequence of steps to instruct the computer as to the tasks that need to be accomplished. This type of language allows developers to perform decision-making logic, such as IF–THEN conditions, within their applications. This procedural language allows the developer to make some decisions within the code, such as checking values entered by the user and determining the shipping cost based on the total number of items ordered.

PL/SQL is a proprietary language of Oracle, and a PL/SQL processing engine is built into the Oracle database and developer tools (such as Forms and Reports). SQL statements can be embedded within the PL/SQL code to combine the data manipulation power of SQL with the procedural powers of PL/SQL. This produces a robust language to create database-driven applications.

PL/SQL is tightly integrated in the Oracle database server product and, therefore, the code processes very efficiently. Because a PL/SQL engine is part of the server, these code modules can be used or called from almost any application development language. You can use Visual Basic or Java to develop an application but still harness the power of PL/SQL with Oracle.

There are other advantages as well. You can create program units with PL/SQL that include SQL statements to handle database manipulation tasks. These program units can be stored within the Oracle database and called from your development tool of choice. It is more efficient to process SQL statements in an application than within the user's machine. In the latter case, the statements need to be transmitted to the database server to be processed.

If you have a piece of code that has the potential of being used by various applications, saving this code on the server allows it to be shared by several applications. In addition, Oracle has built a PL/SQL engine into a number of their developer tools, such as Oracle Forms. This allows an Oracle developer to code an entire application—from client screen logic to database manipulation—using PL/SQL.

History of PL/SQL

PL/SQL was originally modeled after Ada, a programming language built for the U.S. Department of Defense, which was considered conceptually advanced for the time. More than a decade ago, Oracle recognized a need for not only expanding functionality in their database system, but also for improving application portability and database security. At this point, PL/SQL was born and has continually been improved with every Oracle database release. The following list is a summary of just a few of the advantages PL/SQL offers when working with an Oracle database.

- *Tight integration with SQL*—You can leverage your knowledge of SQL because PL/SQL supports the use of SQL data manipulation, transaction control, functions, cursors, operators, and pseudocolumns. In addition, PL/SQL fully supports SQL data types, which reduces the need to convert data passed between your applications and the database.

- *Increased performance*—First, PL/SQL allows blocks of statements to be sent to Oracle in a single transmission. This is important in reducing network communication between your application and Oracle, especially if your application is database intensive. PL/SQL blocks can be used to group SQL statements before sending them to Oracle for execution. Otherwise, each SQL statement must be transmitted individually. Second, PL/SQL code modules, or stored program units, are stored in executable form—making procedure calls very efficient. Third, executable code is automatically cached in memory and shared among users. This can speed processing tremendously with a multiuser application that has repeated calls to modules of code. Fourth, a PL/SQL engine is embedded in Oracle developer tools, so PL/SQL code can be processed on the client machine. This reduces network traffic.

- *Increased productivity*—PL/SQL can be used in many of the Oracle tools and the coding is the same within all. Therefore, your PL/SQL knowledge can be used with many development tools and the code created can be shared across applications.

- *Portability*—PL/SQL can run on any platform that Oracle can run. This is important to enable a developer to easily deploy an application on different platforms.

- *Tighter security*—Database security can be increased with application processing supported by PL/SQL stored program units. Program units can enable the access of database objects to users without the users being granted privileges to access the specific objects. Therefore, the users can access these objects only via the PL/SQL program units.

 A database administrator (DBA) can automate and handle some tasks easier by leveraging the power of PL/SQL. Many PL/SQL scripts developed to handle DBA-type tasks are available free in books, Web sites, and user groups, so it is quite beneficial for a DBA to be familiar with the PL/SQL language.

APPLICATION MODELS

An application model is a general framework or design that describes how the various components of the application will be addressed. An application model has three main components:

- *User interface or screens*—The screens that are presented to the end user to enter information and/or take actions, such as clicking a button on the screen. The user interface component may be developed with tools such as Visual Basic, Java, or Oracle Forms.

- *Program logic or the brains behind the screens*—The programming code that provides the logic of what the application will do. PL/SQL addresses the "brains" or logic portions of an application.

- *Database*—The database management system providing the physical storage structure for data and mechanisms to retrieve, add, change, and remove data. The Oracle server provides the database.

To more clearly see where PL/SQL fits into an application, let's take a closer look at the basic two-tier and three-tier application models in which each tier represents specific components of an application.

Note that the traditional two-tier application model is also referred to as a client/server application. You will find both terms used throughout this book.

Two-Tier or Client/Server Application Model

Review the depiction of the client/server application components in Figure 1–2.

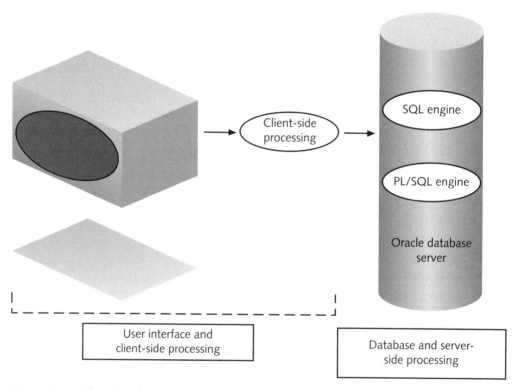

Figure 1-2   The client/server application model

In a client/server model, an executable program or application has been loaded onto the user's computer that contains the user screens and some programming logic. Some processing can take place on the client-side or user's computer. The processing might be verifying that information has been entered into required fields. However, other processing requires transmitting requests to the database server, such as an SQL statement to query requested data from the database. If you are using non-Oracle developer tools such as Visual Basic, client-side code can be VBScript that includes calls to PL/SQL program units stored on the Oracle server. However, if you are using Oracle development tools such as Forms, a PL/SQL engine exists on the client as well as on the database server and, therefore, all the application coding is accomplished using PL/SQL.

In this scenario, PL/SQL code resides on both the client-side and server-side. PL/SQL is saved as part of the Oracle Forms application logic on the client-side and is stored as named program units on the database server. A **named program unit** is simply a block of PL/SQL code that has been named so that it can be saved (stored) and reused. We will create PL/SQL stored procedures, functions, packages, and triggers in this book.

The term **stored** indicates the program unit is saved in the database and, therefore, can be used or shared by different applications. To give you an idea of what each of these different types of program units are, Table 1-1 lists a brief description of each.

Table 1-1   Stored program unit types

Stored Program Unit Type	Description
Procedure	Performs a task. Can receive and return multiple values.
Function	Performs a task. Typically returns only one value. Within certain parameters, it can be used within SQL statements.
Database Trigger	Performs a task automatically when a DML action occurs on the table with which it is associated. Recall DML represents Data Manipulation Language or insert, update, and delete.
Package	Groups related by procedures and functions. By doing so, additional programming features become available.

Application procedures, functions, and triggers are integrated within the Forms application and are considered client-side program units. An example of an application trigger is PL/SQL code that automatically runs when a particular button is clicked on the screen. These are addressed within Oracle Forms development, which is beyond the scope of this text.

Three-Tier Application Model

Let's move from the two-tier model to the three-tier model. The three-tier model is growing in popularity because it attempts to ease application maintenance and enables the support of larger numbers of users. In this scenario, the user interface is typically presented via a Web browser and is often referred to as a thin client. Application code is not loaded onto the client machine; all the application code now resides on an application server, which is also referred to as the middle tier. Figure 1-3 shows the three-tier application components.

Three layers now exist in this model: user interface, application server, and database server. The Oracle application server allows the deployment of Oracle Forms applications via the Web. The user interface is delivered from the application server that contains the logic to respond to user actions and, in turn, sends appropriate code to the database server for processing.

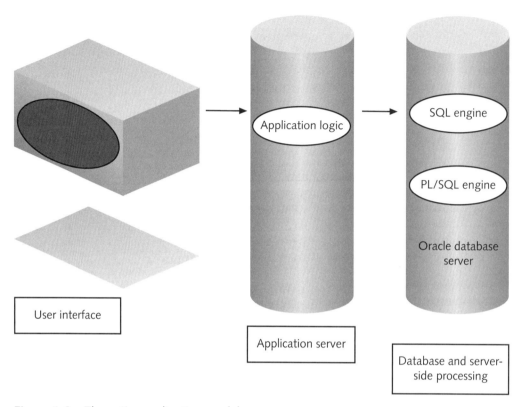

Figure 1-3   Three-tier application model

 In either application model, PL/SQL's role is the same: to provide the logic to instruct the computer as to what to do when an event occurs. (An **event** can range from a user action, such as clicking a button on the screen, to a table update statement that automatically calls a database trigger.) In this book, all the code is placed server-side as stored program units.

DOCUMENTATION ON THE WEB AND ON CD-ROM

As you have probably already discovered, numerous books are available on various Oracle system topics—including PL/SQL. However, another valuable documentation resource is provided free via the Oracle Web site. Oracle has created an Oracle Technology Network (OTN) that contains a variety of useful resources, including documentation, white papers, downloads, and discussion forums. This Web site is located at *http://otn.oracle.com*.

At the site, one of the available links is "Documentation." Within this link, you can view information by product or technologies. If you then choose Oracle9*i* Database, you will discover an HTML and PDF link for general documentation that allows you to list all the reference books available online. Both an SQL and a PL/SQL reference book can be found in this list. When you first visit this site, you need to follow the membership link to set up your new free membership.

 OTN membership is free at the time of this text publication.

In addition, documentation is also included on software installation CDs distributed by Oracle. The documentation is offered in both HTML and PDF formats. If you look at a directory listing of your software CD, you will typically find a directory named doc. In this directory, you will find an index file that outlines the documentation included on the CD. Documentation can also be purchased at the Oracle store located at *http://oraclestore.oracle.com*.

SQL AND PL/SQL TOOLS

Wading through the many software tools that Oracle offers can easily get confusing. In addition, you'll soon learn that many independent companies produce third-party software tools that can assist Oracle developers. This section of the chapter introduces Oracle software tools applicable to this book and PL/SQL development. It then discusses some third-party tools available. The text does not give recommendations, nor is it exhaustive. It merely attempts to expose you to the many options a developer faces when selecting tools.

 Your instructor is a valuable resource in helping you decide which tools are best for you.

Software Tools Used in this Book

This book addresses the newest version of the Oracle database, or Oracle9*i*. PL/SQL version numbering matches the database version numbering starting with Oracle8, so we will be using PL/SQL 9*i*. SQL*Plus will be used throughout this text as this software is available regardless of the developer tools you install. However, you also will be introduced to other developer tools that are available. In Appendix B and Appendix C, Procedure Builder and TOAD will be introduced. Procedure Builder is a tool included in Oracle Developer Forms 6*i* software for creating and editing PL/SQL program units. Procedure Builder is not available in Developer Forms 9*i*, so that is why a third-party tool named TOAD (Tool for Oracle Developers) is also introduced. In both Oracle

Forms 6*i* and 9*i*, the PL/SQL Editor is a window used to enter PL/SQL code. Developer Forms 9*i* is introduced in Appendix D to assist you in visualizing how PL/SQL coding is linked to the application screens that are presented to end users. Appendix D is available at *www.course.com* via the "Student Downloads" link, on the Web page for this book.

SQL*Plus

SQL*Plus is a basic tool available with the Oracle server that allows a user to enter SQL and PL/SQL statements directly to the Oracle database server for processing. Figure 1-4 displays SQL*Plus with a simple query entered.

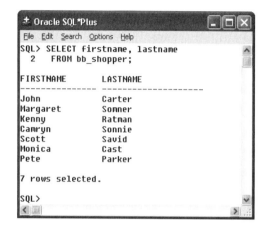

Figure 1-4   SQL*Plus screen

Procedure Builder

Procedure Builder is a utility that is part of the Oracle Developer 6*i* Suite and can be loaded individually or as a part of Oracle Forms. Procedure Builder provides a more graphical user interface (GUI) that allows the developer to explore database objects, quickly review/edit code, and easily create/debug program units. Figure 1-5 shows the main interface screen of Procedure Builder just to give you an idea of what it looks like.

1

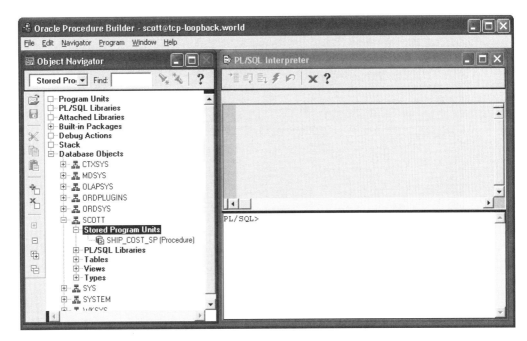

Figure 1-5   Procedure Builder screen

The OTN Web site contains documentation on Procedure Builder. A Getting Started guide is located at *http://technet.oracle.com/doc/pbuilder.htm* as of this text printing.

Even though PL/SQL code can be run in SQL*Plus, Procedure Builder offers features to simplify processes, such as testing or debugging code. Note that this tool is not available in the Oracle Developer 9i Suite, only with 6i. Reason? The debugger is now more closely integrated with forms development, a discussion of which is beyond the scope of this text. To get yourself started, you just need to know this fact: SQL*Plus and third-party tools are now used to test and debug stand-alone PL/SQL program units being developed.

PL/SQL Editor

Regardless of the developer suite version, PL/SQL program units can be created using the PL/SQL Editor. The PL/SQL Editor allows the construction and compilation of program units that will test the syntax. The Editor window looks the same whether using Oracle Developer 6i or 9i, and is shown in Figure 1-6.

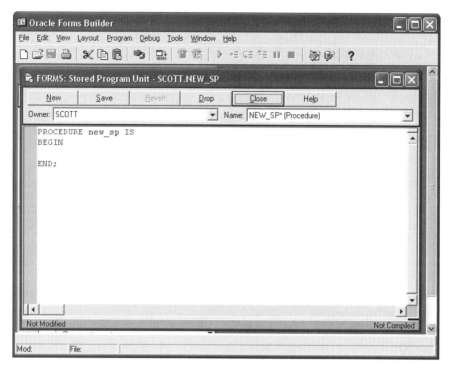

Figure 1-6 PL/SQL Editor

Working with the Tools

To perform the coding activity in this text, you need access to the Oracle9*i* database, which contains the SQL*Plus tool. To work with the appendices, you will need Forms 6*i* Procedure Builder, Forms 9*i*, and TOAD, which is a third-party tool. If you are working in a computer lab setting, then the appropriate software installation should already be prepared for you.

If you wish to install on your own computer, you can refer to the installation instructions for the Oracle9*i* database at *www.course.com/cdkit* (as was already mentioned in the Read This Before You Begin section of this book). For learning purposes, many programmers install Personal Oracle, which is a desktop version of the Oracle database that is available for learning and testing. It is used as a single-user environment. It is not intended or constructed to be used as a production database server. However, it is a powerful tool in that it contains much of the utility of the full database server version.

Third-Party PL/SQL Development Tools

Many other companies have developed software to assist the PL/SQL developer. As you progress through this text and become familiar with creating PL/SQL program units, you may want to take some time to experiment and compare some of the tools available by third-party companies. Table 1-2 lists some of the popular tools and their related

Web sites for your reference. Many of the companies offer a free trial download for you to test the product.

Table 1-2 Third-party PL/SQL development tools

Tool Name	Web Site Address
TOAD	www.quest.com
Rapid SQL	www.embarcadero.com
PL/SQL Developer	www.allroundautomations.nl
SQL-Programmer	www.bmc.com
DevPartner DB	www.compuware.com

TOAD or Tool for Oracle Application Developers by Quest Software is demonstrated in Appendix C.

DATABASES USED IN THIS BOOK

Every programmer quickly realizes that to become a successful application developer, he or she must thoroughly understand the database. Therefore, be sure to review this section carefully to obtain a good understanding of the design of the databases used in this book.

The main database used throughout this book supports a company retailing coffee via the Internet, over the phone, and in one walk-in store. The company has two main areas of products: coffee consumables and brewing equipment. They also hope to add a coffee club feature to entice return shoppers. Figure 1-7 depicts a basic entity relationship diagram (ERD), or a visual representation, of the database.

The BB_DEPARTMENT table contains the identity of the main areas of business. The BB_PRODUCT table contains information on all the individual products, including name, description, price, and sales pricing information. Three tables exist to allow the management of various product options, such as size (for example, 1/2 lb of coffee) and form (for example, ground or whole bean).

The BB_PRODUCTOPTIONCATEGORY table maintains the list of main categories applicable to any of the products, such as size and form. The BB_PRODUCTOPTION-DETAIL table identifies the choices available under each of these categories (such as whole bean for form). The BB_PRODUCTOPTION table links each product to the applicable options for that item. Keep in mind that each product can be associated with many options so we have a one-to-many relationship between the BB_PRODUCT and the BB_PRODUCTOPTION tables.

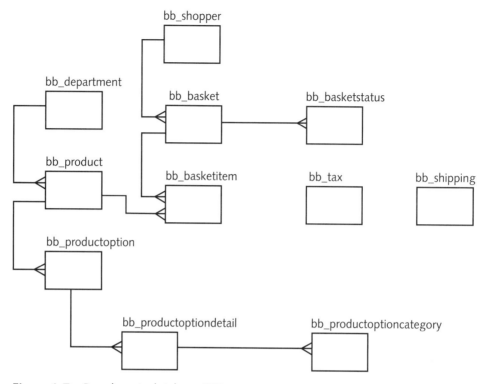

Figure 1-7   Brewbean's database ERD

The BB_SHOPPER table serves as the focal point in identifying customers and it contains the shopper name, address, e-mail address, and logon information. When a customer begins his or her shopping experience, a new basket is created in the BB_BASKET table. The BB_BASKET table is one of the largest tables and it holds order summary, shipping, and billing information. As the shopper selects items, they are inserted into the BB_BASKETITEM table, which holds all the shopper selections by basket number.

The BB_BASKETSTATUS table stores data related to the status of an order. Each status update is recorded as a new row in this table. Possible statuses include order placed, order verified and sent to shipping, order shipped, order cancelled, and order on backorder. The other tables associated with completing an order include BB_TAX and BB_SHIPPING. The company currently calculates shipping based on the quantity of items ordered.

Data listing by table are in Appendix A for your reference. It is important to take time now to become familiar with the database. All the table names for the Brewbean's database start with a prefix of BB_. Let's run the script to create all the necessary objects for the following chapters.

To create the Brewbean's database:

1. Locate the c1Dbcreate file in the Chapter.01 folder to ensure that it exists. This file contains the script to create the database.

2. Open SQL*Plus and connect using the appropriate user id, password, and host string. If you are using Personal Oracle, you can log on as user **Scott**, password **Tiger**, and no host string.

3. First, we create a spool file so that SQL*Plus will keep a copy of all the messages received from running the file that creates the database. If any errors do occur, you can then present this file to your instructor to determine why the errors occurred. On the main menu in SQL*Plus, click **File**, point to **Spool**, and then click **Spool File**. A Select File dialog box appears.

4. Browse to the Chapter.01 folder and in the File name: text box, type **DB_log**, and click **Save**. Now, whatever text is seen in our SQL*Plus session will be saved to this file for future reference.

 Avoid saving any Oracle files in directories whose name includes a blank space. Attempting to use a file in such a directory will typically produce an error.

5. Now let's create the database. In SQL*Plus, enter the following command, which runs all the statements contained in the c1Dbcreate.sql file. Messages verifying the creation and data insertion steps will scroll on the SQL*Plus screen. This takes a couple minutes to complete.

   `@<pathname to PL/SQL files>\Chapter.01\c1Dbcreate`

6. Now, we will turn the spooling off so that we can review the results in the spool file created. On the main menu, click **File**, point to **Spool**, and then click **Spool Off**.

7. Open a text editor.

8. Open the file named **DB_log.lst** in the Chapter.01 folder.

9. Review the messages for any errors.

10. Close all open windows.

A different database is used with the case projects at the end of each chapter. Case 1 has differing themes from chapter to chapter. Case 2 presents a continuous scenario throughout this book using a database that supports a movie rental company. More Movies is a movie rental company that is creating a new application to support their membership and renting processes. The ERD of the database is shown in Figure 1-8.

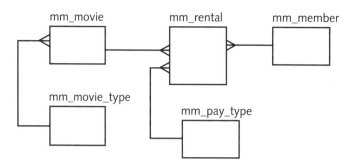

Figure 1-8 More Movies rental company ERD

To create the More Movies database:

1. Locate the **case2_Dbcreate.sql** file in the Chapter.01 folder to ensure that it exists. This file contains the script to create the database.

2. Open or return to SQL*Plus and connect using the appropriate user id, password, and host string. Remember, if you are using Personal Oracle, you log on as **Scott**, **Tiger**, and no host string.

3. First, we create a spool file so that SQL*Plus will keep a copy of all the messages received from running the file that creates the database. If any errors do occur, you can then present this file to your instructor to determine why the errors occurred. On the main menu in SQL*Plus, click **File**, point to **Spool**, and then click **Spool File**. A Select File dialog box appears.

4. Browse to the Chapter.01 folder and in the File name: text box, type **case2_DB_log**, and click **Save**. Now, whatever text is seen in our SQL*Plus session is saved to this file for future reference.

5. Now let's create the database. In SQL*Plus, enter the following command, which runs all the statements contained in the case2_Dbcreate.sql file. Messages verifying the creation and data insertion steps scroll on the SQL*Plus screen. This takes a couple minutes to complete.

   `@<pathname to PL/SQL Files>\Chapter.01\case2_Dbcreate`

6. Now, we will turn the spooling off so we can review the results in the spool file created. In SQL*Plus, click **File**, point to **Spool**, and then click **Spool Off**.

7. Open a text editor.

8. Open the file named **case2_DB_log.lst** in the Chapter.01 folder.

9. Review the messages for any errors.

10. Close all open windows.

CHAPTER SUMMARY

- ❐ A procedural language allows a programmer to code a logical sequence of steps to make decisions and to instruct the computer as to the tasks that need to be accomplished.

- ❐ SQL allows the creation and manipulation of a database, but it is not a procedural language.

- ❐ PL/SQL is a procedural language and the Oracle database and software tools contain a PL/SQL processing engine.

- ❐ PL/SQL improves application portability in that it can be processed on any platform on which Oracle runs.

- ❐ The use of PL/SQL can lead to improved security by not requiring that users be granted direct access to database objects.

- ❐ Applications contain three main components: user interface, program logic, and a database.

- ❐ The two-tier or client/server application model splits the programming logic between the client user machine and the database server.

- ❐ The three-tier application model places much of the application coding on an application server or middle tier.

- ❐ Programming code on the client machine is referred to as client-side code, and code on the database server is referred to as server-side.

- ❐ A named program unit is a block of PL/SQL code that has been saved and assigned a name so that it can be reused.

- ❐ A stored program unit is a named program unit that is saved in the database and can be shared by various applications.

- ❐ A procedure and function are named program units that are called to perform a specific task.

- ❐ A database trigger is PL/SQL code that is processed automatically when a particular DML action occurs.

- ❐ A package is a structure that allows the grouping of functions and procedures into one container.

- ❐ An application event is some activity that occurs, typically a user selecting an item on the screen, which causes some processing to occur.

- ❐ Documentation is available on the Oracle Technology Network (OTN) Web site.

- ❐ SQL*Plus is an Oracle software tool that allows the submission of SQL and PL/SQL statements directly to the Oracle server.

❑ Procedure Builder is a software tool to ease the testing of program units and is included in Developer 6*i*.

❑ The PL/SQL Editor allows the creation of program units and is included in the Oracle Developer 6*i* and 9*i* suites.

❑ Many third-party software tools are on the market to assist in the development of PL/SQL program units.

❑ You need to know the ERD of a database before you can work with it effectively.

REVIEW QUESTIONS

1. What application model uses a thin client that is typically a browser-based front end?

    a. client/server

    b. two-tier

    c. three-tier

    d. thin-tier

2. Which of the following is not a type of stored PL/SQL program unit?

    a. procedure

    b. application trigger

    c. package

    d. database trigger

3. The term "named program unit" indicates _____.

    a. that the PL/SQL block is assigned a name so it can be saved and reused

    b. the type of PL/SQL block

    c. that the PL/SQL block is executable as an anonymous block

    d. that the PL/SQL block is saved client-side

4. Any application model typically addresses what three basic components?

    a. user interface

    b. program logic

    c. coding style

    d. database

5. A software tool that is a part of the Oracle Developer 6*i* Suite (but not the 9*i* Suite) that assists in the construction of PL/SQL program units is called _____.

   a. SQL*Plus

   b. PL/SQL Builder

   c. PL/SQL Creator

   d. Procedure Builder

6. Variable naming conventions are used _____.

   a. because they are required

   b. to assist in the declaration of a variable

   c. to assist the developer in identifying the variable type

   d. to assist the PL/SQL engine to execute the code

7. Which of the following is an Oracle software tool that is included with the Oracle database server and that allows SQL and PL/SQL statements to be submitted to the server?

   a. SQL*Plus

   b. PL/SQL Builder

   c. PL/SQL Creator

   d. Procedure Builder

8. A procedural programming language allows the inclusion of _____.

   a. decision-making logic

   b. inserts

   c. DML

   d. create table statements

9. Application portability refers to the ability to _____.

   a. upload and download

   b. create a small executable

   c. move to various computer platforms

   d. transmit data efficiently

10. A two-tier application model is commonly referred to as _____.

   a. *n*-tier

   b. client/server

   c. double layered

   d. user-database

11. Name the four different types of stored program unit structures and the basic difference among them.

12. Name and briefly describe three advantages of using PL/SQL with an Oracle database.

13. If you are not using Oracle development tools such as Oracle Forms, should you pursue learning PL/SQL? Why or why not?

14. Describe the significant difference between a two- and three-tier application model.

15. Describe what a user interface is and the role a procedural language plays in regards to the user interface.

HANDS-ON ASSIGNMENTS

Assignment 1-1: Review Data in the Brewbean's Database

It is important to become familiar with the Brewbean's database. We refer to this database throughout this text. Let's do a few queries to take a look at the data.

To query the database:

1. Open SQL*Plus.

2. Enter the query shown in Figure 1-9 and check your data against the listing shown in the same figure.

Note If you like, you can set options such as PAGESIZE and LINESIZE in SQL*Plus to control the output appearance. On the main menu of SQL*Plus, click **Options**, and then click **Environment**. The list of items that can be set are listed on the left side of the pane. Start with linesize = **100** and pagesize = **60**.

```
Oracle SQL*Plus
File  Edit  Search  Options  Help
SQL> SELECT idProduct, productname, price, active, type, idDepartment, stock
  2    FROM bb_product;

IDPRODUCT PRODUCTNAME                      PRICE    ACTIVE T IDDEPARTMENT    STOCK
--------- ------------------------- ----------- --------- - ------------ --------
        1 CapressoBar Model #351          99.99         1 E            2       23
        2 Capresso Ultima                129.99         1 E            2       15
        3 Eileen 4-cup French Press        32.5         1 E            2       30
        4 Coffee Grinder                   28.5         1 E            2       26
        5 Sumatra                          10.5         1 C            1       41
        6 Guatamala                          10         1 C            1       42
        7 Columbia                         10.8         1 C            1       61
        8 Brazil                           10.8         1 C            1       53
        9 Ethiopia                           10         1 C            1       54
       10 Espresso                           10         1 C            1       50

10 rows selected.

SQL> |
```

Figure 1-9   Query Brewbean's product information

3. Enter the query shown in Figure 1-10 and check your data against the listing shown in the same figure. Note that ANSI standard joins introduced in Oracle9*i* are being used.

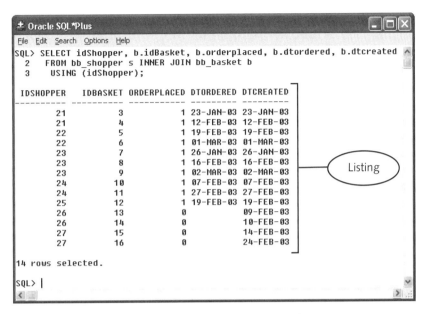

Figure 1-10   Query Brewbean's shopper basket information

4. Enter the query shown in Figure 1-11 and check your data against the listing shown.

Assignment 1-2: Review Third-Party Software Tools

Different third-party PL/SQL software tools were mentioned in this chapter. (See Table 1-2.) Go to the Web site for one of these tools and describe at least two features of the tool that will potentially assist in the development of PL/SQL code.

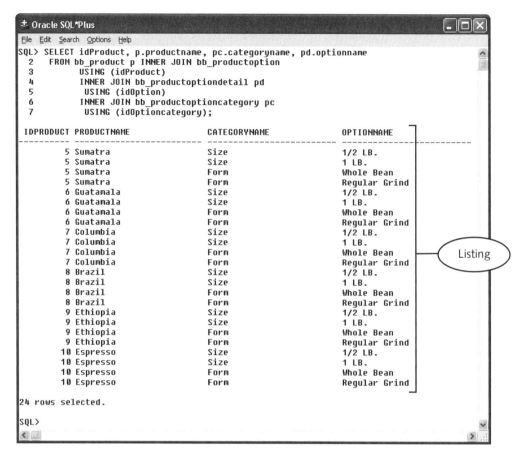

Figure 1-11   Query Brewbean's product option information

Assignment 1-3: Identify Processing Steps

Review the Brewbean's application screen displayed in Figure 1–12. List the logical processing steps that need to occur if the Check Out link is clicked and for the next screen to display the order subtotal, shipping amount, tax amount, and final total. The shipping costs are stored in a database table by number of items. The tax percentage is stored by state in a database table and is based on the billing state of the customer.

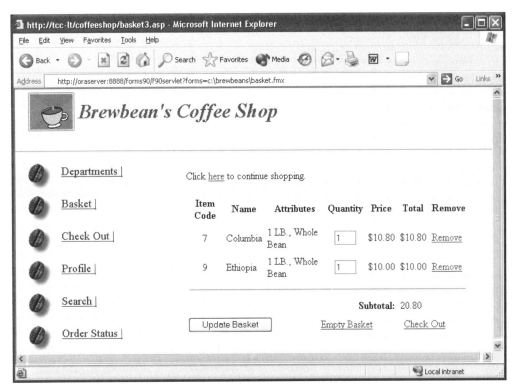

Figure 1-12   Brewbean's application screen

Assignment 1-4: Use OTN Documentation

Go to the Oracle9*i* database documentation section on the OTN Web site. Find the PL/SQL User's Guide and Reference under the list of books available. Within this reference, locate the chapter covering control structures. Briefly explain what a control structure is and provide an example of a specific PL/SQL statement that is considered to be a control structure statement.

CASE PROJECTS

Case 1-1: Review Procedure Builder Documentation

Review the Procedure Builder Getting Started guide to identify at least two features that you believe will be helpful in creating PL/SQL code. Your instructor will identify the location of this documentation.

Case 1-2: The More Movies Database

The entity relationship diagram (ERD) and steps to create the More Movies rental company database were presented in this chapter. This database is used in a continuous manner for the second case in every chapter of this text. In this case project, we perform several queries on this data to become familiar with it and to verify the content. Perform the queries listed in Table 1-3 and compare them with the associated figure to verify your results.

Table 1-3   Query tasks on the More Movies database

Query #	What to Query	Results Figure
1	List the movie id, movie title, movie category id, and movie category name for every movie in the database. What five categories exist for movies?	Figure 1-13
2	List the member id, last name, and suspension code for every member. Are any members suspended at this point?	Figure 1-14
3	List the member last name, rental checkout date, and movie title for all rentals. What checkout data applies to all the recorded rentals?	Figure 1-15

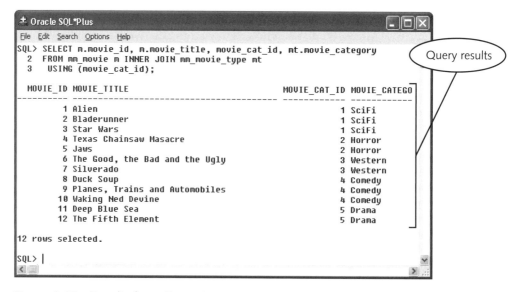

Figure 1-13   Results from Query #1

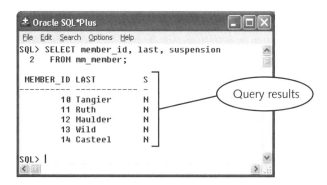

Figure 1-14   Results from Query #2

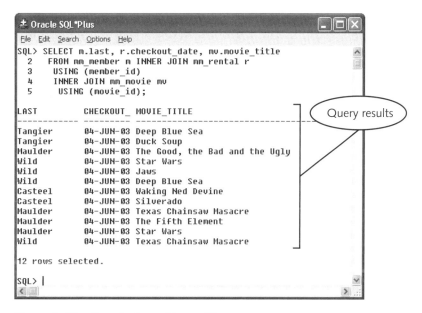

Figure 1-15   Results from Query #3

2

HANDLING DATA IN PL/SQL BLOCKS

In this chapter, you will:

♦ Learn about the PL/SQL block

♦ Define variables

♦ Create scalar variables

♦ Include SQL within PL/SQL

♦ Execute a PL/SQL block

♦ Use host or bind variables

♦ Understand the %TYPE attribute

♦ Use composite data types

♦ Process conditional logic with IF statements and loops

♦ Create collections

♦ Manipulate data with cursors

♦ Identify variable scope

Before we do any processing in PL/SQL, we need to understand the basic block structure of PL/SQL code and how data flows through the block. You probably have a lot of questions about the material in this chapter—What are the parts of a block? How does data get into the block? What forms of data can we use in a block? These questions and others are answered in this chapter.

It is almost impossible to separate the discussion of data handling and statement processing because these two functions are intertwined within a block. Thus, even though Chapter 3 is the chapter that concentrates on statement processing, you will find some of its content introduced in this chapter.

Our first step in this chapter is to identify a current data-handling challenge in the Brewbean's application. In the rest of the chapter, you'll learn what you need to do to address that challenge.

THE CURRENT CHALLENGE IN THE BREWBEAN'S APPLICATION

The Brewbean's application already has certain functionalities. As you can see in Figure 2-1, a shopping cart screen shows a shopper all the items and quantities selected thus far. A number of potential processing tasks could be required when the Check Out link is clicked by the shopper. Those possible tasks include calculating taxes, calculating shipping charges, checking and/or updating product inventory, and determining if the shopper already has credit card information stored.

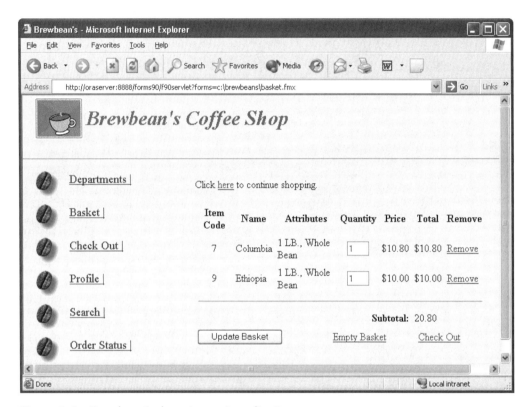

Figure 2-1   Brewbean's shopping cart application screen

So, how do we approach programming the needed tasks? Well, a block is typically written to accomplish a particular task. For example, in the Brewbean's shopping cart scenario, a PL/SQL block may be written to specifically handle the calculation of shipping costs based on the number of items in the order and what type of shipping method is selected. Another block may be developed to calculate the applicable taxes and yet another block for addressing inventory issues. Using this approach makes developing an application and maintaining and reusing code more manageable than having larger, more complex blocks of code, and it accomplishes a variety of tasks.

You construct the block for the shipping calculation to accept inputs such as a basket number from the shopping cart screen. The program uses this input to query the database regarding the number of items in that basket. By accepting input values, the same PL/SQL block is made dynamic in that it can calculate the shipping costs for any order because it queries the database for the total number of items based on the given basket number.

PL/SQL BLOCK STRUCTURE

In this section, we explore the basic construct of a PL/SQL block. To learn the basic syntax of a block, we begin working with **anonymous blocks**, which are blocks of code that are not stored for reuse and, as far as the Oracle server is concerned, no longer exist after being executed. The block code is entered into SQL*Plus, just as we would enter an SQL statement.

As outlined in Table 2-1, PL/SQL code is created in a block structure that contains four main sections: DECLARE, BEGIN, EXCEPTION, and END. A block is always closed with an **END;** statement. The only required sections of a PL/SQL block are the BEGIN and END sections; however, most blocks use all four sections.

Table 2-1   PL/SQL block sections

Block Section	Section Content
DECLARE	Creates variables, cursors, and types
BEGIN	Contains SQL statements, conditional logic processing, loops, and assignment statements
EXCEPTION	Contains error handlers
END	Closes the block

The following sections discuss each of the first three sections (noted in Table 2-1) in detail.

DECLARE Section

The DECLARE section contains code that creates variables, cursors, and types. The block uses these to hold data for processing and manipulation. For example, if a shopper on the Brewbean's Web site clicks the Check Out link, the code needs to retrieve all the item quantities in the shopper's basket to calculate shipping costs. In addition, the code needs to calculate the actual shipping cost. Variables hold or house this data to be used while the block executes.

As you progress in your studies, you'll discover that a number of different variable types exist and that they serve different purposes. For example, a scalar variable can hold only a single value, whereas a composite variable can hold multiple values.

BEGIN Section

The BEGIN section is the heart of the PL/SQL block in that it contains all the processing action, or programming logic. SQL is used for database queries and data manipulation. Conditional logic, such as IF statements, is used to make decisions on what action to take. Loops are used to repeat code, and assignment statements are used to put or change values in variables. In our shipping cost example, we may use IF statements to check the quantity of items and apply the appropriate shipping cost.

EXCEPTION Section

One job of the developer is to anticipate possible errors that could occur and to include exception handlers to provide a user with understandable messages on corrections needed or system problems that must be addressed.

The EXCEPTION section contains handlers that allow you to control what the application will do if an error occurs during the executable statements in the BEGIN section. For example, if your code attempts to retrieve all the items in a basket but there are no items, an Oracle error occurs. You can use exception handlers so that shoppers will not see Oracle system error messages and have the application halt operation. Instead, you can display friendly user messages and allow the application to continue via exception handlers.

WORKING WITH VARIABLES

In most PL/SQL blocks, variables are needed to hold values for use in the program logic. Variables are named memory areas that hold values to allow retrieval and manipulation of values within your programs. For example, if a SELECT statement is included in a block, variables are needed to hold the data retrieved from the database. In addition, if the block contains a calculation, a variable is needed to hold the resulting value.

The type of data that needs to be stored determines the type of variable needed. For example, if only a single value needs to be held, scalar variables are used. However, if multiple values need to be held, such as an entire row from a database table, then a record (which is a composite variable) is needed. Last, if you intend to process a number of rows retrieved with a SELECT statement, you may create a cursor that is a structure specifically suited to processing a group of rows.

We discuss different data types in detail in the following sections of this chapter. Note now, however, that no matter what type of variable is needed, the variable must be declared in the DECLARE segment of the block before you can use it in the BEGIN section.

To declare a variable, you must supply a variable name and data type. Variable names follow the same naming conventions as Oracle database objects:

- Begin with an alpha character
- Contain up to 30 characters
- Can include upper- and lowercase letters, numbers, and the _ , $, #, and special characters

You will encounter naming conventions for variables as you move through the remaining sections of this chapter.

WORKING WITH SCALAR VARIABLES

Scalar variables are variables that can hold a single value. The common data types used for scalar variables include character, numeric, date, and Boolean. Table 2-2 provides a brief description of each.

Table 2-2   Scalar variable data types

Type	How It Is Written in PL/SQL Code	Description
Character	CHAR(n)	Stores alphanumeric data, with n representing the number of characters in length. The variable always stores n number of characters regardless of the actual length of the value it is storing because it pads the data with blanks. If n is not provided, the length defaults to 1. The maximum length is 32,767 characters. Use CHAR for items that will always contain the same number of characters. Otherwise, it is more efficient (in terms of system resources) to use VARCHAR2.
	VARCHAR2(l)	Stores alphanumeric data with l representing the number of characters in length. The variable stores only the number of characters needed to hold the value placed into this variable, regardless of the actual variable length (hence "VAR" for variable). The l value is required. The maximum length is 32,767 characters.
Numeric	NUMBER(p,s)	Stores numeric data with p representing the size or precision and the s representing the scale. The size includes the total number of digits and the scale is the number of these digits that will be to the right of the decimal point. For example, the declaration of NUMBER(2,1) can hold 9.9 but is too small for 19.9. If s is not provided, the variable will not be able to store decimal amounts. If both p and s are not provided, the variable defaults to a size of 40.
Date	DATE	Stores date and time values. The default format with which Oracle can identify a string value as a date is *DD-MON-YY*. For example, 15-NOV-03 is recognized as a date. This setting is in the init.ora file and can be changed by the DBA.
Boolean	BOOLEAN	Stores a value of TRUE, FALSE, or NULL. Typically used to indicate the result of a condition or set of conditions that have been checked. In other words, the Boolean data type provides a variable that represents the state of a logical condition in terms of true or false.

Tip The init.ora file contains settings for a wide array of parameters on the database and is referenced when the database is started up. The DBA is typically responsible for determining and handling any necessary changes to these settings. This file can be found in your database directory and viewed as a text file.

Note that the Boolean data type is used quite a bit in association with logic that checks for the existence of some condition in your program. For example, a Boolean variable may be set to TRUE if the shipping address should be the same as the shopper's address. In your program logic, you can easily check this flag to determine which database fields to use to retrieve the shipping address.

In the next five subsections, you learn how to use scalar variables, starting with their declarations.

Variable Declarations in Code

To give you a picture of what various variable declarations look like, the following code snippet displays four scalar variable declarations in the DECLARE segment of a PL/SQL block. Notice that only one variable declaration per line is allowed.

```
DECLARE
   lv_ord_date DATE;
   lv_last_txt VARCHAR2(25);
   lv_qty_num NUMBER(2);
   lv_shipflag_bln BOOLEAN;
BEGIN
   ---- PL/SQL executable statements ----
END;
```

Each line of code declares a variable by supplying a variable name and data type, which is the minimum information necessary to declare a variable. The LV_ PREFIX indicates these are all scalar variables that are local to (used only within) the block. The suffixes indicate the data type with which the variable is declared. As you review blocks of code, this information will assist you in identifying the types of values that each variable holds.

Each of the variables in the preceding code example will contain a NULL value when the BEGIN section starts execution. However, at times, you may want to have a starting or initial value in a variable. This can be accomplished with initialization in the declaration, as explained in the next section.

Variable Initialization

If desired, a variable can be initialized with a value in the declaration statement. That is, the variable will already contain a value when the BEGIN section of the block starts execution. Let's look at how we could initialize the variables from our earlier example.

```
DECLARE
  lv_ord_date DATE := SYSDATE;
  lv_last_txt VARCHAR2(25) := 'Unknown';
  lv_qty_num NUMBER(2) := 0;
  lv_shipflag_bln BOOLEAN := 'FALSE';
BEGIN
  ---- PL/SQL executable statements ----
END;
```

Notice that := followed by a value is used to assign initial values to the variables within the declaration statements. The keyword DEFAULT can be used in place of the := symbol to achieve the same result. It is typical to initialize a Boolean variable to FALSE and have executable statements in the BEGIN section check for conditions and determine if it should be changed to TRUE. It is also typical to initialize numeric variables that will be used in calculations. This keeps you from scenarios in which you are trying to perform calculations on a NULL value, which is not the same as a 0 value.

In addition to providing initial values, you can set further controls on the variable value with the NOT NULL and CONSTANT options, as discussed in the next section.

NOT NULL and CONSTANT

The NOT NULL option can be added to the variable declaration to require the variable to always contain a particular value within the block. That is, the CONSTANT option can keep the value of the variable from being changed within the block. Let's look at how the NOT NULL and CONSTANT options are included in the variable declaration:

```
DECLARE
  lv_shipcntry_txt VARCHAR2(15) NOT NULL := 'US';
  lv_taxrate_num CONSTANT NUMBER(2,2) := .06;
BEGIN
  ---- PL/SQL executable statements ----
END;
```

Note that in both declarations in the preceding code, the variable is initialized with a value using the := assignment symbol. In the Brewbean's application, the same tax rate applies to all sales; therefore, the variable LV_TAXRATE_NUM is declared as a constant to ensure this value is not mistakenly modified in the executable section of the block. The LV_SHIPCNTRY_TXT contains US because, at this point, all shoppers are from the United States. Also note that the CONSTANT keyword is listed prior to the variable data type, whereas the NOT NULL keywords are listed after the data type.

The Role of Scalar Variables in the BEGIN Section

You can use scalar variables in assignment statements and calculations. An example of this use is shown in the following code:

```
DECLARE
  lv_taxrate_num CONSTANT NUMBER(2,2) := .06;
  lv_taxamt_num NUMBER(4,2);
BEGIN
  lv_taxamt_num := 50 * lv_taxrate_num;
END;
```

The BEGIN section in the preceding block contains one assignment statement that involves a calculation. Both variables declared are used in this statement. The multiplication resolves to 50*.06, or 3, and, therefore, the LV_TAXAMT_NUM variable holds this value. Again, note that the := symbol (beginning with a colon) is used to create the assignment statement. Many developers become accustomed to using an equal sign in SQL and forget the colon needed in PL/SQL—don't be one of them!

You can also use the DBMS_OUTPUT package to display values of variables during the execution of a block to verify proper execution. We discuss that next.

Using DBMS_OUTPUT to Check Values

The DBMS_OUTPUT package is a quick way to check variable values within a PL/SQL block. This is a popular mechanism used in testing and debugging a block of code. Let's run an example to review.

To execute a PL/SQL block using DBMS_OUTPUT:

1. Start SQL*Plus and log on.

2. Type **SET SERVEROUTPUT ON**, and then press **Enter**. This is an SQL*Plus command that enables the display of values from DBMS_OUTPUT statements.

3. Type the following code, and then press **Enter**:

```
DECLARE
 lv_basket_date DATE := SYSDATE;
BEGIN
 DBMS_OUTPUT.PUT_LINE(lv_basket_date);
END;
/
```

4. Compare your results to Figure 2-2. The date displayed below the block should reflect the current date. Of course, your date will differ.

5. Close SQL*Plus.

2

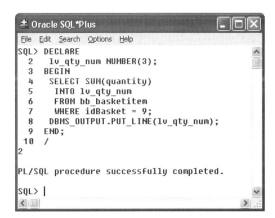

Figure 2-2   Using DBMS_OUTPUT.PUT_LINE to display values

PUT_LINE is a procedure contained within the DBMS_OUTPUT Oracle-supplied package.

INCLUDING SQL WITHIN A PL/SQL BLOCK

Now that we know how to declare variables and check values held by variables, let's look at a couple of examples of blocks using declared variables to hold data retrieved from the database. The first example, shown in Figure 2-3, is a block that contains a SELECT statement to retrieve the total number of items contained in a basket to be used to calculate the shipping cost.

Figure 2-3   A SELECT statement

When included in a PL/SQL block, a SELECT statement is written just as we accomplished in our SQL with one major exception—an INTO clause must be used. The INTO clause follows the SELECT clause and indicates which variables are to hold the values that are retrieved from the database. In the example in Figure 2-3, only a single value is retrieved and, therefore, only one variable is needed. Notice that the WHERE clause ensures that only rows for basket 9 are retrieved and the SUM operation ensures only a single value is returned, which is all a scalar variable can hold. DBMS_OUTPUT.PUT_LINE is used to display the value retrieved, which we can check against the database table to ensure accuracy.

Let's look at a more involved SELECT statement. Review Figure 2-4, which displays an application page that Brewbean's has developed to notify a shopper who logs onto the site if he or she has previously started a shopping cart that has not yet been completed. (It is common for Internet shoppers to be permitted to save shopping carts for completion at a later time.) Right now, we will assume that a shopper can have only one saved, noncompleted basket.

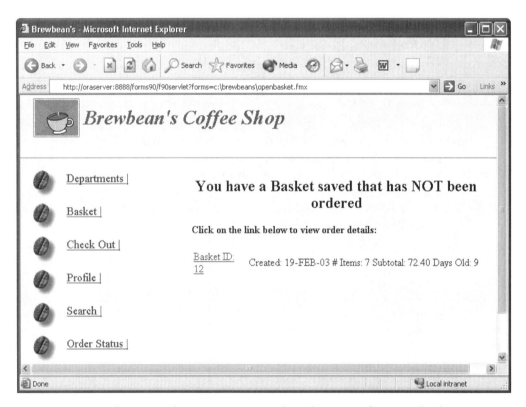

Figure 2-4   Brewbean's application page reminding shoppers of existing baskets

When a shopper logs onto the Brewbean's Web site, programming code is needed to determine if the shopper has an existing noncompleted basket saved. If so, the basket date, amounts, and age need to be retrieved to display on the screen. To accomplish this task, data needs to be queried from the BB_BASKET table and the age needs to be calculated using the date the basket was created. Figure 2-5 displays a PL/SQL block that accomplishes these tasks.

2

```
Oracle SQL*Plus
File  Edit  Search  Options  Help
SQL> DECLARE
  2    lv_basket_num NUMBER(3);
  3    lv_created_date DATE;
  4    lv_qty_num NUMBER(2);
  5    lv_sub_num NUMBER(5,2);
  6    lv_days_num NUMBER(3);
  7    lv_shopper_num NUMBER(3) := 25;
  8  BEGIN
  9    SELECT idBasket, dtcreated, quantity, subtotal
 10      INTO lv_basket_num, lv_created_date, lv_qty_num, lv_sub_num
 11      FROM bb_basket
 12      WHERE idShopper = lv_shopper_num
 13        AND orderplaced = 0;
 14    lv_days_num := SYSDATE - lv_created_date;
 15    DBMS_OUTPUT.PUT_LINE(lv_basket_num||' * '||lv_created_date||' * '||
 16                         lv_qty_num||' * '||lv_sub_num||' * '||lv_days_num);
 17  END;
 18  /
12 * 19-FEB-03 * 7 * 72.4 * 9

PL/SQL procedure successfully completed.

SQL> |
```

Output will differ based on the date on your system

Figure 2-5   A PL/SQL block to retrieve a noncompleted basket

The first action occurs at Line 9, which performs data retrieval with a SELECT command. Notice this command uses a WHERE clause to ensure only one row of data is returned from the SELECT statement. This is critical because we can hold only single values in the scalar variables declared. The first four variables declared are used in the INTO clause to hold the values retrieved in the SELECT statement. The column values retrieved are moved to the variables in a positional manner. In other words, the value retrieved for the first column listed in the SELECT clause (which is IDBASKET) is placed into the first variable listed in the INTO clause (which is LV_BASKET_NUM). Notice again that the SELECT statement retrieves only a single row from the table due to the WHERE clause, so scalar variables are sufficient in this block.

In Line 14, you'll see an assignment statement that calculates the age of the basket or, in other words, the current date of SYSDATE minus the LV_CREATED_DATE variable. (LV_CREATED_DATE contains the value retrieved from the DTCREATED column.) The result of this calculation is held in the LV_DAYS_NUM variable. Remember that assignment statements in PL/SQL use a combination of a colon and an equal sign.

Leaving the colon off the assignment statement is probably one of the most common errors experienced by new PL/SQL programmers. It takes a bit to adjust to using the := symbol. However, do not get this mixed up with your SQL syntax. Notice the WHERE clause of the SQL query in Figure 2-5 uses a plain equal sign just as we always have in our SQL statements. The := syntax applies only to PL/SQL statements.

The last statement in the block uses DBMS_OUTPUT.PUT_LINE to display values on the screen. This is a common method used by developers to check the values of variables during the execution of a block. Each DBMS_OUTPUT statement can display only one item, so the concatenation operator is used in this example to display the values of all the variables with a single DBMS_OUTPUT statement. You can include multiple DBMS_OUTPUT statements throughout your block to test values at any point in the block. In Line 18, notice that a forward slash must be placed at the beginning of the line immediately following the block to instruct SQL*Plus to execute the block.

EXECUTING A PL/SQL BLOCK WITH ERRORS

Let's jump in and experiment with running several versions of the PL/SQL block in Figure 2-5 so that we can get a feel for executing anonymous blocks. We will first run the block as is with no errors. Then, we will run the block with various errors to get accustomed to the error messages encountered with PL/SQL blocks.

To execute an anonymous block with scalar variables:

1. Open or return to SQL*Plus.
2. Type **SET SERVEROUTPUT ON**, and then press **Enter**. This enables the display of values from DBMS_OUTPUT statements.
3. In a text editor, open **scalarA02.txt** from the Chapter.02 folder. Copy all the code.
4. Return to SQL*Plus. Click **Edit** on the menu bar, and then click **Paste**.
5. Press the **Enter** key to execute the block. The results should appear similar to Figure 2-5, which shows information for basket 12.
6. Locate the file **scalarB02.txt** in the Chapter.02 folder.
7. Open the file in a text editor. Note that the same block is used except the assignment statement for LV_DAYS_NUM has been modified to contain an error.
8. Copy all the code.
9. Return to SQL*Plus. Click **Edit** on the menu bar, and then click **Paste**.
10. Press **Enter**, if necessary, to execute the block. Notice the error message should match that shown at the bottom of Figure 2-6, which identifies that an = symbol was encountered when something else was expected, which, in this case, is a PL/SQL assignment of :=.

```
Oracle SQL*Plus
File  Edit  Search  Options  Help
SQL> DECLARE
  2    lv_basket_num NUMBER(3);
  3    lv_created_date DATE;
  4    lv_qty_num NUMBER(2);
  5    lv_sub_num NUMBER(5,2);
  6    lv_days_num NUMBER(3);
  7    lv_shopper_num NUMBER(3) := 25;
  8  BEGIN
  9    SELECT idBasket, dtcreated, quantity, subtotal
 10      INTO lv_basket_num, lv_created_date, lv_qty_num, lv_sub_num
 11      FROM bb_basket
 12      WHERE idShopper = lv_shopper_num
 13        AND orderplaced = 0;
 14    lv_days_num = SYSDATE - lv_created_date;
 15    DBMS_OUTPUT.PUT_LINE(lv_basket_num||' * '||lv_created_date||' * '||
 16                          lv_qty_num||' * '||lv_sub_num||' * '||lv_days_num);
 17  END;
 18  /
  lv_days_num = SYSDATE - lv_created_date;
              *
ERROR at line 14:
ORA-06550: line 14, column 14:
PLS-00103: Encountered the symbol "=" when expecting one of the following:
:= . ( @ % ;
ORA-06550: line 15, column 2:
PLS-00103: Encountered the symbol "DBMS_OUTPUT"
ORA-06550: line 16, column 74:
PLS-00103: Encountered the symbol ";" when expecting one of the following:
. ( , * % & - + / at mod rem <an identifier>
<a double-quoted delimited-identifier> <an exponent (**)> as
from into || bulk

SQL>
```

Identifies that an error was encountered

Figure 2-6 Assignment statement error message

When multiple errors are listed in SQL*Plus, you should resolve the first one listed and rerun the block. Many times, the resolution of the first error corrects the other errors, which usually result from the first error offsetting the statements in the block.

11. Locate the file **scalarC02.txt** in the Chapter.02 folder.

12. Open the file in a text editor. Note that the same block is used, except the LV_BASKET_NUM variable is misspelled in the INTO clause.

13. Copy all the code.

14. Go to SQL*Plus and click **Edit**, and then click **Paste**.

15. Press **Enter**, if necessary, to execute the block. The error message should appear, as shown in Figure 2-7. The error message "identifier 'LV_BASKET_NU' must be declared" is common because it is easy to either forget to declare a variable or to misspell a variable name.

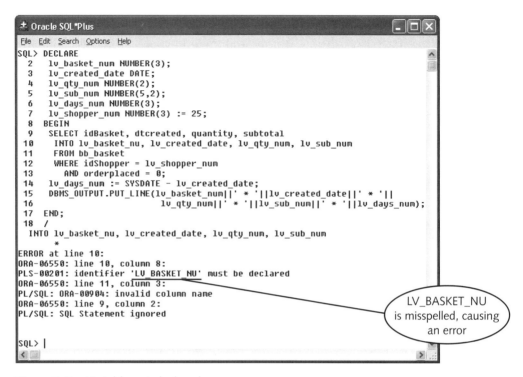

Figure 2-7   Variable not declared error message

16. Locate the file **scalarD02.txt** in the Chapter.02 folder.

17. Open the file in a text editor. Note that the same block is used, except the semicolon is left off the end of the LV_DAYS_NUM assignment statement.

18. Copy all the code.

19. Return to SQL*Plus. Click **Edit**, and then click **Paste**.

20. Press **Enter**, if necessary, to execute the block. The error message shown in Figure 2-8 is displayed. When an error occurs at the beginning of a line that starts a new statement, the cause is often that the previous statement was not terminated with the appropriate semicolon.

21. Locate the file **scalarE02.txt** in the Chapter.02 folder.

22. Open the file in a text editor. Note that the same block is used except the shopper id supplied is 33 and there is no shopper 33 in the database.

23. Copy and paste all the code into SQL*Plus.

24. Press **Enter**, if necessary, to execute the block. The error message shown in Figure 2-9 is displayed. The important part of this error message is the reference to no data found. A SELECT statement that is embedded into a PL/SQL block must return a row or it generates this error.

25. Close SQL*Plus.

```
± Oracle SQL*Plus                                                    □□✕
File  Edit  Search  Options  Help
SQL> DECLARE
  2    lv_basket_num NUMBER(3);
  3    lv_created_date DATE;
  4    lv_qty_num NUMBER(2);
  5    lv_sub_num NUMBER(5,2);
  6    lv_days_num NUMBER(3);
  7    lv_shopper_num NUMBER(3) := 25;
  8  BEGIN
  9    SELECT idBasket, dtcreated, quantity, subtotal
 10     INTO lv_basket_num, lv_created_date, lv_qty_num, lv_sub_num
 11     FROM bb_basket
 12     WHERE idShopper = lv_shopper_num
 13       AND orderplaced = 0;
 14    lv_days_num := SYSDATE - lv_created_date
 15    DBMS_OUTPUT.PUT_LINE(lv_basket_num||' * '||lv_created_date||' * '||
 16                         lv_qty_num||' * '||lv_sub_num||' * '||lv_days_num);
 17  END;
 18  /
  DBMS_OUTPUT.PUT_LINE(lv_basket_num||' * '||lv_created_date||' * '||
  *
ERROR at line 15:
ORA-06550: line 15, column 2:
PLS-00103: Encountered the symbol "DBMS_OUTPUT" when expecting one of the
following:
. ( * @ % & = - + ; < / > at in is mod not rem
<an exponent (**)> <> or != or ~= >= <= <> and or like
between ||
The symbol "." was substituted for "DBMS_OUTPUT" to continue.

SQL>
```

Figure 2-8   "Not closing previous statement" error

```
± Oracle SQL*Plus                                                    □□✕
File  Edit  Search  Options  Help
SQL> DECLARE
  2    lv_basket_num NUMBER(3);
  3    lv_created_date DATE;
  4    lv_qty_num NUMBER(2);
  5    lv_sub_num NUMBER(5,2);
  6    lv_days_num NUMBER(3);
  7    lv_shopper_num NUMBER(3) := 33;
  8  BEGIN
  9    SELECT idBasket, dtcreated, quantity, subtotal
 10     INTO lv_basket_num, lv_created_date, lv_qty_num, lv_sub_num
 11     FROM bb_basket
 12     WHERE idShopper = lv_shopper_num
 13       AND orderplaced = 0;
 14    lv_days_num := SYSDATE - lv_created_date;
 15    DBMS_OUTPUT.PUT_LINE(lv_basket_num||' * '||lv_created_date||' * '||
 16                         lv_qty_num||' * '||lv_sub_num||' * '||lv_days_num);
 17  END;
 18  /
DECLARE
*
ERROR at line 1:
ORA-01403: no data found
ORA-06512: at line 9

SQL> |
```

Figure 2-9   "No data found" error from a SELECT statement

Note that the last example demonstrating the "no data found" error is a run-time error versus a syntax error. That is, the code syntax in the block is correct, but an error occurred due to the results of the execution.

WORKING WITH HOST OR BIND VARIABLES

PL/SQL blocks use host or bind variables to move values from an application environment into the PL/SQL block for processing. In our example concerning the noncompleted basket, the shopper number would not be hard coded into the PL/SQL block. Why? Our goal would be to reuse this block of code so that we can check for a non-completed basket for anyone who logs onto the Brewbean's site. Therefore, the WHERE clause should be written to use a host variable, as shown in the following code:

```
WHERE idShopper = :g_shopper
        AND orderplaced = 0;
```

The host or bind variable (also referred to as a global variable) :G_SHOPPER is simply a value that we can reference in our block but that is coming from our application environment. With them, we can also send values back to screens or the front end to display information on the user's screen. Note that the colon preceding the variable name instructs PL/SQL that this is a host variable, so this variable is not declared within the block.

We use the VARIABLE command to create host variables. Keep in mind the VARIABLE command creates SQL*Plus variables (not PL/SQL variables) that will serve as our host variables because this is the environment we are using to execute our blocks. Let's try using a host variable to provide the shopper number.

To test a host variable in a PL/SQL block:

1. Open or return to SQL*Plus and type **VARIABLE g_shopper NUMBER**, and then press **Enter**. This creates our host variable in SQL*Plus. Notice there is not a semicolon at the end of the VARIABLE statement. No semicolon is needed because the code that you typed is an SQL*Plus command, not an SQL or PL/SQL statement. No SQL*Plus feedback results from this statement.

2. For the block to run properly, a value needs to be placed into the host variable. We accomplish this by using an anonymous PL/SQL block that consists only of an assignment statement. Type the following anonymous block, and then press **Enter**.

```
BEGIN
  :g_shopper := 25;
END;
/
```

3. Locate the file **host02.txt** in the Chapter.02 folder.

4. Open the file in a text editor. Note that the same block is used except the host variable is now used in the WHERE clause of the SELECT statement to provide a shopper number.

5. Copy all the code and paste all the code into SQL*Plus.

6. If necessary, press **Enter** to execute the block. The results should display the information from the DBMS_OUTPUT statement.

7. Close SQL*Plus.

The three main types of host variables we will create in SQL*Plus are NUMBER, CHAR, and VARCHAR2. The DATE and BOOLEAN data type does not exist for SQL*Plus variables. It is important to note that the NUMBER data type does not accept any size information but the character data types do. After you create an SQL*Plus variable, it will exist for your entire session or, in other words, for as long as you are logged on. Therefore, host variables are sometimes referred to as session variables and can be used by multiple blocks that you execute during a session.

USING THE %TYPE ATTRIBUTE

Another important item to mention regarding the example block is that all the variables declared, with the exception of the LV_DAYS_NUM variable, are created specifically for holding values retrieved from database columns. When creating variables to hold database column values, the %TYPE attribute could be used in the variable declaration to provide the data type. The %TYPE attribute tells the system to look up the data type of a database column and use this data type for the declared variable. This is referred to as an **anchored data type**.

The modified block using the %TYPE attribute in the variable declarations would look like the following:

```
DECLARE
  lv_basket_num bb_basket.idBasket%TYPE;
  lv_created_date bb_basket.dtcreated%TYPE;
  lv_qty_num bb_basket.quantity%TYPE;
  lv_sub_num bb_basket.subtotal%TYPE;
  lv_days_num NUMBER(3);
BEGIN
  SELECT idBasket, dtcreated, quantity, subtotal
   INTO lv_basket_num, lv_created_date, lv_qty_num,
        lv_sub_num
   FROM bb_basket
   WHERE idShopper = :g_shopper
     AND orderplaced = 0;
  lv_days_num := SYSDATE - lv_created_date;
  DBMS_OUTPUT.PUT_LINE(lv_basket_num||' * '||
     lv_created_date||' * '||lv_qty_num||' * '||
     lv_sub_num||' * '||lv_days_num);
END;
/
```

The use of the %TYPE in the first variable declaration instructs the system to go to the BB_BASKET TABLE, look up the IDBASKET column, and return the data type. This data type declaration is advantageous for two reasons. First, the programmer does not have to look up the data type to be sure to accommodate the value correctly with the declared variable. Second, if changes are made to the database structure, such as making a column longer in length, the programmer does not have to be concerned with making any changes in all the variable declarations. Minimizing maintenance of code in this fashion is always a goal of the programmer. On the downside, there is a slight performance hit in that the database server must look up the data type from the data dictionary.

WORKING WITH COMPOSITE DATA TYPES

The scalar variables are quite valuable; however, many times, our programs need to handle logical groups of data, such as an entire row from a database table or a number of rows being input from an application screen. To handle these types of situations, PL/SQL offers composite data types that allow the creation of a variable that can hold multiple values with various data types as a single unit to make these tasks easier to manage.

In PL/SQL, developers often use the terms "composite data type" and "collection" interchangeably to indicate a variable that handles multiple values as a single unit. However, in this text, we differentiate between these two terms as follows:

- A **composite data type** is one that can store and handle multiple values of different data types as one unit.

- A **collection** is a variable that can store and handle multiple values of the same data type as one unit.

The main difference between the two is that composite variables hold a variety of data types, whereas a collection holds values that all have the same data type. In this section, we explore the two composite data types.

Record Data Type

A **record** data type is quite similar to the structure of a row in a database table, and a variable declared with a record data type can hold one row of data consisting of a number of column values. A row of data typically includes a number of different fields. For a record variable, we must construct our own data type using a TYPE statement that indicates the fields that will be included in the record and their associated data types. After we have created the data type, we can then declare a variable with that data type. Note that declaring a composite variable is different from declaring a scalar variable in that we must create our own data types.

The syntax layout of the TYPE statement is as follows:

```
TYPE type_name IS RECORD (
    field_1_name data type [NOT NULL] [:= default_value],
    ...
    field_n_name data type [NOT NULL] [:= default_value] );
```

Let's return to our block example we constructed for the Brewbean's application to retrieve a noncompleted basket that a shopper has saved. Previously, we created four scalar variables to hold the values retrieved from the BB_BASKET table for a particular shopper. The following code shows the modifications needed to use a record variable to handle the data in the query. Note that the new code is in bold to make it easy to identify:

```
DECLARE
  TYPE type_basket IS RECORD (
    basket bb_basket.idBasket%TYPE,
    created bb_basket.dtcreated%TYPE,
    qty bb_basket.quantity%TYPE,
    sub bb_basket.subtotal%TYPE);
  rec_basket type_basket;
  lv_days_num NUMBER(3);
  lv_shopper_num NUMBER(3) := 25;
BEGIN
  SELECT idBasket, dtcreated, quantity, subtotal
   INTO rec_basket
   FROM bb_basket
   WHERE idShopper = lv_shopper_num
     AND orderplaced = 0;
  lv_days_num := SYSDATE - rec_basket.created;
END;
/
```

Notice that the TYPE statement (starting in the second line) creates a record data type named TYPE_BASKET that will hold four values—one date and three numbers. The REC_BASKET variable is then declared using the TYPE_BASKET data type we just created. The INTO clause of the SELECT statement now only contains one variable, the REC_BASKET record variable, which will hold all four values returned by the query. Note that the column data values are moved into the record fields in a positional manner. For example, the value of the first column, IDBASKET, is copied to the first field defined in the data type creation, which is the basket field. Be sure that the order of columns in the SELECT statement matches the order of the fields in the record variable.

Because we now only have one variable, how do we reference each of the four different values that it holds? Review the LV_DAY_NUM assignment statement that must reference the date created value returned by the query to calculate the age in days of the basket, as listed in the following code:

```
lv_days_num := SYSDATE - rec_basket.created;
```

Using a *record_variable_name.field_name* notation on the right side of :=, you can reference each individual field of the record variable. In our example, the date created value is referenced as REC_BASKET.CREATED with REC_BASKET being the record variable declared and "created" being one of the fields defined in the record data type by the TYPE statement.

Let's try our example with the addition of the DBMS_OUTPUT statement to verify the values held in the record variable.

To use a record variable:

1. Locate the file **record02.txt** in the Chapter.02 folder.

2. Open the file in a text editor. Notice that the block declares a record variable, as shown in the previous code example. DBMS_OUTPUT statements have been added to print out the values retrieved from the database.

3. Copy and paste all the code into SQL*Plus.

4. Press **Enter**, if necessary, to execute the block. The results from the DBMS_OUTPUT statements will appear as shown in the following example. (Note that the LV_DAYS_NUM variable value is dependent on the current date.)

```
12
19-FEB-03
7
72.4
9
```

5. Close SQL*Plus.

In the preceding example, each of the four fields contained in the record variable are displayed individually with DBMS_OUTPUT statements to verify the correct values were moved to the expected field in the record. In addition, notice that the TYPE statement that creates the record data type uses the %TYPE attribute to define the data types of each field in the record variable. It uses the %TYPE attribute because the variable will be used to hold data retrieved from those database table columns. Any data type we have used for scalar variables could be used here as well.

The Role of the %ROWTYPE Attribute

In our noncompleted basket example, only a few columns from the BB_BASKET table were being retrieved. However, what if we want to retrieve most or all of the columns of a row from a table? We could create a data type with a TYPE statement to accomplish the task. However, we can accomplish the task more efficiently by using the %ROWTYPE attribute that reviews the table structure regarding column names and data types and creates the record data type based on this information. In essence, the system creates the record data type for us based on the table structure information.

Let's consider a screen in the Brewbean's application in which the %ROWTYPE feature can be used. Shoppers on the site will have a profile saved to avoid having to reenter their names, addresses, and contact information. A screen will be available to users that displays all their profile information and allows them to make necessary edits (such as an address change). The profile information displayed includes almost all of the columns from the BB_SHOPPER table. In this case, we could use the %ROWTYPE feature to simplify the code with the creation of a record variable to hold the row of data queried for a shopper.

The following block accomplishes this task:

```
DECLARE
 rec_shopper bb_shopper%ROWTYPE;
BEGIN
 SELECT *
  INTO rec_shopper
  FROM bb_shopper
  WHERE idshopper = :g_shopper;
 DBMS_OUTPUT.PUT_LINE(rec_shopper.lastname);
 DBMS_OUTPUT.PUT_LINE(rec_shopper.address);
 DBMS_OUTPUT.PUT_LINE(rec_shopper.email);
END;
/
```

The preceding block assumes a host variable named :G_SHOPPER exists in the application environment. Recall that in SQL*Plus, we create host variables with the VARIABLE command.

Now, we only have one statement in the DECLARE section to create a record variable. No TYPE statement is needed to create the data type because the %ROWTYPE now instructs the system to automatically do this based on the table structure. Notice that the column names from the table will be used as the field names of the record variable, and a SELECT * statement is used to move an entire row of data from a table to the record. The %ROWTYPE feature can greatly simplify the maintenance of code related to database structure modifications in that we do not have a TYPE statement that needs to be modified as the columns and data types are retrieved from the data dictionary.

Table of Records

A **table of records** is another type of composite data type and it is the same as a record data type except that it can handle more than one record or row of data. To see why we might need a table of records data type, let's look at what happens in a block that retrieves more than one row with a SELECT statement while using scalar variables. Review the execution errors displayed in Figure 2-10.

```
Oracle SQL*Plus
File  Edit  Search  Options  Help
SQL> DECLARE
  2    lv_basket_num NUMBER(3);
  3    lv_created_date DATE;
  4    lv_qty_num NUMBER(2);
  5    lv_sub_num NUMBER(5,2);
  6    lv_days_num NUMBER(3);
  7    lv_shopper_num NUMBER(3) := 26;
  8  BEGIN
  9    SELECT idBasket, dtcreated, quantity, subtotal
 10     INTO lv_basket_num, lv_created_date, lv_qty_num, lv_sub_num
 11     FROM bb_basket
 12    WHERE idShopper = lv_shopper_num
 13      AND orderplaced = 0;
 14    lv_days_num := SYSDATE - lv_created_date;
 15    DBMS_OUTPUT.PUT_LINE(lv_basket_num||' * '||lv_created_date||' * ||
 16                    lv_qty_num||' * '||lv_sub_num||' * '||lv_days_num);
 17  END;
 18  /
DECLARE
*
ERROR at line 1:
ORA-01422: exact fetch returns more than requested number of rows
ORA-06512: at line 9


SQL> |
```

Figure 2-10   Error using scalar variables with multiple row select

The error references Line 9, which contains the SELECT statement. The message "exact fetch returns more than requested number of rows" indicates that the variables in the INTO clause of the query cannot handle multiple rows of data.

Let's think about another part of the Brewbean's coffee ordering application. As a shopper is perusing the site and selecting items to purchase, the Brewbean's application needs a variable to hold information on all the items selected until the shopper indicates a desire to save the cart for later completion or to complete the purchase (at which time the data is inserted into the database). In this scenario, we need a variable that holds data similar to a row of the BB_BASKETITEM table. In addition, the shopper might select more than one item; therefore, the variable needs to be able to handle more than one row of data. A table of records would handle both possibilities.

First, review the application screen that the PL/SQL block will support. Figure 2-11 displays that screen. On the screen, if a user clicks the Add to Cart button to select a product, then the application needs to add this product selection to the table of records variable.

We need to create a number of host variables to use in our block because the product selection data is coming from the application environment. In addition, we need to create a host variable that stores a number to indicate the row number of the table of records that should be added. Keep in mind, we now have a variable that can hold multiple rows and multiple fields. Therefore, when placing data into the table of records variable, we must indicate not only which field, but also which row into which to insert the values. Figure 2-12 displays the needed host variable declaration and initialization.

Figure 2-11   User screen in which shopper can add an item to his or her cart

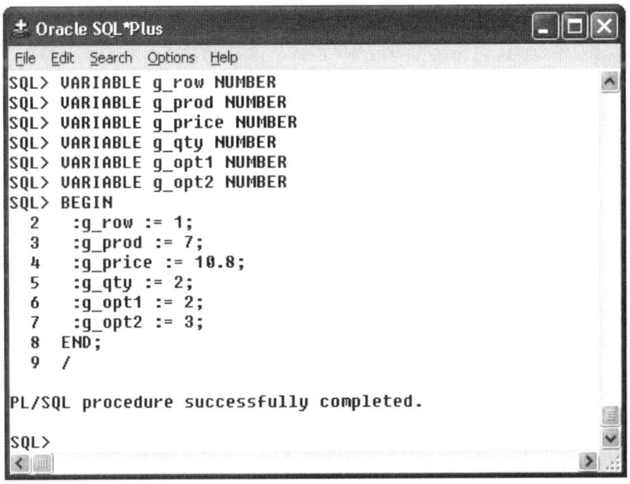

Figure 2-12   Host variable declaration and initialization

Now, let's review the executed block in Figure 2-13 that provides the needed processing to add a selected item to a table of records variable.

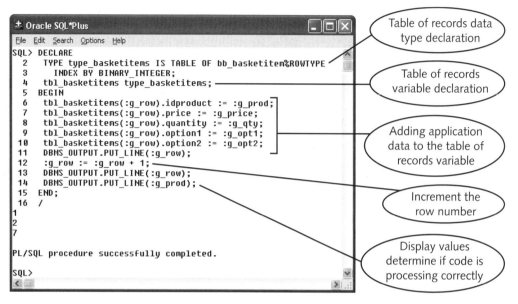

Figure 2-13   Using a table of records variable

First, a variable data type is created with the TYPE command containing the key words IS TABLE OF BB_BASKETITEM%ROWTYPE to provide a record data type—just as we used in creating a record variable. A record data type that we have created could also be used in this statement. The INDEX BY BINARY_INTEGER creates a primary key index that is used to reference the particular rows of the table of records variable. Again, keep in mind that our data type must set up not only different fields of data to handle, but also the ability to handle multiple rows. The table of records variable named TBL_BASKETITEMS is then declared just as we would any variable—by indicating a variable name and data type. In the example, the table of records data type named TYPE_BASKETITEMS is used.

The body or executable section of the block needs to place the application variables into a row of the TBL_BASKETITEMS variable. In the following code, notice the structure of the assignment statements that are used to accomplish this task:

```
tbl_basketitems(:g_row).idproduct := :g_prod;
```

The code :G_ROW indicates in which row of the table of records the data will be added and .IDPRODUCT indicates the field into which the value will be placed. The :G_ROW value is also incremented by one so that the next row added will be placed in the next row of the table of records variable. Right about now, you might be asking "Don't the variables in a PL/SQL block exist only during the execution of the block?

If so, what happens to the table of records variable in this example?" Well, those are good questions. It is true that variables declared within a block last only as long as the block is executing. In other words, these variables do not persist beyond the block execution. So, the question now becomes, "How can we resolve the problem in our example so that the table of records variable will persist for a user session?" To do this, we need to declare the table of records variable in a package specification; this declaration allows variables to persist for a user session.

Discussing packages does raise the important issue of the persistence of the variables. At times, the need will arise within an application to have variables that persist across the execution of a number of PL/SQL blocks. This need is typical of a table of records variable.

PROCESSING WITH IF STATEMENTS AND LOOPING ACTIONS

Your next step is to understand simple IF statements, basic loops, and FOR loops. These are very simple processing statements; you'll need them to help you understand data handling.

Simple IF Statements

In any programming language, you need to be able to test for the existence of various conditions. The IF statement is one feature that allows the developer to check values and determine what actions to take based on the values.

Review the syntax layout of an IF statement:

```
IF condition THEN statement END IF;
```

The IF portion of the statement lists the value to check for and the THEN portion provides the action to take if the IF portion is true.

Let's consider an IF statement we could use in the Brewbean's application. Earlier in the chapter, we created a block that would retrieve a noncompleted basket if a shopper had one saved. What if we needed to check whether the basket was more than 7 days old and, if it was, to send a message that the basket has expired and cannot be used? We can accomplish this task by using an IF statement, as shown in the following code snippet. In the code, assume that the LV_CREATED variable holds the date created value retrieved from the database for the basket.

```
IF   SYSDATE-lv_created > 7   THEN
     lv_flag := 'Expired';
END IF;
```

Loops: Basic and FOR

Loops enable the repeating of a statement or set of statements. One of the most important parts of a loop is ensuring that proper instructions have been included to stop the loop iteration when appropriate. The structure of the basic loop is shown in the following code. Note that the EXIT WHEN clause must provide a condition that at some point will evaluate to TRUE. This evaluation to TRUE is how the loop will stop.

```
LOOP
    - - Statements - -
    EXIT WHEN condition;
END LOOP;
```

The FOR loop completes the same job of iterating; however, this type of loop indicates how many times to loop by providing a range at the beginning of the statement. The structure of the FOR loop is shown in the following code:

```
FOR counter IN lower_bound..upper_bound LOOP
    - - Statements - -
END LOOP;
```

The counter is a variable that holds the value of the current iteration number and is automatically incremented by 1 each time the loop iterates. The counter begins with the value supplied as the lower bound and the iterations continue until the counter reaches the value supplied for the upper bound. Even though any name can be used for the counter, it is typical to use an "i". The lower and upper bounds can be numbers or variables containing numeric values.

Let's turn to a situation in the Brewbean's application and see how a loop may simplify our code. Assume that we have a table of records variable that is holding all the items a shopper has placed into the cart and we need to insert this data into the BB_BASKETITEM table when the shopper elects to check out. The following partial block shows the use of a FOR loop to iterate through the rows in a table of records variable:

```
FOR i IN 1..lv_cnt LOOP
  INSERT INTO bb_basketitem (idBasketitem,idProduct,
                             idBasket, price, quantity)
    VALUES (bi_seq.NEXTVAL, tbl_cart(i).prod,
      tbl_cart(i).bask, tbl_cart(i).price, tbl_cart(i).qty)
  ;
END LOOP;
```

The LV_CNT variable holds the number of rows in the table of records. If LV_CNT contains a value of 5, the loop will iterate 5 times. Note that i will start at 1 and increment by 1 on each loop iteration. When the i is set to the value 5, the statements in the loop will run once more and then the loop ceases iterating. The counter variable is used in the INSERT statement to indicate the table row to be used for the input values.

WORKING WITH COLLECTIONS

2

A **collection** is an ordered group of elements that allows the handling of multiple values of the same data type as a single unit. Collections are similar to arrays used on other languages. A collection may hold many rows of data but only a single field. The values in each row of the collection must be of the same type and an index allows references to individual values or rows within the collection.

We focus on the collection type called index-by tables in this section of the chapter. Other available collections, VARRAYS and nested tables, are briefly mentioned at the end of this section but are beyond the scope of this text.

Index-by Tables

An **index-by table** was commonly called a PL/SQL table in earlier Oracle versions and is a variable that can handle many rows of data but only one field. The index-by table is essentially the same as a table of records except that it holds only a single column of data. Table 2-3 contains a list of the main characteristics of the index-by table.

Table 2-3   Index-by table characteristics

Characteristic	Description
One-dimensional	Can have only one column.
Unconstrained	Rows added dynamically as needed.
Sparse	A row only exists when a value is assigned. Rows do not have to be assigned sequentially.
Homogeneous	All elements have the same data type.
Indexed	Integer index serves as primary key of the table.

Recognize that index-by tables are not physical tables in the database. They are variables used to hold and manipulate data within PL/SQL programs. Therefore, we cannot perform SQL commands on these tables.

Declaring an index-by table data type is quite similar to declaring a table of records data type. First, create a data type defining the table structure that consists of one column and an index to reference the rows. Second, declare a table using the data type.

Table attributes are functions that can be used in conjunction with table variables and allow greater ability to manipulate table values. Table 2-4 lists available table attributes and a description of each.

Table 2-4   PL/SQL Index-by table attributes

Attribute Name	Value Data Type	Description
COUNT	NUMBER	Number of rows in the table
DELETE	None	Removes a row from the table
EXISTS	BOOLEAN	TRUE if specified row does exist
FIRST	BINARY_INTEGER	Index for the first row in the table
LAST	BINARY_INTEGER	Index for the last row in the table
NEXT	BINARY_INTEGER	Index for the next row in the table after the row specified
PRIOR	BINARY_INTEGER	Index for the row in the table before the row specified

The FIRST, LAST, NEXT, and PRIOR attributes are available to enable movement through the table data. COUNT returns the total number of values in the table, which is quite valuable when using a loop. EXISTS checks if a value has been entered for the stated index number. These functions are referenced using a dot notation, as shown in the assignment statement in the following code:

```
v_rows := tbl_orders.COUNT;
```

In this statement, TBL_ORDERS represents an index-by table variable, and the .COUNT attribute notation instructs the system to return the total count of rows that exist in the table variable. This value is held in a PL/SQL scalar variable named V_ROWS.

Let's look at an example of using an index-by table variable. Brewbean's roasts their own coffee beans and, as a quality assurance test to monitor the performance of the roasting equipment, sample weight measurements are taken for each batch. Depending on the size of the batch, an employee records four or five one-cup weight measurements. The average of the measurements is saved to a database table that holds the average weight for each batch. In this scenario, Brewbean's has an application screen that allows an employee to record all the sample measurements. The application then calculates the average and inserts the result into the appropriate table. The PL/SQL block to support this application screen uses an index-by table to hold all the sample measurements for processing.

Figure 2-14 displays the creation and initialization of host variables to represent the weight measurements entered into the application screen. The figure also shows the PL/SQL block that calculates the average measurement.

The first statements of the executable section of the block place the values of the host variables into the table variable. Because the number of measurements entered vary by batch size, the IF statements are used only to add rows to the table variable for host variables that actually contain a value or measurement. Notice that only the row number needs to be indicated when entering a value into an index-by table variable because there is only a single column available in the variable.

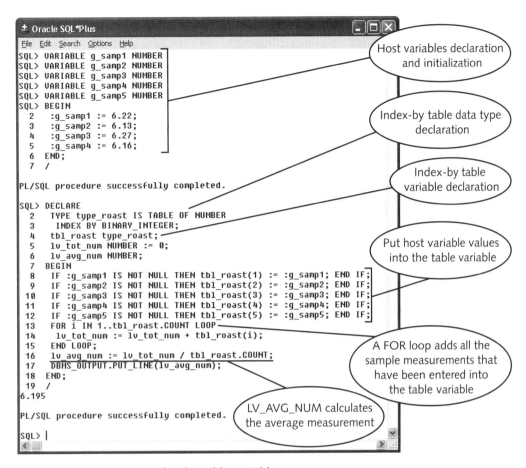

Figure 2-14   Using an index-by table variable

In Figure 2-14, note that the FOR loop iterates through the items in the table variable to sum the values of all the measurements. In addition, because the number of measurements entered may vary, the FOR loop iteration is controlled by using the COUNT attribute of the table variable. At the end, the assignment statement for LV_AVG_NUM calculates the average of the measurements by using the sum calculated in the FOR loop and the COUNT attribute of the table variable. Using the COUNT attribute in this manner makes this block flexible to handle any number of measurements that are entered for the sample.

Let's run this example while adding a few more DBMS_OUTPUT statements to look at some of the other table attribute values.

To create a block using an index-by table:

     1. Open or return to SQL*Plus.

2. Enter the following code to create and initialize the needed host variables:

```
VARIABLE g_samp1 NUMBER
VARIABLE g_samp2 NUMBER
VARIABLE g_samp3 NUMBER
VARIABLE g_samp4 NUMBER
VARIABLE g_samp5 NUMBER
BEGIN
 :g_samp1 := 6.22;
 :g_samp2 := 6.13;
 :g_samp3 := 6.27;
 :g_samp4 := 6.16;
END;
/
```

3. Locate the file **ibtable02.txt** in the Chapter.02 folder.

4. Open the file in a text editor. In the bottom rows of the file, notice a DBMS_OUTPUT statement is used to display the average weight calculated.

5. Copy and paste all the code into SQL*Plus.

6. Press **Enter**, if necessary, to execute the block. The average weight of 6.195 should be displayed on the screen.

7. Close SQL*Plus.

Other Collections: VARRAYS and Nested Tables

With nested tables and VARRAYS, we can now have database columns that hold more than one value. One major advantage with using collections as a part of the physical database is being able to retrieve multiple values with a query of a single column. An example in which we might use a nested table could be a column that is created to hold customer preference choices. A single column on the customer table could be created as a nested table and hold any number of preference values. The main differences between a nested table and a VARRAY are that a VARRAY has a set size upon creation, the order of elements is preserved, and the data is stored inline with the table data.

WORKING WITH CURSORS

A **cursor** represents a work area or section of memory in which an SQL statement is being processed in the Oracle server. This memory area is also referred to as the context area. Cursors provide a powerful mechanism in regards to handling multiple rows of data retrieved with an SQL query.

Implicit and explicit are two types of SQL statement cursors available. **Implicit cursors** are declared automatically for all DML and SELECT statements issued within a PL/SQL block. **Explicit cursors** are declared and manipulated in the PL/SQL block code for handling a set of rows returned by a SELECT statement. In addition, there are also **cursor variables**, which are references or pointers to a work area. We discuss each in turn.

Implicit Cursors

When an SQL statement is executed, the Oracle server creates an implicit cursor automatically. This cursor is a work area in memory where the statement is processed and that contains the results of the SQL statement. Cursor attributes allow the results of an SQL statement to be checked in regards to if the statement affected any rows and how many. The attributes that are available are listed in Table 2-5.

Table 2-5   SQL Cursor attributes

Attribute Name	Data Type	Description
%ROWCOUNT	Number	Number of rows affected by the SQL statement
%FOUND	Boolean	TRUE if at least one row is affected by the SQL statement, otherwise FALSE
%NOTFOUND	Boolean	TRUE if no rows are affected by the SQL statement, otherwise FALSE

If an UPDATE statement is processed, you can check how many rows were affected by referencing SQL%ROWCOUNT, or check if any rows were affected by looking for a TRUE value in SQL%FOUND. For example, the programmer can check the results of the UPDATE statement via cursor attributes to confirm that at least one row was updated—if this is a requirement of the application. Keep in mind that an update may affect zero or many rows and none of these cases raises an Oracle error. Note that the attribute references (ROWCOUNT, FOUND) are preceded by SQL%. This instructs the system to look at the implicit SQL cursor area, which contains information for the most recent SQL statement processed.

Let's look at an example in the Brewbean's application. The Brewbean's application includes an inventory update screen for employees recording product shipments received. The employee can enter a product id and shipment quantity to be added to the stock data. The block supporting this screen includes an UPDATE statement on the BB_PRODUCT table. Figure 2-15 displays the execution of this block.

```
± Oracle SQL*Plus                          [_][□][X]
File  Edit  Search  Options  Help
SQL> BEGIN                                          ▲
  2    UPDATE bb_product                            ▤
  3      SET stock = stock + 25
  4      WHERE idProduct = 15;
  5    DBMS_OUTPUT.PUT_LINE(SQL%ROWCOUNT);
  6    IF SQL%NOTFOUND THEN
  7      DBMS_OUTPUT.PUT_LINE('Not Found');
  8    END IF;
  9  END;
 10  /
0
Not Found

PL/SQL procedure successfully completed.

SQL>                                                ▼
◄ ▥                                      ►  .::
```

Figure 2-15   Using SQL attributes

 Note For example purposes, the product id and quantity are coded as numeric values in the UPDATE statement. In reality, both of these values are provided via host variables.

Notice that two SQL attributes are referenced in the block. The SQL%ROWCOUNT returns the value of zero because no rows were updated (in other words, product id 15 does not exist). The SQL%NOTFOUND evaluated to TRUE, also indicating no rows were affected by the UPDATE statement.

One important item to consider when referencing SQL attributes is that the attribute information always reflects the information from the most recent SQL statement processed. Therefore, when we have multiple SQL statements in a block, we need to consider where we check the attribute values. Let's expand the previous example and add a SELECT statement and verify that the %ROWCOUNT attribute changes to reflect each SQL statement. Notice the block executed in Figure 2-16 demonstrates that the %ROWCOUNT is 1 after the SELECT and 0 after the UPDATE.

Explicit Cursors

You can process multiple rows of data from a database by creating explicit cursors. Explicit cursors are cursors that are declared with the declaration containing a SELECT statement. If we need to process a group of rows retrieved from a database, this is the way to go! Remember how we developed a PL/SQL block that would retrieve noncompleted shopping cart information for a shopper to support the screen in Figure 2-4? Let's review this code and see what happens when more than one noncompleted basket exists in the database for a shopper, as shown in Figure 2-17. Notice an error occurs because the scalar variables cannot handle more than one row being returned from a query. We can resolve this by using an explicit cursor that allows the easy handling of multiple rows returned from a query.

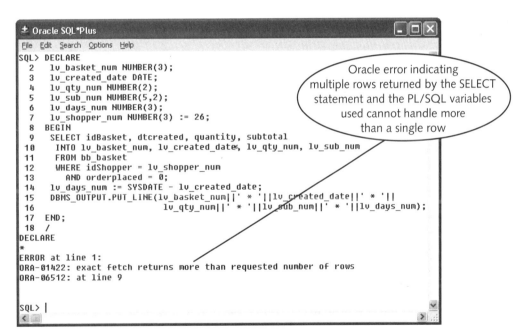

Figure 2-16   The SQL attributes reflect the most recent SQL statement

```
± Oracle SQL*Plus                                          [_][□][X]
File  Edit  Search  Options  Help
SQL> DECLARE
  2    lv_basket_num NUMBER(3);
  3    lv_created_date DATE;
  4    lv_qty_num NUMBER(2);
  5    lv_sub_num NUMBER(5,2);
  6    lv_days_num NUMBER(3);
  7    lv_shopper_num NUMBER(3) := 26;
  8  BEGIN
  9    SELECT idBasket, dtcreated, quantity, subtotal
 10      INTO lv_basket_num, lv_created_date, lv_qty_num, lv_sub_num
 11      FROM bb_basket
 12      WHERE idShopper = lv_shopper_num
 13        AND orderplaced = 0;
 14    lv_days_num := SYSDATE - lv_created_date;
 15    DBMS_OUTPUT.PUT_LINE(lv_basket_num||' * '||lv_created_date||' * ||
 16                         lv_qty_num||' * '||lv_sub_num||' * '||lv_days_num);
 17  END;
 18  /
DECLARE
*
ERROR at line 1:
ORA-01422: exact fetch returns more than requested number of rows
ORA-06512: at line 9

SQL> |
```

> Oracle error indicating multiple rows returned by the SELECT statement and the PL/SQL variables used cannot handle more than a single row

Figure 2-17   Scalar variables cannot handle multiple row queries

Using an explicit cursor involves several steps, as listed in Table 2-6.

The steps include declaring the cursor in the DECLARE section of a block and then OPEN, FETCH, and CLOSE activities occur in the executable or BEGIN section.

Table 2-6   Actions in using an explicit cursor

Step	Step Activity	Activity Description
1	DECLARE	Creates a named cursor identified by a SELECT statement. The SELECT statement does not include an INTO clause. Values in the cursor are moved to PL/SQL variables with the FETCH step.
2	OPEN	Processes the query and create the active set of rows available in the cursor.
3	FETCH	Retrieves a row from the cursor into block variables. Each consecutive FETCH issued retrieves the next row in the cursor until all rows have been retrieved.
4	CLOSE	Clears the active set of rows and frees the memory area used for the cursor.

Let's work through an example to see how an explicit cursor is set up and processed. Brewbean's sells both equipment and coffee and all in-state orders must be taxed, with different tax rates applied to equipment and coffee items. Review the block in Figure 2-18 that uses a cursor to retrieve all the items in a basket and calculate the total tax for the order by applying the correct rates to equipment and coffee items.

Note that not only was a cursor declared, but a record variable was also declared to hold values returned with the FETCH action. The declaration includes a SELECT statement just as we would issue a stand-alone query directly in SQL*Plus with no INTO clause. In addition, notice that a host variable is referenced in the WHERE clause of the cursor. This is typical to make the cursor dynamic or, in other words, a different set of rows may be retrieved each time the SELECT statement executes depending on user input. PL/SQL block variables can also be referenced within the cursor declaration, just be sure the variable values have been set before the OPEN statement executes. The values of host and PL/SQL variables at the time the OPEN is issued will be used to create the active set of rows in the cursor.

The OPEN command prompts the SELECT statement of the cursor to be processed and an active set of rows created within the cursor. The FETCH is used to move one of the rows in the cursor into a record variable to become available for processing in the block. A list of scalar variables could be used instead of a record variable.

The cursor contains a pointer that will automatically keep track of the next row to be returned by a FETCH and it is initially set to the first row returned from the query. The pointer moves to the next row in the active set upon each consecutive FETCH statement. The FETCH is accomplished in a loop so that one row at a time is processed from the cursor and the same processing logic is applied to each row. The EXIT WHEN clause instructs the loop to stop after all the rows of the cursor have been fetched. Notice this checks the cursor attribute %NOTFOUND that would be TRUE if no more records remain in the cursor. So, in this case, we are processing all the rows in the cursor. Then, we are finished with the cursor, so the CLOSE command is issued to free up associated resources.

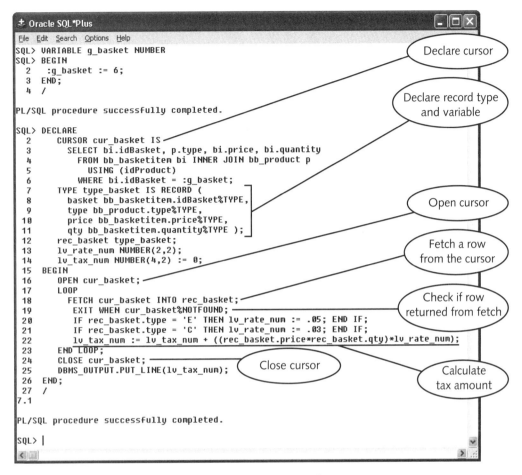

Figure 2-18 Using an explicit cursor to process rows from a query

Using the explicit cursors described allows complete control over every aspect of processing the cursor. For example, what if a maximum order tax amount existed? In this case, a statement could be added in the loop to exit after the LV_TAX_NUM variable reached this maximum. At times, this could mean that not all the rows in the cursor need to be processed, only the ones needed. Let's run an example that implements a maximum tax amount of $5.

Use an explicit cursor to calculate tax amount:

1. Open or return to SQL*Plus.

2. Add another item in basket 6 by typing the following and then pressing **Enter**:

```
INSERT INTO bb_basketitem
    VALUES (44, 8, 10.80, 1, 6, 2, 3);
```

3. Check the items in basket 6 by typing the following code and then pressing **Enter**. Three rows or items should be displayed, as shown in Figure 2-19.

```
SELECT bi.idBasket, p.type, bi.price, bi.quantity
  FROM bb_basketitem bi INNER JOIN bb_product p
   USING (idProduct)
  WHERE bi.idBasket = 6;
```

Figure 2-19   List of items in basket 6

4. Locate the file **excurloop02.txt** in the Chapter.02 folder.

5. Open the file in a text editor. Notice an IF statement is added to the bottom of the loop to check if the maximum $5 limit has been reached.

6. Copy and paste all the code into SQL*Plus.

7. Press **Enter**, if necessary, to execute. Two amounts will be output. First, a 2 is displayed to indicate the cursor ROWCOUNT. Second, a 5 is displayed to indicate the maximum tax amount was reached.

8. Close SQL*Plus.

To check the ROWCOUNT attribute of the explicit cursor, the reference uses cursor name followed by a percent symbol rather than SQL%, which is used with implicit cursors. Notice this cursor attribute was checked with a DBMS_OUTPUT statement before the CLOSE cursor statement was issued. If we attempt to check an explicit cursor attribute after the CLOSE, an Oracle error is raised stating that the cursor does not exist. In this case, the ROWCOUNT was 2 indicating that was the last row returned with a FETCH statement before closing the cursor. This confirms that only two of the three rows for basket 6 were processed in this block because the $5 maximum tax amount was met after only two items. Therefore, keep in mind that ROWCOUNT of an implicit cursor always returns the total rows in the cursor, whereas ROWCOUNT of an explicit cursor reflects how many rows have been fetched.

Now that we know what implicit and explicit cursors are and how they work, let's look at a couple more cursor considerations including CURSOR FOR loops and using subqueries and parameters with cursors.

CURSOR FOR Loop

In many instances, you will need to retrieve rows from a database and perform processes on each of the rows including data manipulation commands. An example of this might be a process for calculating new salaries in which each row of the EMPLOYEE table needs to be reviewed for job class and years employed to apply the correct percentage raise and then perform an UPDATE to store the new salary in the table. A CURSOR FOR loop is particularly suited to handle this very task quite easily.

The CURSOR FOR loop makes this task easier in that it handles many of the explicit cursor actions automatically. These actions include creating a record variable, opening the cursor, looping through one row at a time until the last row is retrieved from the cursor, and then closing the cursor automatically.

In the new salary calculation case, we also need to perform an UPDATE on each row to change the salary in the database. However, we are looking at rows from a cursor, so how do we perform the update on the database? Well, the CURSOR FOR loop contains a feature that allows you to instruct the Oracle server to keep track of which physical database row corresponds to each row in the cursor. This feature is invoked by using the FOR UPDATE clause in the cursor declaration. This instructs the system to lock the rows retrieved with the SELECT statement because you intend to issue updates via the cursor, and to keep track of the physical row of the database table to which each row in the cursor corresponds. In addition, the WHERE CURRENT OF cursor clause will be added to the UPDATE statement to instruct the system to update the physical row of the table that corresponds to the row of the cursor that is currently being processed. The FOR UPDATE and WHERE CURRENT OF clauses work together to simplify UPDATE activity via a cursor.

Let's look at an example with the Brewbean's application. The BB_PRODUCT table contains a column named SALEPRICE that holds prices on associated sale items. The Brewbean's manager wants to do some analysis on sales pricing and has requested that a sale price be loaded for every product. For coffee products, which have a type of C, the sale price should be set 10 percent below the regular price. For equipment products, which have a type of E, the sale price should be set five percent below the regular price. A sale price should be entered only for the products that are currently active; currently active products have a code of 1 in the active column. Review the following code, which uses a CURSOR FOR loop to calculate and update the sale price (bold is used on the new code):

```
DECLARE
  CURSOR cur_prod IS
    SELECT type, price
    FROM bb_product
    WHERE active = 1
    FOR UPDATE NOWAIT;
  lv_sale bb_product.saleprice%TYPE;
```

```
BEGIN
  FOR rec_prod IN cur_prod LOOP
    IF rec_prod.type = 'C' THEN lv_sale :=
                                rec_prod.price * .9;
    END IF;
    IF rec_prod.type = 'E' THEN lv_sale :=
                                rec_prod.price * .95;
    END IF;
    UPDATE bb_product
      SET saleprice = lv_sale
      WHERE CURRENT OF cur_prod;
  END LOOP;
  COMMIT;
END;
/
```

The first statement in the declaration section creates a cursor named CUR_PROD. Notice the last line in the cursor declaration is a FOR UPDATE clause, which instructs the server to track the physical database row that relates to each row in the cursor to be able to achieve the UPDATE action through the cursor. The FOR UPDATE clause ends with the NOWAIT option, which controls what will occur if the rows being retrieved by the cursor are already locked by another session. If you do not use the NOWAIT option, the statement would wait indefinitely for the rows to unlock and become available.

Using the NOWAIT option raises the Oracle error shown in the following code example, if the desired rows are currently locked. This enables the application code to let the user know that the rows are currently not available and allows the user to choose to continue with other tasks. This is valuable when dealing with databases that have many users potentially vying to perform DML activities on the same rows.

```
ORA-00054 resource busy and acquire with NOWAIT specified
```

The first line of the executable section sets up the CURSOR FOR loop. Oracle automatically creates a record variable with the name supplied as the second item in this clause or REC_PROD. Notice that we did not declare a record variable. Oracle does this for us by using the table information from the database for the columns being selected in the cursor. The record variable is created to hold one row of the cursor at a time through the loop. The execution also automatically exits the loop after all the rows have been retrieved from the cursor. The UPDATE statement ends with the WHERE CURRENT OF CUR_PROD clause, which instructs the system that this update should actually occur on the physical database row that corresponds to the current cursor row. After we complete the transaction, the last action in the block is a COMMIT to save the DML actions permanently.

Let's try running this example and checking the results.

To execute the CURSOR FOR loop block:

1. Locate the file **curloop02.txt** in the Chapter.02 folder.

2. Open the file in a text editor.

3. Copy and paste all the code into SQL*Plus.

4. Press **Enter**, if necessary, to execute the block.

5. Verify the block execution by querying the BB_PRODUCT table, as shown in Figure 2-20, and reviewing the sale prices.

6. Close SQL*Plus.

```
± Oracle SQL*Plus                                    _ □ X
File  Edit  Search  Options  Help
SQL> SELECT idproduct, type, saleprice, price
  2     FROM bb_product
  3     WHERE active = 1;

 IDPRODUCT T  SALEPRICE        PRICE
---------- - ---------- ----------
         1 E      94.99        99.99
         2 E     123.49       129.99
         3 E      30.88         32.5
         4 E      27.08         28.5
         5 C       9.45         10.5
         6 C          9           10
         7 C       9.72         10.8
         8 C       9.72         10.8
         9 C          9           10
        10 C          9           10

10 rows selected.

SQL> |
```

Figure 2-20   Check CURSOR FOR loop results

We should always attempt to minimize row locking in an application in an effort to minimize user waits for rows. Therefore, it is important to be aware that the FOR UPDATE clause in a cursor causes all the table rows related to the cursor rows to be locked upon the opening of the cursor. If the cursor involves multiple tables but the UPDATES do not affect all of the tables, an OF option can be used to minimize the rows locked. For example, review the cursor declaration in the following code that includes an OF option in the FOR UPDATE clause (bold is used to show the new code):

```
CURSOR cur_prod IS
   SELECT s.idShopper, s.promo, b.total
    FROM bb_shopper s INNER JOIN bb_basket b
     USING (idShopper)
   FOR UPDATE OF s.promo NOWAIT;
```

The OF S.PROMO part of the FOR UPDATE clause instructs the system that only the PROMO column is needed for the UPDATE action and, therefore, only the BB_SHOPPER table rows need to be locked. This avoids also locking the BB_BASKET table rows, which are not needed for the UPDATE. If more than one column is needed in the UPDATE, a list of columns separated by commas can be used in the OF option.

Cursors with Subqueries and Parameters

Cursors can also utilize subqueries and parameters to be more dynamic. Subqueries are included in the SELECT statement in the cursor declaration just as they are normally added to an SQL statement. Parameters are values passed into the cursor when opened and used in the SELECT statement of the cursor to determine what data the cursor will contain. The partial block in the following code displays a cursor that uses a parameter, P_BASKET, to pass in the basket number to the SELECT statement in a cursor:

```
DECLARE
   CURSOR cur_order (p_basket NUMBER) IS
     SELECT idBasket, idProduct, price, quantity
       FROM bb_basketitem
       WHERE idBasket = p_basket;
BEGIN
   OPEN cur_order(:g_bask_1);
   ...
   OPEN cur_order(:g_bask_2);
   ...
END;
```

Notice the parameter name of P_BASKET, created in the first line of the cursor statement along with a data type that contains no size information. This parameter is then referenced in the WHERE clause of the SELECT statement. Therefore, this cursor could retrieve a different set of rows every time it is opened, as shown in the executable section of this block. This example uses host variables to supply the P_BASKET parameter with a value; however, this could come from other sources such as local variables in the block. Only one parameter is used in this case; however, many parameters are allowed and included by listing them in the cursor declaration separated by commas.

If more than one parameter is used, the values passed in when the cursor is opened work in a positional fashion. The first value listed in the open cursor command is put into the first parameter listed in the cursor declaration, and so forth. Passing parameters can also be accomplished with a CURSOR FOR loop by listing the values to pass into the parameters after the cursor name in the FOR LOOP statement. The FOR LOOP statement in the following code shows the host variable :G_BASK_1 will be supplied to the cursor parameter when the cursor is opened:

```
FOR rec_order IN cur_order(:g_bask_1) LOOP
```

Cursor Variables

An explicit cursor assigns a name to a work area holding a specific result set, whereas a cursor variable is simply a pointer to a work area in which a query can be processed. Implicit and explicit cursors are considered static because they are associated with specific queries. Because the cursor variable is just a pointer to a work area, this work area could be used for different queries. One of the important benefits of the cursor variable is the ability to more efficiently pass the result sets of queries. This method passes the pointer—rather than all the data—to the cursor.

Creating a cursor variable includes a TYPE statement for a REF CURSOR and a variable declaration using this data type. The processing is handled like an explicit cursor using the OPEN, FETCH, and CLOSE statements. However, the OPEN statement for a cursor variable provides the query to be processed.

Assume that Brewbean's had one table in the database for coffee products and another table for equipment products. The shopper indicates which product line he or she wants to peruse, and a host variable is used to identify this selection within a PL/SQL block. A different query would be processed in the cursor depending on the product line the shopper chose. The following code demonstrates the use of a cursor variable to accomplish this task:

```
DECLARE
   TYPE type_curvar IS REF CURSOR;
   cv_prod type_curvar;
   rec_coff  bb_coffee%ROWTYPE;
   rec_equip bb_equip%ROWTYPE;
BEGIN
   IF :g_choice = 1 THEN
     OPEN cv_prod FOR SELECT * FROM bb_coffee;
     FETCH cv_prod INTO rec_coff;
     . . .
   END IF;
   IF :g_choice = 2 THEN
     OPEN cv_prod FOR SELECT * FROM bb_equip;
     FETCH cv_prod INTO rec_equip;
     . . .
   END IF;
END;
```

In the declaration section, the REF CURSOR data type named TYPE_CURVAR is first declared followed by a variable named CV_PROD of that type. The other two declarations create record variables that hold the rows fetched from the cursor. In the executable section, the OPEN statement now indicates not only a cursor name but also the query to be processed for the cursor. Notice that IF statements are used to determine the correct OPEN statement to run based on the user choice of which products to view. The same cursor is used in both OPEN statements; however, it holds different data depending on which OPEN is executed because the query is different in each.

WORKING WITH VARIABLE SCOPE

Blocks can be nested in PL/SQL, which raises issues regarding variable scope. **Variable scope** is the area of a program block that can identify a particular variable. A nested block can be placed in a block wherever an executable statement is allowed, which includes the BEGIN and EXCEPTION sections of a block. A block first looks for a variable within itself. If the variable cannot be located and it is a nested block, the block searches the enclosing block for the referenced variable. Figure 2-21 displays a block containing a nested block.

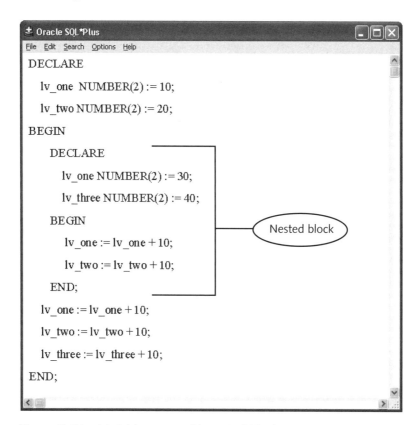

Figure 2-21   Variable scope with nested blocks

A nested block can look up into the enclosing block; however, the reverse is not true. The enclosing block cannot look down into the nested block.

Let's run the variable scope example block with DBMS_OUTPUT statements to check how the assignment statements in the executable section resolve variable references. Notice that a variable named LV_ONE is declared in both the enclosing and the nested blocks. The variable LV_TWO is declared only in the enclosing block and LV_THREE only in the nested block. Figure 2-22 displays the first attempted execution that produces an error.

```
± Oracle SQL*Plus                                              [_][□][X]
File  Edit  Search  Options  Help
SQL> DECLARE
  2       lv_one  NUMBER(2) := 10;
  3       lv_two NUMBER(2) := 20;
  4   BEGIN
  5         DECLARE
  6             lv_one NUMBER(2) := 30;
  7             lv_three NUMBER(2) := 40;
  8         BEGIN
  9             lv_one := lv_one + 10;
 10             lv_two := lv_two + 10;
 11           DBMS_OUTPUT.PUT_LINE('Nested lv_one = '||lv_one);
 12           DBMS_OUTPUT.PUT_LINE('Nested lv_two = '||lv_two);
 13           DBMS_OUTPUT.PUT_LINE('Nested lv_three = '||lv_three);
 14         END;
 15       lv_one := lv_one + 10;
 16       lv_two := lv_two + 10;
 17       lv_three := lv_three + 10;
 18       DBMS_OUTPUT.PUT_LINE('Enclosing lv_one = '||lv_one);
 19       DBMS_OUTPUT.PUT_LINE('Enclosing lv_two = '||lv_two);
 20       DBMS_OUTPUT.PUT_LINE('Enclosing lv_three = '||lv_three);
 21   END;
 22   /
     lv_three := lv_three + 10;
     *
ERROR at line 17:
ORA-06550: line 17, column 5:
PLS-00201: identifier 'LV_THREE' must be declared
ORA-06550: line 17, column 5:
PL/SQL: Statement ignored
ORA-06550: line 20, column 51:
PLS-00201: identifier 'LV_THREE' must be declared
ORA-06550: line 20, column 5:
PL/SQL: Statement ignored

SQL>
```

Figure 2-22   *Variable reference error*

The error occurs in the enclosing block at the point where the LV_THREE variable is referenced. A variable named LV_THREE was declared only in the nested block and the enclosing block cannot see the variables within the nested block, so it is considered undeclared. Let's comment out the LV_THREE variable reference to examine the results of the other variable references, as shown in Figure 2-23.

Table 2-7 reviews each of the assignment statements in Figure 2-23 to explain how each instance of the variable scope works.

Figure 2-23   Variable scope example

Table 2-7   Discussion of variable scope in Figure 2-23

Block	Assignment Statement	Result
Nested	`lv_one := lv_one + 10;`	LV_ONE variable declared in the nested block is used and now has a value of 40. LV_ONE in the enclosing block was not affected.
Nested	`lv_two := lv_two + 10;`	A variable named LV_TWO does not exist in the nested block. The block looks at the enclosing block and does the calculation on the LV_TWO in the enclosing block. It does not move a copy of this variable to the nested block; it just uses the variable in the enclosing block. Therefore, now the enclosing block, as well as the nested block, will see a value of 30 in the LV_TWO variable.
Enclosing	`lv_one := lv_one + 10;`	LV_ONE is declared in the main block and now has a value of 20.

Table 2-7   Discussion of variable scope in Figure 2-23 (continued)

Block	Assignment Statement	Result
Enclosing	`lv_two := lv_two + 10;`	LV_TWO exists in the main block and has a value of 30 from the change the nested block did in the earlier assignment statement. This statement now makes the value 40.
Enclosing	`lv_three := lv_three + 10;`	LV_THREE does not exist in the enclosing block and the enclosing block cannot see the variables in a nested block, so this statement raises an error.

CHAPTER SUMMARY

- A PL/SQL block can contain DECLARE, BEGIN, EXCEPTION, and END sections. The BEGIN and END sections are required.

- Variables are named memory areas that hold values to allow retrieval and manipulation of values within our programs.

- Variables to hold and manipulate values within the block are created in the DECLARE section.

- Scalar variables can hold a single value. Common data types are VARCHAR2, NUMBER, DATE, and BOOLEAN.

- Variable naming rules are the same as those of database objects, including beginning with an alpha character, and can contain up to 30 characters.

- At a minimum, variable declaration must include a name and data type.

- Variables can be initialized with a value in the DECLARE section using the PL/SQL assignment operator.

- A NOT NULL option can be used in a variable declaration to require the variable to always contain a value. These variables must be initialized.

- A CONSTANT option can be used in a variable declaration to enforce that the value cannot be changed in the block. These variables must be initialized.

- The DBMS_OUTPUT.PUT_LINE statement is used to display values to the screen.

- A SELECT statement within a PL/SQL block must return at least one row of data or the error "no data found" will occur.

- Host or bind variables are used as a mechanism to move values from our application environment into the PL/SQL block for processing. They are also used to return values to the application environment.

❑ When variables are used to hold data from database tables, the %TYPE attribute can be used to retrieve the appropriate data type from the data dictionary based on the table and column indicated.

❑ Composite data types allow the creation of a variable that can hold multiple values with various data types as a single unit to make these tasks easier to manage.

❑ A record is a composite data type that can hold a row of data; a table of records can hold multiple rows of data.

❑ The %ROWTYPE attribute can be used to declare a record variable data type based on a table.

❑ A collection is a variable that can store and handle multiple values of the same data type as one unit.

❑ An index-by table is a collection that can handle many rows of data but only one field.

❑ A cursor represents a work area or section of memory in which an SQL statement is being processed in the Oracle server.

❑ Implicit cursors are declared automatically for all DML and SELECT statements issued within a PL/SQL block.

❑ Explicit cursors are cursors that are declared with the declaration containing a SELECT statement. Four processing steps are used: DECLARE, OPEN, FETCH, and CLOSE.

❑ CURSOR FOR loops simplify processing a set of rows from the database. If an UPDATE action is necessary, the FOR UPDATE and WHERE CURRENT OF clauses will assist.

❑ Parameters can make cursors more dynamic. Parameters are values passed into the cursor when opened and used in the SELECT statement of the cursor to determine what data the cursor will contain.

❑ A cursor variable is a pointer to a work area in which a query can be processed and is declared with the REF CURSOR clause.

❑ With nested PL/SQL blocks, the nested block can reference variables in the enclosing block; however, the enclosing block cannot see the variables in the nested block.

REVIEW QUESTIONS

1. Which of the following variable declarations is illegal?

    a. lv_junk NUMBER(3);

    b. lv_junk NUMBER(3) NOT NULL;

    c. lv_junk NUMBER(3) := 11;

    d. lv_junk NUMBER(3) DEFAULT 11;

2

2. Which of the following is a possible value for a BOOLEAN variable? (Choose all that apply.)

   a. TRUE

   b. FALSE

   c. BLANK

   d. NULL

3. What type of variable can store only one value?

   a. implicit cursor

   b. scalar

   c. %ROWTYPE

   d. explicit cursor

4. What does the %TYPE attribute instruct the server to do?

   a. retrieve the database column data type for this variable

   b. copy a variable

   c. retrieve data from the database

   d. use a BOOLEAN data type

5. Which item is a valid reference to a value in a record variable?

   a. `rec_junk(1)`

   b. `rec_junk(1).col`

   c. `rec_junk.col`

   d. `rec_junk.col(1)`

6. A table of records variable can hold _____.

   a. only one row and many columns of data

   b. many columns and only one row of data

   c. many rows and many columns of data

   d. none of the above

7. Which statement allows you to check the number of rows affected by an UPDATE statement?

   a. `SQL%FOUND`

   b. `SQL%NOTFOUND`

   c. `SQL%COUNT`

   d. `SQL%ROWCOUNT`

8. When should you use the %ROWTYPE attribute in creating a record variable?

   a. when using most of the columns from a table

   b. when using only a small portion of the columns from a table

   c. when creating a record variable

   d. none of the above

9. The FOR UPDATE clause in a cursor declaration instructs the server to

   _____ .

   a. execute the cursor

   b. issue an SQL UPDATE statement

   c. keep track of the physical database row that is related to each row in the cursor

   d. allow transaction control statements

10. Which of the following is a valid reference to a value in an index-by table?

    a. `tbl_junk(1).col`

    b. `tbl_junk.col`

    c. `tbl_junk(1)`

    d. `tbl_junk.col(1)`

11. What are variables and why are they needed?

12. In what way is a SELECT statement different when issued in a PL/SQL block?

13. Describe the differences between implicit and explicit cursors.

14. Describe how a CURSOR FOR loop makes cursor processing easier.

15. Define a composite data type and name two available in PL/SQL.

ADVANCED REVIEW QUESTIONS

1. Review the following block. What value will be displayed by the DBMS_OUTPUT statement?

```
DECLARE
   lv_junk1 CHAR(1) := 'N';
   lv_junk2 CHAR(1) := 'N';
BEGIN
   lv_junk2 := 'Y';
   DECLARE
      lv_junk2 := 'N';
   BEGIN
      lv_junk1 := 'Y';
   END;
   DBMS_OUTPUT.PUT_LINE(lv_junk2);
END;
```

a. Y

b. N

c. NULL

d. This block would raise an error.

2. A DECLARE section of a block is shown in the following code. What type of variable is V_JUNK?

```
DECLARE
    TYPE type_junk IS TABLE OF CHAR(1)
        INDEX BY BINARY_INTEGER;
    v_junk type_junk;
```

a. scalar

b. table of records

c. index-by table

d. cursor

3. In a CURSOR FOR loop, which command is used to open the cursor?

a. FETCH

b. OPEN

c. FOR loop

d. The cursor is opened implicitly by a CURSOR FOR loop.

4. Which of the following is not true regarding CURSOR FOR loops?

a. A record variable must be declared to hold a row of the cursor.

b. Fetching rows is handled implicitly by the loop.

c. Opening the cursor is handled implicitly by the loop.

d. No exit condition is needed to end the looping action.

5. What would be used in the data type creation of a record variable that needs to hold all the column values from the shopper table?

a. %TYPE

b. %ROWTYPE

c. list of columns

d. %ROWCOUNT

HANDS-ON ASSIGNMENTS

Assignment 2-1: Use Scalar Variables

An application screen is being developed for Brewbean's that will allow employees to enter a basket number and view the shipping information for that order to include date, shipper, and shipping number. An IDSTAGE value of 5 in the BB_BASKETSTATUS table indicates the order has been shipped. Let's create the block using scalar variables to hold the data retrieved from the database.

To create a block to check shipping information:

1. Open SQL*Plus.

2. Type **SET SERVEROUTPUT ON** and press **Enter**.

3. Create and initialize a host variable by typing the following code and then pressing **Enter**:

```
VARIABLE g_basket NUMBER
BEGIN
 :g_basket := 3;
END;
/
```

4. Locate the file **assignment02-01.txt** in the Chapter.02 folder.

5. Open the file in a text editor. Review the block of code and note the use of scalar variables to hold the values retrieved in the SELECT statement.

6. Copy and paste all the code into SQL*Plus.

7. Press **Enter**, if necessary, to execute the block. The screen should appear similar to Figure 2-24.

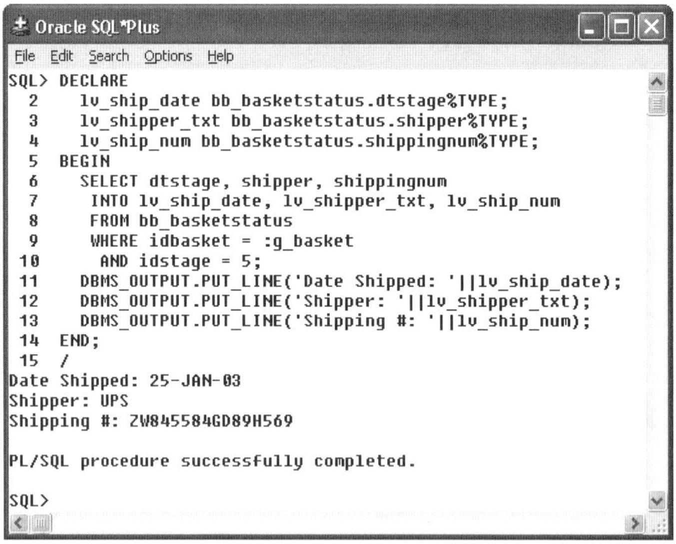

```
SQL> DECLARE
  2     lv_ship_date bb_basketstatus.dtstage%TYPE;
  3     lv_shipper_txt bb_basketstatus.shipper%TYPE;
  4     lv_ship_num bb_basketstatus.shippingnum%TYPE;
  5  BEGIN
  6    SELECT dtstage, shipper, shippingnum
  7     INTO lv_ship_date, lv_shipper_txt, lv_ship_num
  8     FROM bb_basketstatus
  9     WHERE idbasket = :g_basket
 10       AND idstage = 5;
 11    DBMS_OUTPUT.PUT_LINE('Date Shipped: '||lv_ship_date);
 12    DBMS_OUTPUT.PUT_LINE('Shipper: '||lv_shipper_txt);
 13    DBMS_OUTPUT.PUT_LINE('Shipping #: '||lv_ship_num);
 14  END;
 15  /
Date Shipped: 25-JAN-03
Shipper: UPS
Shipping #: ZW845584GD89H569

PL/SQL procedure successfully completed.

SQL>
```

Figure 2-24   Using scalar variables to retrieve shipping information

8. Now let's try to run this same block using a basket id that has no shipping information recorded. In SQL*Plus, type the following code to change the value of the host variable, and then press **Enter**.

```
BEGIN
  :g_basket := 7;
END;
/
```

9. Return to the text editor and copy the block of code that you copied in Step 6. (Note that this code will likely still be in memory. The step is repeated for clarity.)

10. Return to SQL*Plus and paste the code into the window.

11. Press **Enter**, if necessary, to execute the block. The screen should appear similar to Figure 2-25.

12. Close SQL*Plus.

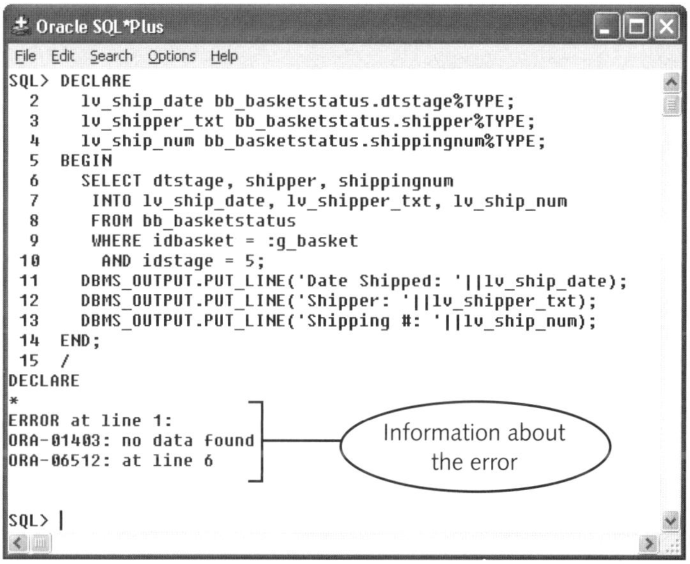

Figure 2-25   No data found error

Assignment 2-2: Use a Record Variable

An application screen is being developed for Brewbean's that will allow employees to enter a basket number and view the shipping information for that order. The screen needs to display all the column values from the BB_BASKETSTATUS table. An IDSTAGE value of 5 in the BB_BASKETSTATUS table indicates the order has been shipped. Let's create the block using a record variable to hold the row of data retrieved from the database.

To create a block with a record variable:

1. Open SQL*Plus.

2. Type **SET SERVEROUTPUT ON** and press **Enter**.

3. Create and initialize a host variable by typing the following code and then pressing **Enter**:

```
VARIABLE g_basket NUMBER
BEGIN
 :g_basket := 3;
END;
/
```

4. Locate the file **assignment02-02.txt** in the Chapter.02 folder.

5. Open the file in a text editor. Review the block of code and note the use of a record variable to hold the values retrieved in the SELECT statement. Also note how individual values of the record variable are referenced in the · DBMS_OUTPUT statements.

6. Copy and paste all the code into SQL*Plus.

7. Press **Enter**, if necessary, to execute the block. The screen should appear similar to Figure 2-26.

8. Close SQL*Plus.

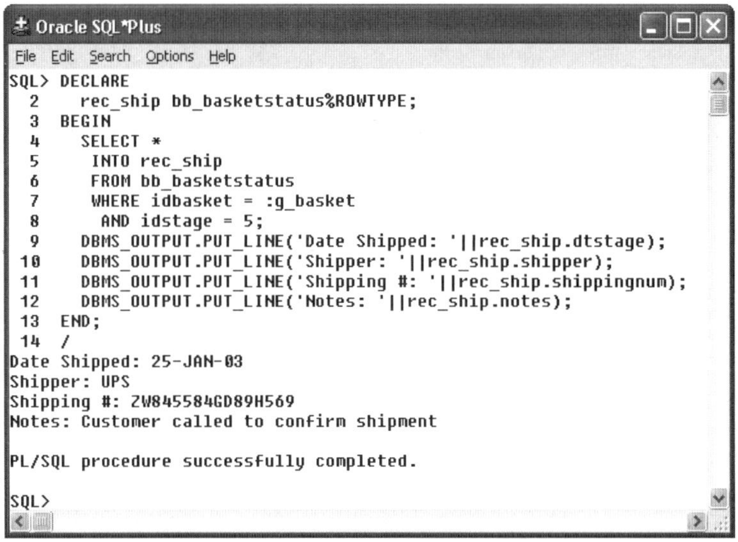

Figure 2-26   Using a record variable to retrieve shipping information

Assignment 2-3: Use an Explicit Cursor

In the Brewbean's application, a customer can request to check if all items in his or her cart are in stock. Create a block that uses an explicit cursor to retrieve all the items in the basket and determine if all items are in stock by comparing the item quantity to the product stock amount. If all items are in stock, send a message "All items in stock!" to the screen. If not, send a message "All items NOT in stock!" to the screen. The basket number will be provided via a host variable.

To use an explicit cursor to check product stock:

1. Open SQL*Plus.
2. Type **SET SERVEROUTPUT ON** and press **Enter**.
3. Create and initialize a host variable by typing the following code and then pressing **Enter**:

   ```
   VARIABLE g_basket NUMBER
   BEGIN
    :g_basket := 6;
   END;
   /
   ```

4. Locate the file **assignment02-03.txt** in the Chapter.02 folder.
5. Open the file in a text editor. Review the block of code and note both a cursor and a record variable is created in the DECLARE section. Also, the cursor must be manipulated with explicit actions of OPEN, FETCH, and CLOSE. A local variable named LV_FLAG_TXT is used to hold the status of the stock check.
6. Copy and paste all the code into SQL*Plus.
7. Press **Enter**, if necessary, to execute the block. The screen should appear similar to Figure 2-27.
8. Close SQL*Plus.

```
Oracle SQL*Plus
File  Edit  Search  Options  Help
SQL> DECLARE
  2      CURSOR cur_basket IS
  3        SELECT bi.idBasket, bi.quantity, p.stock
  4          FROM bb_basketitem bi INNER JOIN bb_product p
  5            USING (idProduct)
  6          WHERE bi.idBasket = :g_basket;
  7      TYPE type_basket IS RECORD (
  8        basket bb_basketitem.idBasket%TYPE,
  9        qty bb_basketitem.quantity%TYPE,
 10        stock bb_product.stock%TYPE);
 11      rec_basket type_basket;
 12      lv_flag_txt CHAR(1) := 'Y';
 13  BEGIN
 14      OPEN cur_basket;
 15      LOOP
 16        FETCH cur_basket INTO rec_basket;
 17          EXIT WHEN cur_basket%NOTFOUND;
 18          IF rec_basket.stock < rec_basket.qty THEN lv_flag_txt := 'N'; END IF;
 19      END LOOP;
 20      CLOSE cur_basket;
 21      IF lv_flag_txt = 'Y' THEN DBMS_OUTPUT.PUT_LINE('All items in stock!'); END IF;
 22      IF lv_flag_txt = 'N' THEN DBMS_OUTPUT.PUT_LINE('All items NOT in stock!'); END IF;
 23  END;
 24  /
All items in stock!

PL/SQL procedure successfully completed.

SQL>
```

Figure 2-27   Using an explicit cursor to verify items in stock

Assignment 2-4: Use a CURSOR FOR loop

Brewbean's wants to send a promotion via e-mail to shoppers. If a shopper has purchased more than $50 at the site, the shopper will receive a $5 off coupon for his or her next purchase over $25. If the shopper has spent more than $100, he or she will receive a free shipping coupon.

The BB_SHOPPER table contains a column named PROMO, which needs to hold the appropriate promotion code. Follow the steps in this assignment to create a block using a CURSOR FOR loop to check the total spent by each shopper and update the PROMO column in the BB_SHOPPER table accordingly. Note that the cursor SELECT statement contains a subquery in the FROM clause to retrieve the shopper totals. This is done because a cursor using a GROUP BY statement cannot use the FOR UPDATE clause because its results are summarized data versus individual rows from the database.

To create a block with a CURSOR FOR loop:

1. Open SQL*Plus.
2. Locate the file **assignment02-04.txt** in the Chapter.02 folder.
3. Open the file in a text editor. Review the block of code and note the SELECT in the cursor declaration that uses a subquery. Also note that because an UPDATE is to be accomplished, the FOR UPDATE and WHERE CURRENT OF clauses are used.
4. Copy and paste all the code to SQL*Plus.
5. Press **Enter**, if necessary, to execute the block.
6. Run a query, as shown in Figure 2-28, to check the results.
7. Close SQL*Plus.

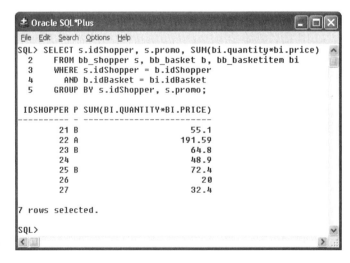

Figure 2-28   Query the BB_SHOPPER table to check the PROMO column

Assignment 2-5: Use Implicit Cursors

The BB_SHOPPER table in the Brewbean's database contains a column named PROMO that holds code to indicate promotions to send to shoppers. This column needs to be cleared after the promotion has been sent. First, open the file named **assignment02–05.txt** in the Chapter.02 folder. Run the UPDATE and COMMIT statements at the top of this file (not the anonymous block at the end). Modify the anonymous block so that it will display the number of rows updated by the block on the screen. Run the block.

Assignment 2-6: Use Variable Scope

The anonymous block in the file named **assignment02–06.txt** contains a nested block. Add DBMS_OUTPUT statements to display the values of the LV_ONE and LV_TWO variables both at the end of the nested block and at the end of the enclosing block. Run the block and explain the results.

Assignment 2-7: Use Scalar Variables for Data Retrieval

The Brewbean's application contains a screen that displays order summary information including IDBASKET, SUBTOTAL, SHIPPING, TAX, and TOTAL columns from the BB_BASKET table. Create a PL/SQL block that will use scalar variables to retrieve this data and then display it on the screen. A host variable should provide the IDBASKET value. Test the block using IDBASKET 12.

Assignment 2-8: Use a Record Variable for Data Retrieval

The Brewbean's application contains a screen that displays order summary information including IDBASKET, SUBTOTAL, SHIPPING, TAX, and TOTAL columns from the BB_BASKET table. Create a PL/SQL block that will use a record variable to retrieve this data and display it on the screen. A host variable should provide the IDBASKET value. Test the block using IDBASKET 12.

CASE PROJECTS

Case 2-1: Variable Types

The Brewbean's manager has just hired another programmer to assist you in developing the application code for their coffee business. Explain to the new employee the difference between scalar, record, and table variables.

Case 2-2: More Movie Rentals

As business is becoming strong and the movie stock is growing for the More Movie Rentals, the manager wants to do more inventory evaluations. One item of interest concerns any movie for which the company is holding $75 or more in value. The manager wants to focus on these movies in regards to their revenue generation to ensure the stock level is warranted. To make these stock queries more efficient, the application team decides that a column should be added to the MM_MOVIE table named STK_FLAG that will hold an '*' if the stock value is $75 or more. Otherwise, the value should be NULL. Add the needed column and create an anonymous block containing a CURSOR FOR loop to accomplish this task. The company will run this program monthly to update the STK_FLAG column before the inventory evaluation.

3

PL/SQL PROCESSING

In this chapter, you will:

♦ Use control structures to make decisions

♦ Use looping structures to repeat code

♦ Work with the GOTO statement

♦ Manage errors with exception handlers

♦ Address exception-handling issues, such as RAISE_APPLICATION_ERROR and propagation

♦ Document code with comments

This chapter introduces procedural statements available in PL/SQL that enable statement-processing flow. We begin with IF and CASE statements to address decision-making tasks. In addition, loops are explored to handle situations in which we need to repeat the same group of statements. This chapter also covers the basic, FOR, and WHILE looping structures.

Further in the chapter, you learn how GOTO statements can control the flow of execution in a block. We wrap up the chapter with the realization that no matter how talented we are, there is always potential for errors to occur in an application. Thus, this chapter introduces exception handling to manage application errors within our code.

THE CURRENT CHALLENGE IN THE BREWBEAN'S APPLICATION

Now that we know how to handle various forms of data within a PL/SQL block, we need to be able to perform some logical processing with the data. For example, Figure 3-1 displays a screen from the Brewbean's application in which a shopper can view all of his or her basket items. When the shopper clicks the Check Out link on this screen, the application needs to perform some decision-making tasks, such as identifying the tax rate to be applied based on the shipping state. To accomplish this, you need logical processing statements that allow decision making to occur within the code. You learn about logical processing statements—and more—throughout this chapter.

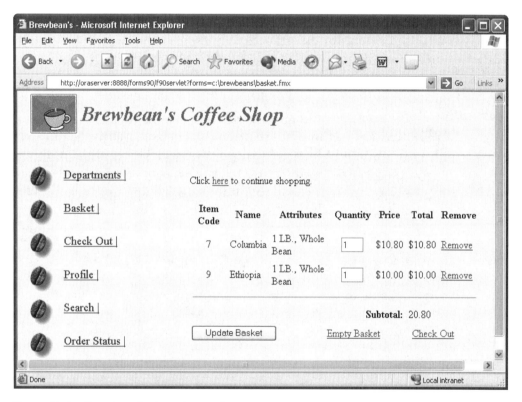

Figure 3-1   Brewbean's shopping cart screen

REBUILDING YOUR DATABASE

Before starting each chapter from this chapter forward, you need to execute a script to rebuild the database to ensure the database is in the proper state to achieve the coding accomplished in the chapter.

To rebuild the Brewbean's database:

1. Open SQL*Plus.

2. You first need to create a spool file so that SQL*Plus keeps a copy of all the messages received from running the file that created the database. If any errors do occur, you can then present this file to your instructor to determine why the errors occurred. On the main menu in SQL*Plus, click **File**, point to **Spool**, and then click **Spool File**. A Select File dialog box appears.

3. Browse to a folder used to contain your working files and in the File name text box, type **DB_log3**, and then click **Save**. Now, whatever text is seen in your SQL*Plus session is saved to this file for future reference.

4. Now let's create the database. In SQL*Plus, type the following code, which runs all the statements contained in the c3Dbcreate.sql file. Messages verifying the creation and data insertion steps will scroll on the SQL*Plus screen. This scrolling takes a couple minutes to complete.

   ```
   @<pathname to PL/SQL files>\Chapter.03\c3Dbcreate.sql
   ```

5. Now, you need to turn the spooling off so that you can review the results in the spool file created. On the main menu, click **File**, point to **Spool**, and then click **Spool Off**.

6. Open a text editor.

7. Open the file named **DB_log3.LST** in your working folder.

8. Review the messages for any errors.

9. Close the file, and then close the text editor.

CONTROL STRUCTURES

The statements used to control the flow of logic processing in our programs are commonly referred to as **control structures**. The control structures provide the capability to perform conditional logic to determine which statements should be run, how many times the statements should be run, and the overall sequence of events.

There are different types of control structures; this section of the chapter discusses each in turn.

IF Statement Logic

The IF statement is a mechanism that allows the checking of a condition to determine if statements should or should not be processed. These statements are commonly used to check values of variables and direct the processing flow to the particular statement(s) that need to be processed based on the values checked.

IF Statements in the Brewbean's Application

In Chapter 2, we used simple IF statements. Let's look at them again in the context of the Brewbean's application. Consider that the PL/SQL block that calculates taxes on online orders may need to determine if the order is from a state requiring taxes. Such a scenario starts with the following assumptions:

- Taxes should be applied only in the company's home state of Virginia.

- We need to check for only one condition: whether the shipping state is Virginia. If it is, the tax amount is calculated using the order subtotal retrieved from the database.

- If the shipping state is not Virginia, no calculation is performed. The tax amount is then 0 because the LV_TAX_NUM variable is initialized to 0 in the declaration segment, as shown in Line 7 of Figure 3-2.

```
Oracle SQL*Plus
File  Edit  Search  Options  Help
SQL> DECLARE
  2    TYPE type_order IS RECORD (
  3      basket bb_basket.idBasket%TYPE,
  4      sub bb_basket.subtotal%TYPE,
  5      state bb_basket.shipstate%TYPE);
  6    rec_order type_order;
  7    lv_tax_num NUMBER(4,2) :=0;
  8  BEGIN
  9    SELECT idBasket, subtotal, shipstate
 10      INTO rec_order
 11      FROM bb_basket
 12      WHERE idBasket = 6;
 13    IF rec_order.state = 'VA' THEN
 14      lv_tax_num := rec_order.sub * .06;
 15    END IF;
 16    DBMS_OUTPUT.PUT_LINE('State = '||rec_order.state);
 17    DBMS_OUTPUT.PUT_LINE('Subtotal = '||rec_order.sub);
 18    DBMS_OUTPUT.PUT_LINE('Tax amount = '||lv_tax_num);
 19  END;
 20  /
State = VA
Subtotal = 149.99
Tax amount = 9

PL/SQL procedure successfully completed.

SQL>
```

Figure 3-2   Simple IF statement using a shipping state of VA

When an IF statement checks only one condition and performs actions only if the condition is TRUE, it is referred to as a simple IF condition. In Figure 3-2, a basket that contains a shipping state of VA was retrieved; therefore, the tax amount is calculated using 6%. What if the shipping state were not VA? The IF condition would resolve to FALSE and the tax calculation would not take place.

As application developers, our testing must include a variety of data situations, so let's examine the same block using a basket with a shipping state of NC (North Carolina), as shown in Figure 3-3. The output shown in the figure confirms that the tax amount remains at 0 because the shipping state is not VA. The IF condition resolves to FALSE and, therefore, the LV_TAX_NUM assignment statement is not processed.

```
± Oracle SQL*Plus                                    _ □ X
File  Edit  Search  Options  Help
SQL> DECLARE
  2    TYPE type_order IS RECORD (
  3      basket bb_basket.idBasket%TYPE,
  4      sub bb_basket.subtotal%TYPE,
  5      state bb_basket.shipstate%TYPE);
  6    rec_order type_order;
  7    lv_tax_num NUMBER(4,2) :=0;
  8  BEGIN
  9    SELECT idBasket, subtotal, shipstate
 10      INTO rec_order
 11      FROM bb_basket
 12      WHERE idBasket = 4;
 13    IF rec_order.state = 'VA' THEN
 14      lv_tax_num := rec_order.sub * .06;
 15    END IF;
 16    DBMS_OUTPUT.PUT_LINE('State = '||rec_order.state);
 17    DBMS_OUTPUT.PUT_LINE('Subtotal = '||rec_order.sub);
 18    DBMS_OUTPUT.PUT_LINE('Tax amount = '||lv_tax_num);
 19  END;
 20  /
State = NC
Subtotal = 28.5
Tax amount = 0

PL/SQL procedure successfully completed.

SQL> |
```

Figure 3-3   Simple IF statement using a shipping state of NC

IF/THEN/ELSE Statements in the Brewbean's Application

The simple IF statement performs an action only if the condition is TRUE. What if we need to perform one action if the condition is TRUE and a different action if the condition is FALSE? Let's change the scenario so that if the shipping state is VA, a tax rate of 6% is applied, and if the shipping state is anything other than VA, the tax rate of 4% is applied. The code in Figure 3-4 reflects this expanded scenario of adding an ELSE clause to the IF statement.

This IF statement uses a 6% tax if the condition checking the shipping state resolves to TRUE. If the condition resolves to FALSE, a 4% tax rate is used. In this scenario, the state is NC; therefore, a tax rate of 4% is applied. Note that by using an IF/THEN/ELSE statement, one tax amount calculation always occurs, regardless of the value of the shipping state.

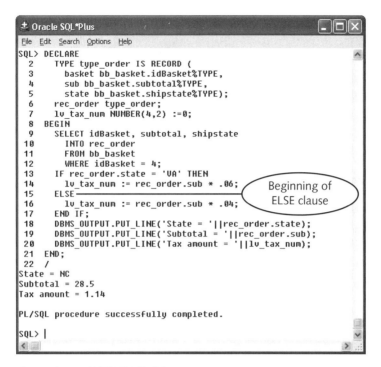

Figure 3-4   IF/THEN/ELSE statement

IF/THEN Versus IF

In Chapter 2, instead of using the ELSE clause, we created the IF condition using two simple IF statements, as shown in the following code:

```
IF rec_order.state = 'VA' THEN
   lv_tax_num := rec_order.sub * .06;
END IF;
IF  rec_order.state <> 'VA' THEN
   lv_tax_num := rec_order.sub * .04;
END IF;
```

Will this code accomplish the same as the IF/THEN/ELSE statement? Both do indeed accomplish the same task and generate the same results. Which method should be used? Well, one method operates more efficiently than the other. In the preceding code containing two simple IF statements, each IF clause is processed regardless of the shipping state value. On the other hand, the ELSE clause in Figure 3-4 processes only one IF clause (checking the value of the shipping state). This is an important difference in regards to processing efficiency because the less code that has to be processed, the faster the program runs.

IF/THEN/ELSIF/ELSE

Let's go a step further with this tax calculation scenario. What if this scenario changes so that the tax rate of 6% should be applied to VA, 5% to ME, 7% to NY, and 4% to all other states? Now, we need to check for the existence of several different values because we no longer have an either/or situation. Review the code in Figure 3-5, which checks for the existence of different values by using ELSIF clauses. Note that each condition in the ELSIF clauses is mutually exclusive, which means that only one of the ELSIF conditions can evaluate to TRUE.

```
± Oracle SQL*Plus                                    [_][□][X]
File  Edit  Search  Options  Help
SQL> DECLARE
  2    TYPE type_order IS RECORD (
  3      basket bb_basket.idBasket%TYPE,
  4      sub bb_basket.subtotal%TYPE,
  5      state bb_basket.shipstate%TYPE);
  6    rec_order type_order;
  7    lv_tax_num NUMBER(4,2) :=0;
  8  BEGIN
  9    SELECT idBasket, subtotal, shipstate
 10      INTO rec_order
 11      FROM bb_basket
 12      WHERE idBasket = 6;
 13    IF rec_order.state = 'VA' THEN
 14      lv_tax_num := rec_order.sub * .06;
 15    ELSIF rec_order.state = 'ME' THEN
 16      lv_tax_num := rec_order.sub * .05;
 17    ELSIF rec_order.state = 'NY' THEN
 18      lv_tax_num := rec_order.sub * .07;
 19    ELSE
 20      lv_tax_num := rec_order.sub * .04;
 21    END IF;
 22    DBMS_OUTPUT.PUT_LINE('State = '||rec_order.state);
 23    DBMS_OUTPUT.PUT_LINE('Subtotal = '||rec_order.sub);
 24    DBMS_OUTPUT.PUT_LINE('Tax amount = '||lv_tax_num);
 25  END;
 26  /
State = VA
Subtotal = 149.99
Tax amount = 9

PL/SQL procedure successfully completed.

SQL> |
```

Figure 3-5   Using an IF/THEN/ELSIF/ELSE statement

The processing begins at the top of the IF statement by checking the shipping state value until it finds a condition that is TRUE. After a TRUE condition is found, the associated program statements are processed and the IF statement is finished. The program then runs the statement immediately following the END IF; line. If no condition resolves to TRUE, the program statements in the ELSE clause are processed. An ELSE clause is not required and if none were provided here, the IF statement could complete without processing a tax calculation.

Because IF clauses are evaluated from the top down, knowing the nature of your data can assist in making the code more efficient. That is, if much of the data processed matches a particular value, then list this value in the first IF condition so that only one IF clause typically has to run.

Do you want to put an 'E' in the ELSIF keyword? This is one of the most common mistakes among PL/SQL programmers. Figure 3-6 displays the error you receive by using ELSEIF rather than ELSIF. Notice the error is raised on the term that follows the ELSEIF. Because this spelling does not represent a keyword, the system thinks the ELSEIF is something else, such as a variable.

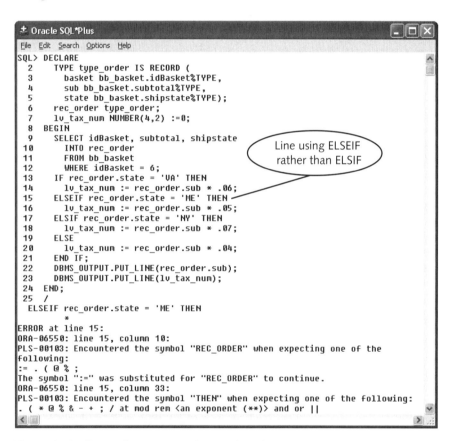

Figure 3-6   Error when using ELSEIF rather than ELSIF

Running an IF/THEN/ELSIF Statement

Let's experiment with running an IF/THEN/ELSIF statement. We will use the code shown in Figure 3-5 and run two different scenarios.

To run an IF/THEN/ELSIF statement:

1. Open or return to SQL*Plus.

2. Enter your user name, password, and host string to connect to the database.

3. Type **SET SERVEROUTPUT ON** and press **Enter**. This enables the display of values from DBMS_OUTPUT statements.

4. In a text editor, open the **ifthen03.txt** file from the Chapter.03 folder. Notice that the SELECT statement is retrieving basket 6, which has a shipping state of VA.

5. Copy and paste all the code into SQL*Plus.

6. Press **Enter**, if necessary, to execute the block. The results should match Figure 3-5 with a tax amount of $9 using the 6% tax rate.

7. Return to the text file and modify the WHERE clause of the SELECT statement (you'll find it on Line 12) to indicate basket 4, which has a shipping state of NC.

8. Copy and paste all the code into SQL*Plus.

9. Press the **Enter** key, if necessary, to execute the block. NC does not match any of the IF/ELSIF conditions; therefore, the ELSE clause processes a tax rate of 4% and a tax amount of $1.14.

10. Return to the text file and remove the ELSE clause (Lines 19 and 20) so that the code appears as shown in Figure 3-7.

11. Copy and paste all the code into SQL*Plus.

12. Press **Enter**, if necessary, to execute the block. NC does not match any of the IF/ELSIF conditions; therefore, no tax calculation is processed. The tax amount remains zero, which was the value given to it during initialization in the DECLARE section.

13. Close SQL*Plus.

```
± Oracle SQL*Plus                                        [_][□][X]
File  Edit  Search  Options  Help
SQL> DECLARE
  2    TYPE type_order IS RECORD (
  3      basket bb_basket.idBasket%TYPE,
  4      sub bb_basket.subtotal%TYPE,
  5      state bb_basket.shipstate%TYPE);
  6    rec_order type_order;
  7    lv_tax_num NUMBER(4,2) :=0;
  8  BEGIN
  9    SELECT idBasket, subtotal, shipstate
 10      INTO rec_order
 11      FROM bb_basket
 12      WHERE idBasket = 4;
 13    IF rec_order.state = 'VA' THEN
 14      lv_tax_num := rec_order.sub * .06;
 15    ELSIF rec_order.state = 'ME' THEN
 16      lv_tax_num := rec_order.sub * .05;
 17    ELSIF rec_order.state = 'NY' THEN
 18      lv_tax_num := rec_order.sub * .07;
 19    END IF;
 20    DBMS_OUTPUT.PUT_LINE('State = '||rec_order.state);
 21    DBMS_OUTPUT.PUT_LINE('Subtotal = '||rec_order.sub);
 22    DBMS_OUTPUT.PUT_LINE('Tax amount = '||lv_tax_num);
 23  END;
 24  /
State = NC
Subtotal = 28.5
Tax amount = 0

PL/SQL procedure successfully completed.

SQL> |
```

Figure 3-7   IF/ELSIF with no ELSE clause

Operators Within an IF Clause

Keep in mind that more than one condition can be checked in each IF clause. The logical operators of OR and AND that are used in SQL are available in PL/SQL as well. For example, the IF clause using an OR operator in the following code uses a tax rate of 6% if the shipping state is either VA or PA:

```
IF rec_order.state = 'VA' OR rec_order.state = 'PA' THEN
   lv_tax_num := rec_order.sub * .06;
ELSE
   lv_tax_num := rec_order.sub * .04;
END IF;
```

Be careful to use complete conditional expressions when using logical operators. A common mistake is to use incomplete conditions in the IF clause, such as occurs in the first line of the following code (note that the first line of the IF statement uses an incomplete condition with the OR):

```
IF rec_order.state = 'VA' OR 'PA' THEN
   lv_tax_num := rec_order.sub * .06;
ELSE
   lv_tax_num := rec_order.sub * .04;
END IF;
```

In regards to being familiar with your data, you also need to consider if the values being checked in the IF clauses could be NULL. If an action must occur when the value being checked contains a NULL value, the IF statement must contain either an explicit check for a NULL value (IS NULL) or an ELSE clause. In the following code, the ELSE clause runs if the state value is NULL or anything but VA or PA.

```
IF rec_order.state = 'VA' OR rec_order.state = 'PA' THEN
   lv_tax_num := rec_order.sub * .06;
ELSE
   lv_tax_num := rec_order.sub * .04;
END IF;
```

However, if we eliminate the ELSE clause but still want some particular action to occur if the state is a NULL value, we must explicitly add an ELSIF to check for a NULL value, as shown in the following code:

```
IF rec_order.state = 'VA' OR rec_order.state = 'PA' THEN
   lv_tax_num := rec_order.sub * .06;
ELSIF rec_order.state IS NULL THEN
   lv_tax_num := rec_order.sub * .04;
END IF;
```

This IF statement calculates a 6% tax amount for states of VA or PA and 4% if the state is a NULL value. No activity occurs if the state is any other value.

The IN operator is another method to check several conditions in one IF clause by providing a list of values. The following example uses the IN operator to check for three state values in the first IF clause:

```
IF rec_order.state IN ('VA','PA','ME') THEN
   lv_tax_num := rec_order.sub * .06;
ELSE
   lv_tax_num := rec_order.sub * .04;
END IF;
```

Next, we cover CASE statements, which provide another method to accomplish conditional checking.

CASE Statements

Oracle9*i* introduced the CASE statement feature to process conditional logic in a manner that is similar to how the IF statement processes conditional logic. CASE statements are available in most programming languages and many developers are familiar with this type of code.

The addition of the CASE statement does not really add functionality in PL/SQL programming; however, advocates of the CASE statement claim it leads to more compact and understandable coding. In addition, the CASE statement can be more efficient in accomplishing IF clause logic that requires a number of ELSIF clauses. The choice between IF statements versus CASE statements really becomes a preference of the programming shop, which should establish standard coding practice rules to maintain consistency of coding throughout the operation.

Structure of the CASE Statement

The CASE statement begins with the key word CASE followed by a selector that is typically a variable name. This is followed by WHEN clauses to determine which statements to run given the value of the selector. The code in Figure 3-8 accomplishes the same task as the code in Figure 3-7, except that it uses the CASE statement.

```
Oracle SQL*Plus
File  Edit  Search  Options  Help
SQL> DECLARE
  2    TYPE type_order IS RECORD (
  3      basket bb_basket.idBasket%TYPE,
  4      sub bb_basket.subtotal%TYPE,
  5      state bb_basket.shipstate%TYPE);
  6    rec_order type_order;
  7    lv_tax_num NUMBER(4,2) :=0;
  8  BEGIN
  9    SELECT idBasket, subtotal, shipstate
 10      INTO rec_order
 11      FROM bb_basket
 12      WHERE idBasket = 6;
 13    CASE rec_order.state
 14      WHEN 'VA' THEN lv_tax_num := rec_order.sub * .06;
 15      WHEN 'ME' THEN lv_tax_num := rec_order.sub * .05;
 16      WHEN 'NY' THEN lv_tax_num := rec_order.sub * .07;
 17      ELSE lv_tax_num := rec_order.sub * .04;
 18    END CASE;
 19    DBMS_OUTPUT.PUT_LINE('State = '||rec_order.state);
 20    DBMS_OUTPUT.PUT_LINE('Subtotal = '||rec_order.sub);
 21    DBMS_OUTPUT.PUT_LINE('Tax amount = '||lv_tax_num);
 22  END;
 23  /
State = VA
Subtotal = 149.99
Tax amount = 9

PL/SQL procedure successfully completed.

SQL>
```

Figure 3-8   Using the CASE statement with a selector

The CASE statement evaluates the same way the IF statement does in that it works from the top down until finding a condition that is TRUE. However, notice that the value of the state variable is evaluated only once at the top of the CASE statement rather than with every condition clause in the IF statement.

Refer back to Figure 3-7. If the ELSE clause is omitted in an IF statement, the IF statement could run and potentially not execute any code. If no matches are found in the IF/ELSIF clauses, the IF statement ends successfully without running any code. This is not what happens when the ELSE clause is omitted from the CASE statement in Figure 3-8. If no TRUE conditions are found in the WHEN clauses and an ELSE clause is not included, the CASE statement includes an implicit ELSE clause that raises an Oracle error, which we look at in the following set of steps:

To generate a NO CASE FOUND error:

1. Open or return to SQL*Plus.

2. Click **File** on the menu bar, and then click **Open**.

3. Navigate to and select the **case_err03.sql** file from the Chapter.03 folder.

4. Click **Open** and a copy of the file contents appears in the SQL*Plus screen. Notice that the SELECT statement is retrieving basket 4, which has a shipping state of NC.

5. To run the code and create the procedure, click **File**, and then click **Run**. The results should match Figure 3-9. The error is there because no WHEN clause addresses NC and the ELSE clause has been removed.

```
Oracle SQL*Plus
File  Edit  Search  Options  Help
SQL> DECLARE
  2     TYPE type_order IS RECORD (
  3       basket bb_basket.idBasket%TYPE,
  4       sub bb_basket.subtotal%TYPE,
  5       state bb_basket.shipstate%TYPE);
  6     rec_order type_order;
  7     lv_tax_num NUMBER(4,2) :=0;
  8   BEGIN
  9     SELECT idBasket, subtotal, shipstate
 10       INTO rec_order
 11       FROM bb_basket
 12       WHERE idBasket = 4;
 13     CASE rec_order.state
 14       WHEN 'VA' THEN lv_tax_num := rec_order.sub * .06;
 15       WHEN 'ME' THEN lv_tax_num := rec_order.sub * .05;
 16       WHEN 'NY' THEN lv_tax_num := rec_order.sub * .07;
 17     END CASE;
 18     DBMS_OUTPUT.PUT_LINE('State = '||rec_order.state);
 19     DBMS_OUTPUT.PUT_LINE('Subtotal = '||rec_order.sub);
 20     DBMS_OUTPUT.PUT_LINE('Tax amount = '||lv_tax_num);
 21   END;
 22   /
DECLARE
*
ERROR at line 1:
ORA-06592: CASE not found while executing CASE statement
ORA-06512: at line 13

SQL> |
```

Figure 3-9   CASE not found while executing CASE statement error

Searched CASE Statement

A second form of the CASE statement, called a **Searched CASE** statement, is available. It does not use a selector but individually evaluates conditions that are placed in the WHEN clauses. The conditions checked in the WHEN clauses must evaluate to a Boolean value of TRUE or FALSE. This allows different items, such as the shipping state and zip code, to be checked within the same CASE statement.

To illustrate the use of the Searched CASE statement, let's assume that Brewbean's tax calculation must consider not only the state but also special rates applied by some localities that can be identified via zip codes. Figure 3-10 displays the use of a CASE statement to apply a 6% tax rate to all VA residents except zip code 23321, which uses a rate of 2%.

```
± Oracle SQL*Plus                                        _ □ X
File  Edit  Search  Options  Help
SQL> DECLARE
  2      TYPE type_order IS RECORD (
  3        basket bb_basket.idBasket%TYPE,
  4        sub bb_basket.subtotal%TYPE,
  5        state bb_basket.shipstate%TYPE,
  6        zip bb_basket.shipzipcode%TYPE);
  7      rec_order type_order;
  8      lv_tax_num NUMBER(4,2) :=0;
  9   BEGIN
 10      SELECT idBasket, subtotal, shipstate, shipzipcode
 11        INTO rec_order
 12        FROM bb_basket
 13        WHERE idBasket = 6;
 14      CASE
 15      WHEN rec_order.zip = '23321' THEN
 16        lv_tax_num := rec_order.sub * .02;
 17      WHEN rec_order.state = 'VA' THEN
 18        lv_tax_num := rec_order.sub * .06;
 19      ELSE
 20        lv_tax_num := rec_order.sub * .04;
 21      END CASE;
 22      DBMS_OUTPUT.PUT_LINE('State = '||rec_order.state);
 23      DBMS_OUTPUT.PUT_LINE('State = '||rec_order.zip);
 24      DBMS_OUTPUT.PUT_LINE('Subtotal = '||rec_order.sub);
 25      DBMS_OUTPUT.PUT_LINE('Tax amount = '||lv_tax_num);
 26   END;
 27   /
State = VA
State = 23321
Subtotal = 149.99
Tax amount = 3

PL/SQL procedure successfully completed.

SQL> |
```

Figure 3-10   Using a Searched CASE statement

This block processes basket 6, which has a shipping state of VA and a shipping zip code of 23321. Keep in mind, the conditional statements are evaluated from the top down and end when a TRUE condition is discovered. Therefore, processing basket 6 finds a TRUE condition with the first WHEN clause and uses a tax rate of 2%.

The Use of the CASE Expression

The CASE keyword can also be used as an expression rather than a statement, as we have seen thus far. A **CASE expression** evaluates conditions and returns a value in an assignment statement. For example, in our tax calculation, the end result has been putting a value for the tax amount into the variable named LV_TAX_NUM. We can also use a CASE expression to accomplish this task, as shown in Line 13 of Figure 3-11.

```
± Oracle SQL*Plus
File  Edit  Search  Options  Help
SQL> DECLARE
  2    TYPE type_order IS RECORD (
  3      basket bb_basket.idBasket%TYPE,
  4      sub bb_basket.subtotal%TYPE,
  5      state bb_basket.shipstate%TYPE);
  6    rec_order type_order;
  7    lv_tax_num NUMBER(4,2) :=0;
  8  BEGIN
  9    SELECT idBasket, subtotal, shipstate
 10      INTO rec_order
 11      FROM bb_basket
 12      WHERE idBasket = 6;
 13    lv_tax_num := CASE rec_order.state
 14      WHEN 'VA' THEN rec_order.sub * .06
 15      WHEN 'ME' THEN rec_order.sub * .05
 16      WHEN 'NY' THEN rec_order.sub * .07
 17      ELSE rec_order.sub * .04
 18    END;
 19    DBMS_OUTPUT.PUT_LINE('State = '||rec_order.state);
 20    DBMS_OUTPUT.PUT_LINE('Subtotal = '||rec_order.sub);
 21    DBMS_OUTPUT.PUT_LINE('Tax amount = '||lv_tax_num);
 22  END;
 23  /
State = VA
Subtotal = 149.99
Tax amount = 9

PL/SQL procedure successfully completed.

SQL> |
```

Figure 3-11   Using a CASE expression

The CASE clause in the figure actually serves as the value expression of an assignment statement. If the goal of our conditional checking is to assign a value to a variable, this format makes this intention clear as it begins with the assignment statement.

 Notice that a CASE expression, even though constructed much the same as a basic CASE statement, has some subtle syntax differences: The WHEN clauses do not end with semicolons, and the statement does not end with END CASE; (instead, it ends with END;).

LOOPING CONSTRUCTS

Conditional IF/THEN and CASE statements are invaluable in creating programming logic; however, additional constructs are needed. What if we are calculating the tax amount for an order and need to apply different tax rates on coffee (as a food item) and equipment items? We would need to look at each detailed item line to determine the appropriate tax amount, yet we would process the same logic on each line or item. **Looping constructs** allow us to repeat processing of a desired portion of code to handle this type of situation efficiently.

Loops are used for situations in which we need to repeat a line or lines of code within our block. In every loop, the system must be instructed as to which statements should be repeated and when to end the repeating action or stop the loop. Three forms of PL/SQL looping structures are covered in this part of the chapter: basic, WHILE, and FOR.

Basic Loop

The **basic loop** uses the LOOP and END LOOP markers to begin and end the loop code, which includes any statements that are to be repeated. It also uses the EXIT WHEN clause to indicate when the looping action should stop.

Examine the code in Figure 3-12. It contains a basic loop that loops five times. The only statement run is the DBMS_OUTPUT action that displays the value of the counter to the screen. The counter is established in the loop to provide a mechanism to instruct the loop when to stop execution. The LV_CNT_NUM variable holds the value of one on the first iteration of the loop. The numbers 1 to 5 are displayed to the screen and then the loop stops because the LV_CNT_NUM variable now holds a value of 5 and the EXIT WHEN condition evaluates to TRUE.

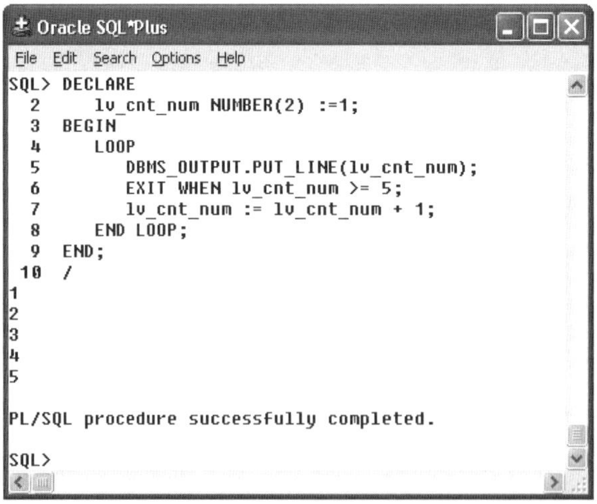

```
SQL> DECLARE
  2      lv_cnt_num NUMBER(2) :=1;
  3  BEGIN
  4      LOOP
  5          DBMS_OUTPUT.PUT_LINE(lv_cnt_num);
  6          EXIT WHEN lv_cnt_num >= 5;
  7          lv_cnt_num := lv_cnt_num + 1;
  8      END LOOP;
  9  END;
 10  /
1
2
3
4
5

PL/SQL procedure successfully completed.

SQL>
```

Figure 3-12   Using a basic loop

Tip If the EXIT WHEN clause is not included in the code in Figure 3-12, the result is the programmer's nightmare of the **infinite loop**, which is a loop that is never instructed when to stop. Thus, it continues looping indefinitely, disrupting the ability of the code to continue with any processing beyond the loop.

3

The DBMS_OUTPUT.PUT_LINE clause allows us to display values within our block to the screen. This is an invaluable tool to use in testing the block code. Used effectively, this allows the developer to determine values of variables at different points in the block, as well as assist in determining which statements actually get processed.

Let's now look at an example that uses a basic loop to process through rows in an explicit cursor. This block retrieves the items in a basket and uses a basic loop to calculate the tax amount using different tax rates for coffee and equipment items.

To use a basic loop to process rows from an explicit cursor:

1. Open or return to SQL*Plus.

2. Create a host variable by typing **VARIABLE g_basket NUMBER**, and then press **Enter**.

3. Initialize g_basket to 6 by typing and running the following anonymous block:
```
BEGIN
   :g_basket := 6;
END;
/
```

4. Click **File** on the menu bar, and then click **Open**.

5. Navigate to and select the **bloop03.sql** file from the Chapter.03 folder.

6. Click **Open** and a copy of the file contents appears in the SQL*Plus screen. Notice that the SELECT statement in the cursor is retrieving the items in basket 6 based on the host variable :G_BASKET.

7. To run the code and create the procedure, click **File**, and then click **Run**. The results should match Figure 3-13.

8. Close SQL*Plus.

Notice the loop provides a mechanism for us to move through each of the rows in the cursor for processing. The EXIT WHEN clause determines when the looping action ceases, which, in this case, is when all the rows of the cursor have been retrieved.

```
Oracle SQL*Plus
File  Edit  Search  Options  Help
SQL> DECLARE
  2      CURSOR cur_basket IS
  3        SELECT bi.idBasket, p.type, bi.price, bi.quantity
  4          FROM bb_basketitem bi INNER JOIN bb_product p
  5            USING (idProduct)
  6          WHERE bi.idBasket = :g_basket;
  7      TYPE type_basket IS RECORD (
  8        basket bb_basketitem.idBasket%TYPE,
  9        type bb_product.type%TYPE,
 10        price bb_basketitem.price%TYPE,
 11        qty bb_basketitem.quantity%TYPE );
 12      rec_basket type_basket;
 13      lv_rate_num NUMBER(2,2);
 14      lv_tax_num NUMBER(4,2) := 0;
 15  BEGIN
 16      OPEN cur_basket;
 17      LOOP
 18        FETCH cur_basket INTO rec_basket;
 19        EXIT WHEN cur_basket%NOTFOUND;
 20        IF rec_basket.type = 'E' THEN
 21            lv_rate_num := .05;
 22        ELSIF rec_basket.type = 'C' THEN
 23            lv_rate_num := .03;
 24        END IF;
 25        lv_tax_num := lv_tax_num + ((rec_basket.price*rec_basket.qty)*lv_rate_num);
 26        DBMS_OUTPUT.PUT_LINE('Type = '||rec_basket.type||' and tax rate = '||lv_rate_num);
 27      END LOOP;
 28      CLOSE cur_basket;
 29      DBMS_OUTPUT.PUT_LINE('Total Tax amount = '||lv_tax_num);
 30  END;
 31  /
Type = C and tax rate = .03
Type = E and tax rate = .05
Total Tax amount = 7.1

PL/SQL procedure successfully completed.

SQL>
```

Figure 3-13   Loop through cursor rows

WHILE Loop

The **WHILE loop** differs from the other types of loops in that it includes a condition to check at the top of the loop in the LOOP clause itself. For each iteration of the loop, this condition is checked and if it is TRUE, the loop continues. If the condition is FALSE, the looping action stops. You should ensure that whatever is being checked in the loop condition actually changes value as the loop iterates so that at some point, the condition resolves to FALSE and, therefore, ends the looping action. Figure 3-14 shows a rewrite of the previous basic loop example shown in Figure 3-12, except that it uses the WHILE loop format.

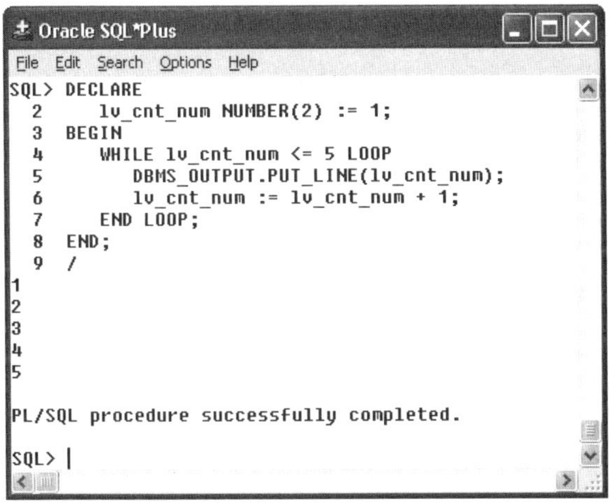

Figure 3-14   Using WHILE loop processing

Notice that this loop iterates only five times, at which time the LV_CNT_NUM variable holds a value of 6. At this point, the WHILE clause runs and determines that the condition evaluates to FALSE (6 is not <= 5); therefore, the looping action stops. Keep in mind that the condition is evaluated at the top of the loop, which means that there is no guarantee that the code inside the loop will run at all. If LV_CNT_NUM had a value of 11 before reaching the loop statement, the WHILE condition would evaluate to FALSE and the looping action would never run.

Other languages offer variations on loop statements. An example of the variation is the LOOP UNTIL, which evaluates a condition at the bottom of the loop. This bottom-of-the-loop evaluation guarantees that the loop always iterates at least once. PL/SQL does not offer this type of loop statement. However, the basic loop with an EXIT WHEN clause can be used to ensure the loop runs at least one time, as shown in the following code:

```
DECLARE
   lv_cnt_num NUMBER(2) :=1;
BEGIN
   LOOP
      DBMS_OUTPUT.PUT_LINE(lv_cnt_num);
      lv_cnt_num := lv_cnt_num + 1;
      EXIT WHEN lv_cnt_num >= 5;
   END LOOP;
END;
/
```

FOR Loop

By indicating a numeric range, the **FOR loop** dictates exactly how many times the loop should run in the opening LOOP clause. Figure 3-15 shows a FOR loop, which achieves the same results as the counter examples used with the basic and WHILE loop. The range is indicated in Line 2.

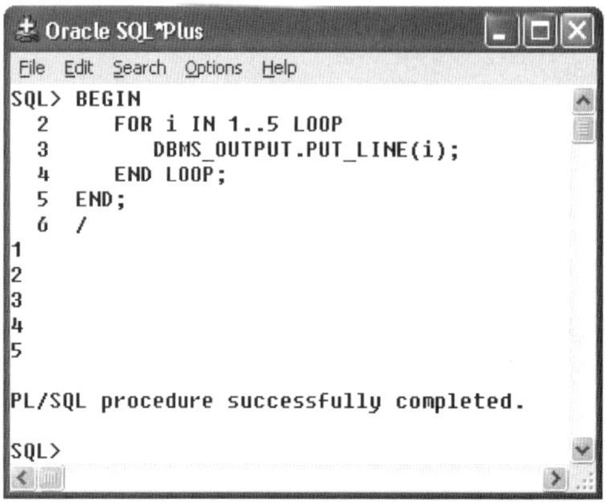

Figure 3-15   Using a FOR loop

Several tasks are accomplished in the FOR clause that opens the loop in Figure 3-15. First, the FOR clause sets up a counter variable automatically. In this example, the counter is named i, which is typical; however, this does not have to be named i. The second task is setting up a range of values that the counter variable takes on. The range controls the number of times the loop runs (1..5). In this example, the counter (i) holds the value of 1 on the first iteration of the loop. Each iteration of the loop increments the counter by 1 automatically, and the loop stops after running five times.

The loop range must indicate a lower bound and upper bound value that determines the number of times the loop iterates. The values can be provided in the form of numbers, variables, or expressions evaluating to numeric values. By default, the counter starts at the lower bound value and is incremented by 1 for each iteration of the loop. The loop in this example always iterates five times because the range is indicated with numeric values. However, a variable could be used in the range to make the number of iterations more dynamic. That is, each time the block containing the loop is run, the loop may iterate a different number of times. For example, the number of iterations of the FOR loop in the following block is not determined until the SELECT statement is executed:

```
DECLARE
   lv_upper_num NUMBER(3);
BEGIN
  SELECT COUNT(idBasket)
     INTO lv_upper_num
     FROM bb_basket;
   FOR i IN 1..lv_upper_num LOOP
     DBMS_OUTPUT.PUT_LINE(i);
   END LOOP;
END;
/
```

If the SELECT statement returns a value of 14 for the count, the loop iterates 14 times. Because the number of rows in the BB_BASKET table could change, the upper bound value may be different when this block is run at another time. Notice the value of the counter can be referenced inside the loop and become an integral value in your processing. In this example, the i value is output to the screen. Another example is if the loop contained an INSERT statement, the counter could be used as a value in the insert.

The counter variable can be referenced in the loop, but cannot be assigned a value because the loop controls the value of the counter. In addition, an option of REVERSE is available to force the counter to begin with the upper bound value of the range and increment by -1 in each iteration of the loop until the lower bound value is reached. The REVERSE key word should be included in the FOR LOOP clause immediately prior to the range, as shown in the following code:

```
FOR i IN REVERSE 1..5 LOOP
```

Many languages allow the programmer to indicate the increment value in the loop. For example, an increment value of 2 would cause the counter variable to increase by 2 instead of by 1 for each iteration of the loop. This feature is not available as an option (such as a step option) on the FOR LOOP statement, which forces the increment value to be 1. However, incrementing by values greater than 1 can be achieved by adding appropriate code into the logic of the loop. Let's say we want a loop that inserts values into the BB_BASKETITEM table but instead of wanting to insert 10 rows, we want to insert two rows with the values of 5 and 10. This task could be easily accomplished if the counter could be incremented by 5; however, we need to manipulate the i value in PL/SQL. In the following block, we add a multiplier in the statements within the loop. The loop runs only twice; however, the insert values are 5 and 10 because a multiplier of 5 is used in the VALUES clause of the INSERT statement:

```
BEGIN
  FOR i IN 1..2 LOOP
    INSERT INTO bb_basketitem (idBasketitem)
      VALUES (i*5);
  END LOOP;
END;
```

Common Errors While Using Looping Statements

Now that we have looked at all three loop constructs, let's discuss a couple items you need to consider while creating loops.

EXIT Clause in Loops

One caution concerns the use of the EXIT clause in loops. Even though it can be used in any type of loop to stop the looping action, it is considered good form to use the EXIT clause only in basic loops. The WHILE and FOR loops are constructed with conditions in the LOOP statement to determine when the looping action begins and ends.

Using an EXIT clause can circumvent these conditions and stop the looping at a different point in the processing. This type of loop is both hard to read and debug.

Static Statements

A second caution is to remember that loops execute all the contained statements for each iteration of the loop. To keep code efficient and minimize statement processing, any statements that are static in nature should be placed outside a loop. For example, let's say we have a loop that contains an insert, as shown in the following code:

```
DECLARE
  lv_today_date DATE;
BEGIN
  FOR i IN 17..21 LOOP
    lv_today_date := SYSDATE;
    INSERT INTO bb_basket (idBasket, dtordered)
      VALUES (i, lv_today_date);
  END LOOP;
END;
/
```

Because the LV_TODAY_DATE assignment statement is inside the loop in this example, this statement is processed five times. Because the current date value remains the same, this repeated processing is not necessary, and the operation should be completed outside the loop.

GOTO STATEMENT

The **GOTO statement** is sometimes called a jumping control in that it instructs the program to "jump to" some specific area of the code. It is used to branch logic in a manner where only certain portions of the code in the block are processed based on some condition.

The GOTO statement can branch to a label that marks a particular spot in the code and that must precede an executable statement. The statement instructs the system to skip any executable statements until reaching the label referenced in the GOTO statement. Then, the statements from that point on are processed. The << >> markers are used in PL/SQL to identify a label.

The following code shows an incomplete executable section of a block that uses a GOTO statement. The GOTO statement instructs the processing to jump to the INSERT_ROW label:

```
BEGIN
  If lv_rows_num = 0 THEN
    GOTO insert_row;
  End If;
  . . .
```

```
   . . .
   <<insert_row>>
   INSERT INTO bb_basket (idBasket)
     VALUES (bb_basket_seq.NEXTVAL);
   . . .
   . . .
 END;
```

3

Even though the GOTO statement exists, most developers believe this should be used very sparingly and only if no other method can be used to accomplish the task at hand. Why? The GOTO action interrupts the flow of execution, making it very difficult to understand and maintain the code. However, it is important to be aware of this statement as you may find yourself supporting existing code that includes GOTO statements.

EXCEPTION HANDLERS

An exception handler is a mechanism to trap an error that occurs in processing. Its code handles the error in a user-friendly manner and allows the application to continue. The section of a lock begins with the EXCEPTION keyword and follows the BEGIN section.

The EXCEPTION section addresses two situations: either an Oracle error is raised or a user-defined error is raised. (In Oracle, "raised" means occurs.) The Oracle error happens automatically if a problem—for example, a DELETE statement is issued but results in a foreign key error—in the executable code exists. On the other hand, there may be times that an Oracle error does not occur but we want an error to be raised. For example, suppose we run an UPDATE statement and no rows are updated. An Oracle error will not occur. Why? It is legal for an UPDATE to affect no rows. However, as developers, we may recognize that we have an invalid value being supplied in the WHERE clause of the UPDATE if no rows are updated. Therefore, a user-defined error needs to be created so that an error is raised if no rows are updated.

Some common Oracle errors, referenced as predefined Oracle errors, already have associated exception names within the Oracle server. All other Oracle errors are considered non-predefined and must have a name assigned within the code. In the following sections, we explore three types of errors: predefined Oracle, non-predefined Oracle, and user-defined.

Predefined Oracle Errors

Oracle supplies a set of **predefined exceptions** or names associated with common Oracle errors. Table 3-1 lists some of the more common predefined errors. The predefined exceptions are declared in the STANDARD package, which makes these globally available on the system.

Table 3-1   Partial list of Oracle predefined errors

Exception Name	Description
NO_DATA_FOUND	A SELECT statement in a PL/SQL block retrieves no rows or a nonexistent row of an index-by table that has been referenced.
TOO_MANY_ROWS	A SELECT statement in a PL/SQL block retrieves more than one row.
CASE_NOT_FOUND	No WHEN clause in the CASE statement is processed.
ZERO_DIVIDE	An attempted division by zero causes this exception.
DUP_VAL_ON_INDEX	An attempted violation of a unique or primary key column constraint causes this exception.

For an exhaustive list of Oracle predefined errors, refer to Oracle database documentation at the OTN Web site at *http://otn.oracle.com*.

A SELECT statement within a PL/SQL block has the possibility of raising the NO_DATA_FOUND or TOO_MANY_ROWS exception. In PL/SQL, if a SELECT statement returns no rows, an Oracle error is raised, which is already associated with the NO_DATA_FOUND exception name. This is a departure from running SQL statements stand-alone, such as directly in SQL*Plus where no error is raised if a SELECT statement returns no rows. This is an issue with a SELECT statement embedded in PL/SQL because we now are moving the data returned by the SELECT statement into variables. On the other hand, if more than one row is returned by a SELECT statement but the variables in the IN clause are either scalar variables or a record variable, the TOO_MANY_ROWS exception occurs. Let's use some examples to explore these exceptions further.

Error Examples

If the INTO clause of a SELECT statement contains either scalar or record variables and if more than one row is selected, the TOO_MANY_ROWS error is raised because multiple rows cannot be handled with these variables. As the basis for the discussion of these exceptions, examine the Brewbean's application screen in Figure 3-16.

The following code example is one of the blocks of code that supports this screen. It completes a SELECT statement on the BB_BASKET table to display a noncompleted shopping basket if one exists for a user who logs onto the Web site:

```
DECLARE
  TYPE type_basket IS RECORD (
    basket bb_basket.idBasket%TYPE,
    created bb_basket.dtcreated%TYPE,
    qty bb_basket.quantity%TYPE,
    sub bb_basket.subtotal%TYPE);
  rec_basket type_basket;
  lv_days_num NUMBER(3);
  lv_shopper_num NUMBER(3) := 25;
```

```
BEGIN
 SELECT idBasket, dtcreated, quantity, subtotal
  INTO rec_basket
  FROM bb_basket
  WHERE idShopper = lv_shopper_num
    AND orderplaced = 0;
 lv_days_num := SYSDATE - rec_basket.created;
 DBMS_OUTPUT.PUT_LINE(rec_basket.basket);
 DBMS_OUTPUT.PUT_LINE(rec_basket.created);
 DBMS_OUTPUT.PUT_LINE(rec_basket.qty);
 DBMS_OUTPUT.PUT_LINE(rec_basket.sub);
 DBMS_OUTPUT.PUT_LINE(lv_days_num);
END;
 /
```

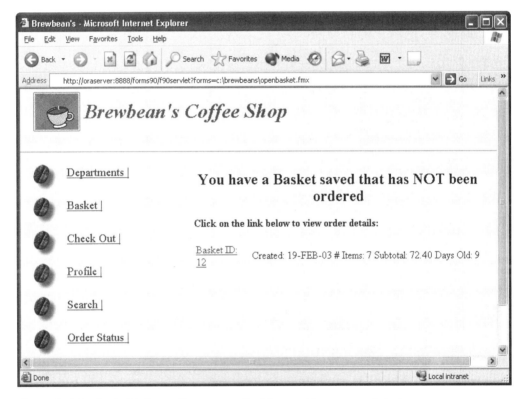

Figure 3-16   Basket information screen that has not been completed

If there is a potential that no rows or more than one row will be retrieved by this SELECT statement, we need to include exception handlers to manage these situations. Let's first run this block using different shoppers to see the types of errors that occur.

To see Oracle errors associated with a SELECT statement:

1. In a text editor, open the **ex_select03.txt** file from the Chapter.03 folder. Notice that the SELECT statement is currently set to retrieve noncompleted basket information for Shopper 22 as the LV_SHOPPER_NUM variable is initialized to this value.

2. Copy and paste all the code into SQL*Plus.

3. Press **Enter**, if necessary, to execute the block. The results should match Figure 3-17, which shows a "no data found" error message because Shopper 22 has no noncompleted baskets saved in the database.

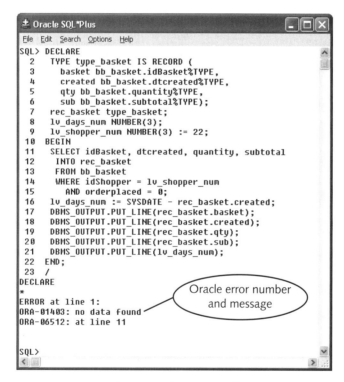

Figure 3-17   No data found error raised

 Some of the Oracle error messages are not as straightforward or understandable. The Oracle database documentation available at *http://otn.oracle.com* contains a section that lists all the Oracle errors in numerical order and with brief descriptions.

4. Return to the text file and change the LV_SHOPPER_NUM variable declaration (in Line 9) to an initialized value of Shopper 26.

5. Copy and paste all the code into SQL*Plus.

6. Press **Enter**, if necessary, to execute the block. The results should match Figure 3-18, which shows an "exact fetch returns more than requested number of rows" error message because shopper 26 has more than one noncompleted basket in the database.

7. Close SQL*Plus.

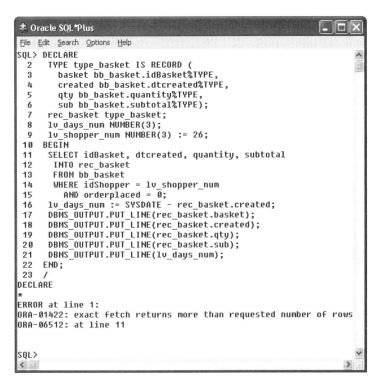

```
Oracle SQL*Plus
File  Edit  Search  Options  Help
SQL> DECLARE
  2    TYPE type_basket IS RECORD (
  3      basket bb_basket.idBasket%TYPE,
  4      created bb_basket.dtcreated%TYPE,
  5      qty bb_basket.quantity%TYPE,
  6      sub bb_basket.subtotal%TYPE);
  7    rec_basket type_basket;
  8    lv_days_num NUMBER(3);
  9    lv_shopper_num NUMBER(3) := 26;
 10   BEGIN
 11     SELECT idBasket, dtcreated, quantity, subtotal
 12       INTO rec_basket
 13       FROM bb_basket
 14       WHERE idShopper = lv_shopper_num
 15         AND orderplaced = 0;
 16     lv_days_num := SYSDATE - rec_basket.created;
 17     DBMS_OUTPUT.PUT_LINE(rec_basket.basket);
 18     DBMS_OUTPUT.PUT_LINE(rec_basket.created);
 19     DBMS_OUTPUT.PUT_LINE(rec_basket.qty);
 20     DBMS_OUTPUT.PUT_LINE(rec_basket.sub);
 21     DBMS_OUTPUT.PUT_LINE(lv_days_num);
 22   END;
 23   /
DECLARE
*
ERROR at line 1:
ORA-01422: exact fetch returns more than requested number of rows
ORA-06512: at line 11

SQL>
```

Figure 3-18   "Exact fetch returns more than requested number of rows" error message

As we have seen, when an Oracle error occurs in the executable section of a PL/SQL block, the processing stops and an error message is displayed. This is definitely not what we want to occur in our applications! We need to anticipate and handle errors in our code so that an application can continue to operate for an end user.

Exception Handler Coding

To create code that can anticipate and handle errors, let's modify the block we just ran to contain exception handlers for the two errors associated with the SELECT statement. Our task is simplified because we are dealing with predefined Oracle errors in this case. Exception names are already associated with these two errors, so we do not need to declare exceptions in the DECLARE section of the block. In addition, both errors are Oracle errors that are raised automatically by the Oracle server, so we do not need to explicitly raise the errors in the executable or BEGIN section of the block. The only task we need to complete is adding exception handlers in the EXCEPTION section of the block to contain instructions on what should be done in the event this error is raised.

To test predefined exception handlers:

1. In a text editor, open the **ex_predef03.txt** file from the Chapter.03 folder. Notice that the SELECT statement is currently set to retrieve noncompleted basket information for shopper 22. Also, review the added EXCEPTION section (which starts eight lines from the bottom). It contains two exception handlers addressing each of our two previous errors.
2. Copy and paste all the code into SQL*Plus.
3. Press **Enter**, if necessary, to execute the block. The results should match Figure 3-19, in which the block traps the "no data found" error and displays the appropriate message to the screen.
4. Return to the text file and change the LV_SHOPPER_NUM variable declaration (which is on Line 9) to an initialized value of shopper 26.
5. Copy and paste all the code into SQL*Plus.
6. Press **Enter**, if necessary, to execute the block. The results should match Figure 3-20, in which the block traps the "too many rows" error and displays a message to the screen.
7. Close SQL*Plus.

```
Oracle SQL*Plus
File  Edit  Search  Options  Help
SQL> DECLARE
  2    TYPE type_basket IS RECORD (
  3      basket bb_basket.idBasket%TYPE,
  4      created bb_basket.dtcreated%TYPE,
  5      qty bb_basket.quantity%TYPE,
  6      sub bb_basket.subtotal%TYPE);
  7    rec_basket type_basket;
  8    lv_days_num NUMBER(3);
  9    lv_shopper_num NUMBER(3) := 22;
 10  BEGIN
 11    SELECT idBasket, dtcreated, quantity, subtotal
 12     INTO rec_basket
 13     FROM bb_basket
 14     WHERE idShopper = lv_shopper_num
 15       AND orderplaced = 0;
 16    lv_days_num := SYSDATE - rec_basket.created;
 17    DBMS_OUTPUT.PUT_LINE(rec_basket.basket);
 18    DBMS_OUTPUT.PUT_LINE(rec_basket.created);
 19    DBMS_OUTPUT.PUT_LINE(rec_basket.qty);
 20    DBMS_OUTPUT.PUT_LINE(rec_basket.sub);
 21    DBMS_OUTPUT.PUT_LINE(lv_days_num);
 22  EXCEPTION
 23    WHEN NO_DATA_FOUND THEN
 24      DBMS_OUTPUT.PUT_LINE('You have no saved baskets!');
 25    WHEN TOO_MANY_ROWS THEN
 26      DBMS_OUTPUT.PUT_LINE('A problem has ocurred in retrieving your saved basket.');
 27      DBMS_OUTPUT.PUT_LINE('Tech Support will be notified and contact you via email.');
 28  END;
 29  /
You have no saved baskets!

PL/SQL procedure successfully completed.

SQL>
```

Message text in the code

Message text displayed to the end user

Figure 3-19   Code displays the appropriate message to the end user

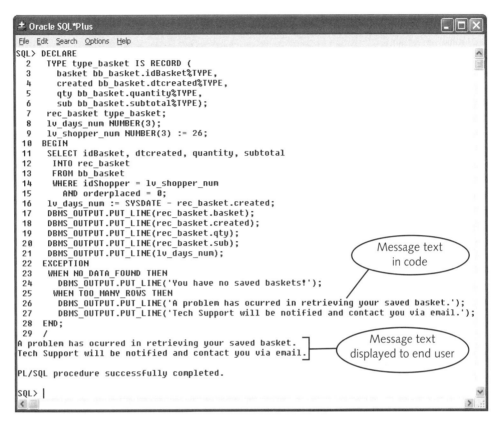

Figure 3-20   Exception handler traps the TOO_MANY_ROWS exception

Keep in mind that as soon as an error occurs in the executable section of the block, the execution moves to the EXCEPTION section and searches for a matching exception handler. If a matching handler is found, the statements in the handler are executed and this block is finished. Execution does not return to the next statement in the executable section of the block. If this block is enclosed in another block (or called from another block as we will see later with named program units), the control returns to the enclosing block.

In the code in the preceding step sequence, notice that each handler displays a different message so that we can determine specifically which handler executed when we are testing this block. The use of DBMS_OUTPUT in this manner is quite useful for testing purposes. However, in an application, we would ultimately provide text that would be assigned to host variables to be displayed on the user application screen or to values assigned to variables to be passed back to the application.

Non-Predefined Oracle Errors

What if an exception handler is needed for a specific Oracle error that does not have a predefined exception (sometimes called unnamed Oracle errors)? In this case, an exception needs to be declared and an Oracle error number needs to be associated with this exception. These two tasks are accomplished in the declaration segment of the block. Figure 3-21 illustrates a block including an error handler for a foreign key violation. The DELETE statement is attempting to eliminate a basket that still has associated item detail rows in the BB_BASKETITEM table.

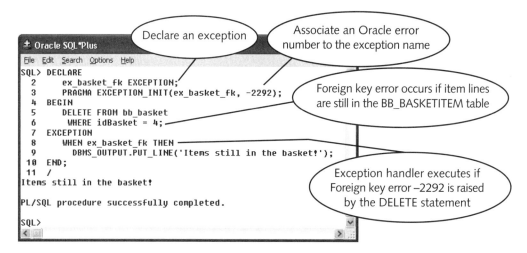

Figure 3-21   Exception handler for a non-predefined Oracle error

First, an exception named EX_BASKET_FK is declared and then the PRAGMA statement associates Oracle error number -2292 with the exception. A **PRAGMA** statement instructs Oracle to use some additional information provided when compiling and executing the block.

Two arguments are required in a PRAGMA EXCEPTION_INIT statement. First, it needs the exception name, which, in this code example, was just declared in the previous line. Second, the statement needs the Oracle error number that should be associated with the exception name. Now, the exception name EX_BASKET_FK is associated with Oracle error ORA-02292. After we have an exception name assigned, the only action that needs to be added is an exception handler that references this error by exception name just as we did to create the predefined Oracle error handler.

In regards to determining non-predefined Oracle error numbers, many developers run SQL statements stand-alone to test which Oracle errors may need to be handled in the block. For example, Figure 3-22 displays the execution of the DELETE statement from our block. This test confirms what kind of Oracle error can occur with this statement.

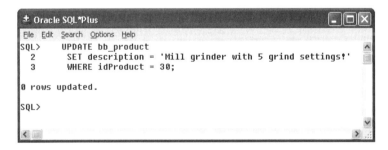

Figure 3-22   DELETE statement Oracle error

This testing assists in developing a list of potential Oracle errors that could be generated by the SQL statement. Keep in mind when dealing with non-predefined Oracle errors that the Oracle error number must be provided in the PRAGMA EXCEPTION_INIT statement.

User-Defined Exception

A **user-defined exception** is an exception that a developer explicitly raises in the block to enforce a business rule. For example, the Brewbean's application has a screen that allows employees to update the description of a product. The list of products is growing, so the employees requested that the screen allow them to type in the product id rather than selecting from a list. They think that the former is faster. However, this also raises the possibility of an invalid product id being entered. How can this situation be handled? Our first step is to examine the code in Figure 3-23 to confirm that the UPDATE of no rows does not raise an Oracle error.

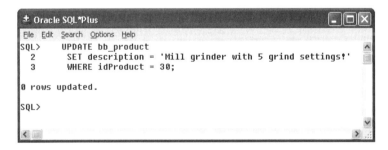

Figure 3-23 UPDATE of no rows does not raise an Oracle error

Because Oracle does not raise an error if an UPDATE does not affect any rows, we must include specific instructions in the block to raise an error if no rows are updated. To accomplish this, three items need to be added to the block. First, an exception needs to be declared. Second, we must explicitly instruct the system when to raise this error in the executable section. Third, an exception handler is needed. Figure 3-24 illustrates the implementation of all three items.

The exception must be explicitly raised in the executable section of the block using the RAISE command. An exception that has been declared must be referred to in the RAISE statement or a PL/SQL error will occur. Do you remember the SQL attributes that allow us to check the results of an SQL statement such as SQL%ROWCOUNT and SQL%NOTFOUND? In the code in Figure 3-24, the SQL cursor attribute of NOTFOUND is used to test (behind the scenes) whether the UPDATE statement affected any rows and whether the exception needs to be raised.

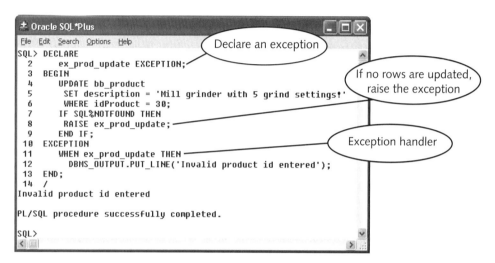

Figure 3-24   User-defined exception

Note that an exception does not have to be related to an SQL statement. Another business rule that may need to be enforced in the Brewbean's application could be a check to ensure the quantity ordered of an item does not exceed the quantity in stock, as shown in Figure 3-25. In this case, we simply compare the two amounts and raise a user-defined exception if the request exceeds the stock amount.

```
± Oracle SQL*Plus                                    [_][□][X]
File  Edit  Search  Options  Help
SQL> DECLARE                                                    ▲
  2      lv_ordqty_num NUMBER(2) := 20;
  3      lv_stock_num NUMBER(4);
  4      ex_prod_stk EXCEPTION;
  5   BEGIN
  6      SELECT stock
  7       INTO lv_stock_num
  8       FROM bb_product
  9       WHERE idProduct = 2;
 10      IF lv_ordqty_num > lv_stock_num THEN
 11       RAISE ex_prod_stk;
 12      END IF;
 13   EXCEPTION
 14      WHEN ex_prod_stk THEN
 15        DBMS_OUTPUT.PUT_LINE('Request quantity beyond stock level');
 16        DBMS_OUTPUT.PUT_LINE('Request qty = '||lv_ordqty_num);
 17        DBMS_OUTPUT.PUT_LINE('Stock qty = '||lv_stock_num);
 18   END;
 19   /
Request quantity beyond stock level
Request qty = 20
Stock qty = 15

PL/SQL procedure successfully completed.

SQL>                                                           ▼
◄ ▥                                                      ►
```

Figure 3-25   Checking the quantity of stock on hand

ADDITIONAL EXCEPTION CONCEPTS

The previous discussion of predefined, non-predefined, and user-defined exceptions are critical in handling anticipated error situations. However, what if unexpected errors occur? We not only want to trap this error to handle it and allow the application to continue, but we also want to document the error number to determine if any code revisions are warranted. This task can be handled using the WHEN OTHERS handler in conjunction with the SQLCODE and SQLERRM functions. All are covered in this section of the chapter. This part of the chapter also covers a mechanism to create your own Oracle error numbers using the RAISE_APPLICATION_ERROR procedure. Last, the flow of exception handling propagation is introduced.

WHEN OTHERS, SQLCODE, and SQLERRM

Another feature available in regards to the exception handlers is the WHEN OTHERS clause in the EXCEPTION segment of the block. It is used to trap any errors that were not specifically addressed in one of the exception handlers. The WHEN OTHERS handler should always be the last handler listed in the EXCEPTION section of a block.

You cannot always anticipate every error that may occur in an application but if an unexpected error does occur, we need the application to handle it so that the application is not abruptly ended. Keep in mind that if an error is raised in the block, the processing jumps right to the exception area of the block and begins looking for a matching handler. If no handler is found for the error, the error is then propagated to the application environment. In other words, an ugly error message is likely to be displayed to the user screen and the application may lock up. To keep this from happening, the WHEN OTHERS handler offers a catchall handler that any error finding no matching handler executes.

Let's return to the example block that retrieves a noncompleted basket for a shopper who logs onto the site. Assume we did not anticipate the TOO_MANY_ROWS error. Figure 3-26 shows the block with a WHEN OTHERS exception handler clause added in the EXCEPTION section. Notice that we do not have a WHEN TOO_MANY_ROWS handler.

```
Oracle SQL*Plus
File  Edit  Search  Options  Help
SQL> DECLARE
  2    TYPE type_basket IS RECORD (
  3      basket bb_basket.idBasket%TYPE,
  4      created bb_basket.dtcreated%TYPE,
  5      qty bb_basket.quantity%TYPE,
  6      sub bb_basket.subtotal%TYPE);
  7    rec_basket type_basket;
  8    lv_days_num NUMBER(3);
  9    lv_shopper_num NUMBER(3) := 26;
 10  BEGIN
 11    SELECT idBasket, dtcreated, quantity, subtotal
 12      INTO rec_basket
 13      FROM bb_basket
 14      WHERE idShopper = lv_shopper_num
 15        AND orderplaced = 0;
 16    lv_days_num := SYSDATE - rec_basket.created;
 17    DBMS_OUTPUT.PUT_LINE(rec_basket.basket);
 18    DBMS_OUTPUT.PUT_LINE(rec_basket.created);
 19    DBMS_OUTPUT.PUT_LINE(rec_basket.qty);
 20    DBMS_OUTPUT.PUT_LINE(rec_basket.sub);
 21    DBMS_OUTPUT.PUT_LINE(lv_days_num);
 22  EXCEPTION
 23    WHEN NO_DATA_FOUND THEN
 24      DBMS_OUTPUT.PUT_LINE('You have no saved baskets!');
 25    WHEN OTHERS THEN
 26      DBMS_OUTPUT.PUT_LINE('A problem has ocurred.');
 27      DBMS_OUTPUT.PUT_LINE('Tech Support will be notified and contact you via email.');
 28  END;
 29  /
A problem has ocurred.
Tech Support will be notified and contact you via email.

PL/SQL procedure successfully completed.

SQL>
```

Figure 3-26   Using the WHEN OTHERS exception handler

In this block, the fetch too many rows error occurs, but no specific handler is included in the EXCEPTION section for this error. Therefore, the last handler of WHEN OTHERS is executed. Keep in mind that if the WHEN OTHERS handler is executed, it probably means an unexpected error occurred in the application and, as developers, we need to examine this error to determine if application modifications are needed. So, how do we capture information regarding the error so that it can be researched?

Our research begins with two functions: The SQLCODE function returns the Oracle error number and the SQLERRM function returns the Oracle error message. Using these functions to save the error information to a transaction log table is quite useful to be able to identify what errors without handlers are occurring in a block. It is typical to include other types of information in this log file, such as user ID, date, and some identification as to what application screen the user was on when the error occurred. Figure 3-27 displays the same block with added statements in the WHEN OTHERS handler to capture the error information using these functions and insert it into a table.

```
Oracle SQL*Plus
File  Edit  Search  Options  Help
SQL> DECLARE
  2    TYPE type_basket IS RECORD (
  3      basket bb_basket.idBasket%TYPE,
  4      created bb_basket.dtcreated%TYPE,
  5      qty bb_basket.quantity%TYPE,
  6      sub bb_basket.subtotal%TYPE);
  7    rec_basket type_basket;
  8    lv_days_num NUMBER(3);
  9    lv_shopper_num NUMBER(3) := 26;
 10    lv_errmsg_txt VARCHAR2(80);
 11    lv_errnum_txt VARCHAR2(10);
 12  BEGIN
 13    SELECT idBasket, dtcreated, quantity, subtotal
 14      INTO rec_basket
 15      FROM bb_basket
 16      WHERE idShopper = lv_shopper_num
 17        AND orderplaced = 0;
 18    lv_days_num := SYSDATE - rec_basket.created;
 19    DBMS_OUTPUT.PUT_LINE(rec_basket.basket);
 20    DBMS_OUTPUT.PUT_LINE(rec_basket.created);
 21    DBMS_OUTPUT.PUT_LINE(rec_basket.qty);
 22    DBMS_OUTPUT.PUT_LINE(rec_basket.sub);
 23    DBMS_OUTPUT.PUT_LINE(lv_days_num);
 24  EXCEPTION
 25    WHEN NO_DATA_FOUND THEN
 26      DBMS_OUTPUT.PUT_LINE('You have no saved baskets!');
 27    WHEN OTHERS THEN
 28      lv_errmsg_txt := SUBSTR(SQLERRM,1,80);
 29      lv_errnum_txt := SQLCODE;
 30      INSERT INTO bb_trans_log (shopper, appaction, errcode, errmsg)
 31        VALUES(lv_shopper_num, 'Get saved basket',lv_errnum_txt,lv_errmsg_txt);
 32      DBMS_OUTPUT.PUT_LINE('A problem has ocurred.');
 33      DBMS_OUTPUT.PUT_LINE('Tech Support will be notified and contact you via email.');
 34  END;
 35  /
A problem has ocurred.
Tech Support will be notified and contact you via email.

PL/SQL procedure successfully completed.

SQL> |
```

Figure 3-27   Using the SQLCODE and SQLERRM functions

The values of SQLCODE and SQLERRM are assigned to variables, as these functions cannot be used directly in an SQL statement. Notice that the SUBSTR function is used on the SQLERRM value. The error message can be up to 512 characters in length, whereas the column used in the BB_TRANS_LOG table to hold this data is only 80 characters in size. Therefore, the SUBSTR function is used to retrieve the first 80 characters of the error message.

Figure 3-28 displays the data that is inserted into the BB_TRANS_LOG table when the WHEN OTHERS handler executes. In this case, the TOO_MANY_ROWS error was raised.

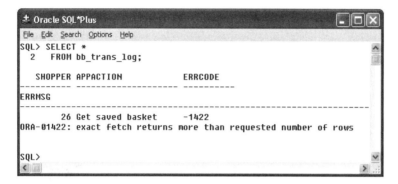

Figure 3-28   Results of the SQLCODE and SQLERRM functions

We'll now test the exception handling process.

To test the exception handling process:

1. In a text editor, open the **ex_test03.txt** file from the Chapter.03 folder. Notice that the SELECT statement is currently set to retrieve noncompleted basket information for Shopper 22, which results in no rows selected. Also, review the EXCEPTION section that contains two exception handlers.

2. Copy and paste all the code into SQL*Plus.

3. Press **Enter**, if necessary, to execute the block. The results should match Figure 3-29, verifying that the NO_DATA_FOUND exception handler was executed.

```
± Oracle SQL*Plus                                    [_][□][X]
File  Edit  Search  Options  Help
SQL> DECLARE
  2    TYPE type_basket IS RECORD (
  3      basket bb_basket.idBasket%TYPE,
  4      created bb_basket.dtcreated%TYPE,
  5      qty bb_basket.quantity%TYPE,
  6      sub bb_basket.subtotal%TYPE);
  7    rec_basket type_basket;
  8    lv_days_num NUMBER(3);
  9    lv_shopper_num NUMBER(3) := 22;
 10  BEGIN
 11    SELECT idBasket, dtcreated, quantity, subtotal
 12     INTO rec_basket
 13     FROM bb_basket
 14     WHERE idShopper = lv_shopper_num
 15       AND orderplaced = 0;
 16    lv_days_num := SYSDATE - rec_basket.created;
 17    DBMS_OUTPUT.PUT_LINE(rec_basket.basket);
 18    DBMS_OUTPUT.PUT_LINE(rec_basket.created);
 19    DBMS_OUTPUT.PUT_LINE(rec_basket.qty);
 20    DBMS_OUTPUT.PUT_LINE(rec_basket.sub);
 21    DBMS_OUTPUT.PUT_LINE(lv_days_num);
 22  EXCEPTION
 23    WHEN NO_DATA_FOUND THEN
 24      DBMS_OUTPUT.PUT_LINE('You have no saved baskets!');
 25    WHEN OTHERS THEN
 26      DBMS_OUTPUT.PUT_LINE('Unexpected error');
 27      DBMS_OUTPUT.PUT_LINE('Error Code = '||SQLCODE);
 28      DBMS_OUTPUT.PUT_LINE('Error Message = '||SQLERRM);
 29  END;
 30  /
You have no saved baskets!

PL/SQL procedure successfully completed.

SQL> |
```

Figure 3-29   NO_DATA_FOUND handler is processed

4. Return to the text file and change the shopper number initialization in the DECLARE section (which is on Line 9) to 26.

5. Copy and paste all the code into SQL*Plus.

6. Press **Enter**, if necessary, to execute the block. The results should match Figure 3-30, verifying that the WHEN OTHERS exception handler was executed. The DBMS_OUTPUT statements display the error number and message.

7. Close SQL*Plus.

RAISE_APPLICATION_ERROR

The RAISE_APPLICATION_ERROR is an Oracle built-in procedure that allows the developers to associate their own error number and message to an error. This procedure is available to use only in stored program units. Stored program units are introduced in Chapter 4, and the discussion of using RAISE_APPLICATION_ERROR is addressed later in this book.

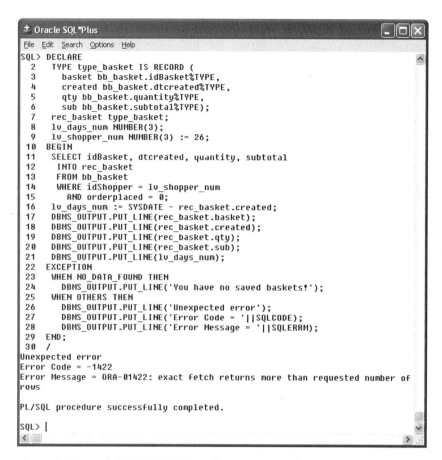

Figure 3-30   WHEN OTHERS handler is processed

Exception Propagation

You know that an error raised in the executable section of a block immediately moves processing to the EXCEPTION section of the block in search of an exception handler. If a matching handler is not found, the error is propagated to the application environment, which is SQL*Plus in our examples. Note, however, that many blocks are called or enclosed by other blocks, which alters the search for a matching exception handler. An exception raised in the executable section first searches the EXCEPTION section of the block for a handler and then, if no handler is found, moves to the enclosing block EXCEPTION section to continue the search for a handler.

Let's take a look at an example of this exception propagation. Figure 3-31 displays a nested block in which a NO_DATA_FOUND error is raised and a matching handler in the nested block is found and executed.

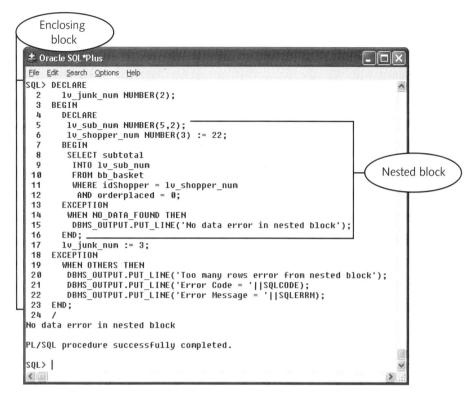

Figure 3-31   A nested block that handles the error

In this case, the error occurs within the nested block so the processing moves to the EXCEPTION section of the nested block. The EXCEPTION section contains an exception handler (WHEN NO_DATA_FOUND) for this error and, therefore, runs and displays the associated message with a DBMS_OUTPUT statement.

What if a matching handler was not found in the nested block? The exception then propagates to the enclosing block and searches for a handler in that block's EXCEPTION section. Figure 3-32 displays the same nested block executed with a TOO_MANY_ROWS error.

Notice the message displayed verifies that the WHEN OTHERS exception handler of the enclosing block executed. The SELECT statement in the nested block retrieves more than one row so an error is raised. The processing first moves to the EXCEPTION section of the nested block; however, no TOO_MANY_ROWS handler is contained in this section. Next, the processing moves to the EXCEPTION section of the enclosing block and processes the WHEN OTHERS exception handler.

If the error is raised in the DECLARE section of a nested block, the exception is immediately propagated to the enclosing block. In such a case, the EXCEPTION section of the nested block is not referenced. Figure 3-33 displays a nested block demonstrating this exception propagation.

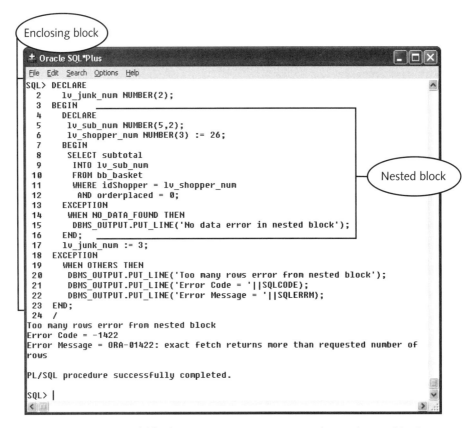

Figure 3-32   A nested block exception propagates to the enclosing block

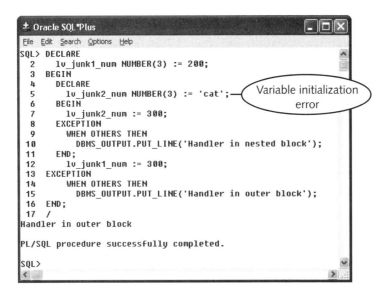

Figure 3-33   DECLARE section error propagates immediately

The LV_JUNK2_NUM variable is declared in the nested block as a number, yet we attempt to initialize with a character string that raises an error. Because the error occurs in the DECLARE section, the process moves immediately to the EXCEPTION section of the enclosing block rather than the EXCEPTION section of the nested block. This process flow is confirmed by the output displayed. Note that both EXCEPTION sections included a WHEN OTHERS handler that processes any errors.

An error can also occur in the EXCEPTION section of a block, which can be an Oracle error raised implicitly or a user-defined error raised explicitly. In this case, the exception propagates to the enclosing block when the error occurs. Figure 3-34 displays a nested block with an error in the executable section and an error in the matching exception handler. Notice that the WHEN OTHERS exception handler in the enclosing block is executed.

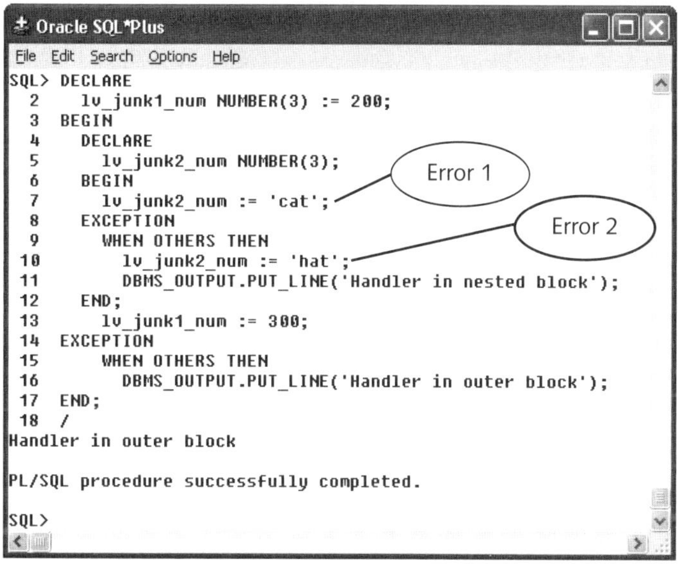

Figure 3-34   Error in exception handler

In this example, an error is first raised in the nested block's BEGIN section by assigning a character value to a numeric variable. The processing then moves to the EXCEPTION section of the nested block and runs the WHEN OTHERS handler. The first statement in the handler raises a second error with another attempt of assigning character data to a numeric variable. Processing then moves to the EXCEPTION section of the enclosing block, which displays the resulting message.

This propagation does not, however, preclude statements in the first exception handler from executing. Let's look at the same example with the exception handler modified to contain a DBMS_OUTPUT statement before the error is raised in the exception handler. Figure 3-35 shows that lines from exception handlers in both the nested and enclosing blocks were processed.

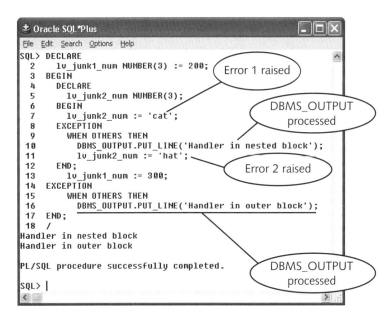

Figure 3-35   Timing of error propagation

In this case, the first statement in the nested block's WHEN OTHERS exception handler is a DBMS_OUTPUT that processes and displays the first message. However, the second statement in this handler raises an error so that the process moves to the EXCEPTION section of the enclosing block and processes the WHEN OTHERS handler.

COMMENTING CODE

A good programming practice is to include comments in your code to document what is being accomplished throughout the program. Note that comment text is not executed; it is included only for the benefit of developers working with the code.

In PL/SQL, you can include both single and multiline comments. A double dash or "--" is used for single line comments. Everything after the "--" is ignored, which means it can be included following code on the same line. Multiline comments open using "/*" and close using "*/". Do not include comments within SQL statements.

The following code demonstrates using both types of comments within a block:

```
DECLARE
   ex_prod_update EXCEPTION;   --For UPDATE of no rows
exception
BEGIN
  /* This block is used to update product descriptions
     Constructed to support the Prod_desc.frm app screen
       Exception raised if no rows updated  */
```

```
     UPDATE bb_product
      SET description = 'Mill grinder with 5 grind settings!'
      WHERE idProduct = 30;
     --Check if any rows updated
     IF SQL%NOTFOUND THEN
      RAISE ex_prod_update;
     END IF;
  EXCEPTION
     WHEN ex_prod_update THEN
        DBMS_OUTPUT.PUT_LINE('Invalid product id entered');
  END;
```

CHAPTER SUMMARY

❐ IF/THEN is one of the basic control structures allowing decisions to be made in the program. Using the ELSIF and ELSE clauses can extend the number of conditions checked. The IF/THEN statement is evaluated from the top down and after a TRUE condition is found, the statement is completed.

❐ CASE statements are now also available to provide decision-making capabilities. The basic CASE statement states a single value or variable to check, which is called a selector. The Searched CASE statement checks different conditions within the WHEN clauses. When many conditions are to be checked, the CASE statement is typically used rather than the IF/THEN.

❐ The loop construct enables us to repeat lines of code. This is especially useful when we need to run a set of logic on each row of a group of rows returned from a database. There is a variety of looping styles, including the basic, WHILE, and FOR loops. The main difference is how the loop is ended and when the condition is checked. The most important item to remember is ensuring that the loop ends.

❐ The GOTO statement allows the logic of the program to jump to a different section of the block.

❐ SQL queries (SELECT) and DML statements (INSERT, UPDATE, and DELETE) can be included in a PL/SQL block.

❐ Errors occurring in the PL/SQL block should be addressed with exception handlers. When handling an Oracle error, you have predefined and non-predefined handlers.

❐ Predefined Oracle exceptions are names already declared in the system for common errors such as NO_ROWS_FOUND, which occurs when a SELECT statement returns no rows.

❐ An exception must be declared for non-predefined Oracle errors. A compiler directive of PRAGMA EXCEPTION_INIT is used to associate an Oracle error number to a handler.

❑ A user-defined error is a situation in which no Oracle error occurs, yet you want to raise an exception to enforce a business rule. In this case, you must explicitly tell the system when the error occurs with the RAISE command.

❑ You can use the WHEN OTHERS handler in conjunction with the SQLCODE and SQLERRM functions in situations such as wanting to trap an error to handle it and to allow the application to continue, but also to document the error number to determine if any code revisions are warranted.

❑ Commenting code assists the developer by documenting what is being accomplished throughout the program. Comment text is not executed; it is included only for the benefit of developers working with the code.

REVIEW QUESTIONS

1. What keyword is used to check multiple conditions with an IF statement?

   a. ELSE IF

   b. ELSEIF

   c. ELSIF

   d. ELSIFS

2. What type of statement should be avoided in structured programming?

   a. CASE

   b. GOTO

   c. PRAGMA

   d. IF/THEN

3. The _____ symbol is used to indicate a single-line comment within PL/SQL code.

4. When does a WHILE loop evaluate the condition that determines if the looping action will continue?

   a. at the beginning of the loop

   b. somewhere within the loop

   c. at the end of the loop

   d. all of the above

5. What type of loop can be used if there is potential that the loop might not need to execute under certain circumstances?

   a. FOR

   b. WHILE

   c. basic

   d. all of the above

6. How is the looping action of a basic loop stopped?

   a. It is stopped when the condition in the LOOP statement is FALSE.

   b. This type of loop has a predetermined number of loops to complete.

   c. The condition in an EXIT WHEN statement is FALSE.

   d. The condition in an EXIT WHEN statement is TRUE.

7. The PRAGMA EXCEPTION_INIT statement accomplishes what task?

   a. It associates an Oracle error number with an exception name.

   b. It associates a user-defined error with an exception name.

   c. It associates a predefined exception name to an Oracle error.

   d. It raises an error in the executable section of a block.

8. The _____ type of exception must use a RAISE statement.

9. If the number of desired loop iterations needed is known, what type of loop should be used?

   a. FOR

   b. WHILE

   c. basic

   d. none of the above

10. What happens first to program execution when an error is raised in the executable section of a PL/SQL block?

    a. The program execution moves to the EXCEPTION section of that block.

    b. The program execution moves to the EXCEPTION section of the enclosing block.

    c. The program execution propagates to the application environment.

    d. none of the above

11. Explain the purpose of exception handlers.

12. User-defined exceptions require three tasks to be accomplished within a block. Describe the three tasks.

13. Describe the two types of exception handlers that exist to manage Oracle errors.

14. Three main types of loop structures exist in PL/SQL. Name each and describe the difference in how each determines how many times the loop executes.

15. Some developers claim benefits exist by using CASE statements instead of IF statements. Describe these benefits.

ADVANCED REVIEW QUESTIONS

1. Review the following IF statement. What is the resulting value of LV_SHIP_NUM if LV_AMT_NUM has a value of 1200?

```
IF lv_amt_num > 500 THEN
    lv_ship_num := 5;
ELSIF lv_amt_num > 1000 THEN
    lv_ship_num := 8;
ELSIF lv_amt_num > 1700 THEN
    lv_ship_num := 10;
ELSE
    lv_ship_num := 13;
END IF;
```

a. 5

b. 8

c. 10

d. 13

2. Review the following block. How many times does the FOR loop process?

```
DECLARE
   lv_cnt_num NUMBER(3);
BEGIN
   FOR I IN 1..7 LOOP
      lv_cnt_num := lv_cnt_num + 2;
   END LOOP;
END;
```

a. 3

b. 4

c. 6

d. 7

3. An Oracle PRAGMA statement _____.

a. associates an Oracle error to an exception name

b. provides additional instructions to be used when the code is executed

c. forces the code to not compile until runtime

d. creates Oracle predefined exceptions

4. Review the following block. What type of exception handler is EX_LIMIT_HIT?

```
DECLARE
    lv_amt_num NUMBER(7,2);
    ex_limit_hit EXCEPTION;
BEGIN
   SELECT amount
      INTO lv_amt_num
```

```
      FROM customer
      WHERE cust_no = :g_cust;
   IF lv_amt_num > 1000 THEN
      RAISE ex_limit_hit;
   END IF;
 EXCEPTION
   WHEN ex_limit_hit THEN
      DBMS_OUTPUT.PUT_LINE('Limit Exceeded!');
 END;
```

   a. predefined Oracle

   b. non-predefined Oracle

   c. user-defined

   d. PRAGMA

5. Which of the following is a valid, predefined Oracle exception?

   a. DATA_NOT_FOUND

   b. TOO_MANY_ROWS

   c. NO_CASE_FOUND

   d. ZERO_DIVISION

HANDS-ON ASSIGNMENTS

Assignment 3-1: Use IF Statements

The Brewbean's application needs a block that determines if a customer is rated high, mid, or low based on his or her total purchases. The block needs to select the total amount of orders for a specified customer, determine the rating, and then display the results to the screen. The code rates the customer HIGH if total purchases are greater than $200, MID if greater than $100, and LOW if $100 or lower. You will use a host variable to provide the shopper ID.

1. Open SQL*Plus.

2. Type **SET SERVEROUTPUT ON** and press **Enter**. This allows you to use DBMS_OUTPUT to display values to the screen.

3. Create and initialize a host variable using the following code:

```
VARIABLE g_shopper NUMBER
BEGIN
 :g_shopper := 22;
END;
/
```

4. Click **File** on the menu bar, and then click **Open**.

5. Navigate to and select the **assignment03-01.sql** file from the Chapter.03 folder.

6. Click **Open**, and a copy of the file contents appears in the SQL*Plus screen. Confirm that this block contains an IF statement to check the total purchase amount of the shopper and to display the correct rating to the screen.

7. To run the code and create the procedure, click **File**, and then click **Run**. The results should match Figure 3-36.

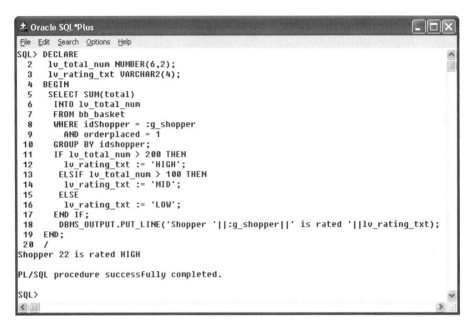

Figure 3-36   Using an IF statement

8. Type and run the following code to confirm the total for this shopper is indeed greater that $200:

```
SELECT SUM(total)
  FROM bb_basket
  WHERE idShopper = 22
    AND orderplaced = 1
  GROUP BY idshopper;
```

9. Close SQL*Plus.

Assignment 3-2: Use Searched CASE Statements

The Brewbean's application needs a block that determines whether a customer is rated high, mid, or low based on total purchases. The block needs to select the total amount of orders for a specified customer, determine the rating, and then display the results to the screen. The code will rate the customer HIGH if total purchases are greater than $200, MID if greater than $100, and LOW if $100 or lower. You will use a host variable to provide the shopper ID.

1. Open SQL*Plus.
2. Type **SET SERVEROUTPUT ON** and press **Enter**.
3. Create and initialize a host variable using the following code:

```
VARIABLE g_shopper NUMBER
BEGIN
 :g_shopper := 22;
END;
/
```

4. Click **File** on the menu bar, and then click **Open**.
5. Navigate to and select the **assignment03-02.sql** file from the Chapter.03 folder.
6. Click **Open**, and a copy of the file contents appears in the SQL*Plus screen. Confirm that this block contains a Searched CASE statement to check the total purchase amount of the shopper and to display the correct rating to the screen.
7. To run the code and create the procedure, click **File**, and then click **Run**. The results should match Figure 3-37.

```
SQL> DECLARE
  2   lv_total_num NUMBER(6,2);
  3   lv_rating_txt VARCHAR2(4);
  4  BEGIN
  5   SELECT SUM(total)
  6    INTO lv_total_num
  7    FROM bb_basket
  8    WHERE idShopper = :g_shopper
  9      AND orderplaced = 1
 10    GROUP BY idshopper;
 11   CASE
 12     WHEN lv_total_num > 200 THEN lv_rating_txt := 'HIGH';
 13     WHEN lv_total_num > 100 THEN lv_rating_txt := 'MID';
 14     ELSE lv_rating_txt := 'LOW';
 15   END CASE;
 16    DBMS_OUTPUT.PUT_LINE('Shopper '||:g_shopper||' is rated '||lv_rating_txt);
 17  END;
 18  /
Shopper 22 is rated HIGH

PL/SQL procedure successfully completed.

SQL>
```

Figure 3-37   Using a Searched CASE statement

8. Type and run the following code to confirm the total for this shopper is indeed greater that $200.

```
SELECT SUM(total)
  FROM bb_basket
  WHERE idShopper = 22
    AND orderplaced = 1
  GROUP BY idshopper;
```

9. Close SQL*Plus.

Assignment 3-3: Use a WHILE Loop

Brewbean's wants to include a feature in their application that calculates the total number of an item that can be purchased with a given amount of money. A WHILE loop is used to increment the cost of the item until the dollar value is met.

1. Open SQL*Plus.

2. Type **SET SERVEROUTPUT ON** and press **Enter**.

3. Create and initialize host variables using the following code:

```
VARIABLE g_total NUMBER

VARIABLE g_prod NUMBER
BEGIN
 :g_total := 100;
 :g_prod := 4;
END;
/
```

4. In a text editor, open the **assignment03-03.txt** file from the Chapter.03 folder. Review the first block that contains a WHILE loop to determine the quantity that can be purchased with the amount available. The condition in the loop statement instructs the system to continue looping only if the LV_AMT_NUM variable is less than the total amount that can be spent.

5. Copy and paste just the first block into SQL*Plus.

6. Press **Enter**, if necessary, to execute the block. The screen should appear as Figure 3-38. Is this correct? No, the total amount calculated for four items is $114, which is greater than the $100 limit. What happened? The WHILE condition is checked at the top of the loop and we have already added the fourth item when this condition is checked and stops the loop.

7. Let's correct this problem. Return to the text file and review the second block in the file. Notice the condition in the WHILE loop now checks if the price of the item is larger than the amount remaining to be spent. This proactively stops the looping before the spending limit is exceeded. Copy and paste the second block into SQL*Plus.

8. Press **Enter**, if necessary, to execute the block. The screen should appear as Figure 3-39. Is this correct? Yes, after three items, at a cost of $28.50 each, are purchased, not enough funds remain to purchase another.

9. Close SQL*Plus.

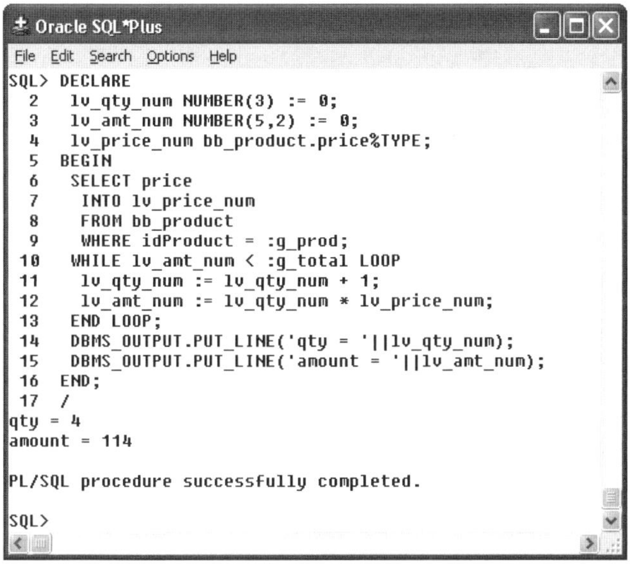

Figure 3-38 Using a WHILE loop improperly

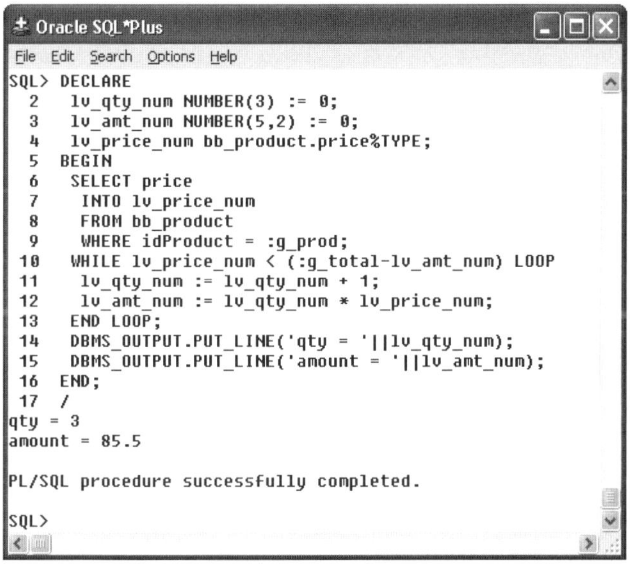

Figure 3-39 Using a WHILE loop

Assignment 3-4: Use Exception Handling

We can test a block containing a CASE statement for errors. Then, we can add an appropriate exception handler for the potential error. The error involved is a predefined Oracle error that has an exception name automatically assigned.

1. Open SQL*Plus.

2. Type **SET SERVEROUTPUT ON** and press **Enter**.

3. Create and initialize host variables using the following code:

```
VARIABLE g_state CHAR(2)
BEGIN
 :g_state := 'NJ';
END;
/
```

4. In a text editor, open the **assignment03-04.txt** file from the Chapter.03 folder. Review the first block that contains a CASE statement and no exception handlers.

5. Copy and paste just the first block into SQL*Plus.

6. Press **Enter**, if necessary, to execute the block. The screen should appear as Figure 3-40. An error is raised, as the state of NJ is not included in the CASE statement; recall that a CASE statement must find a matching case.

```
Oracle SQL*Plus
File  Edit  Search  Options  Help
SQL> DECLARE
  2    lv_tax_num NUMBER(2,2);
  3  BEGIN
  4    CASE :g_state
  5      WHEN 'VA' THEN lv_tax_num := .04;
  6      WHEN 'NC' THEN lv_tax_num := .02;
  7      WHEN 'NY' THEN lv_tax_num := .06;
  8    END CASE;
  9    DBMS_OUTPUT.PUT_LINE('tax rate = '||lv_tax_num);
 10  END;
 11  /
DECLARE
*
ERROR at line 1:
ORA-06592: CASE not found while executing CASE statement
ORA-06512: at line 4

SQL>
```

Figure 3-40   Raising an error with a CASE statement

7. Let's correct this problem. Return to the text file and review the second block in the file. Notice an EXCEPTION section is added to the block and a CASE_NOT_FOUND handler included. Copy and paste the second block into SQL*Plus.

8. Press **Enter**, if necessary, to execute the block. The screen should appear as Figure 3-41. Now, the error is handled in the EXCEPTION section of the block.

9. Close SQL*Plus.

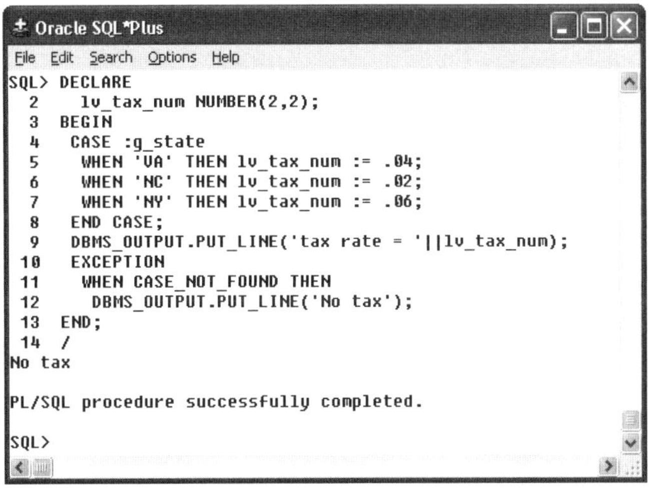

```
± Oracle SQL*Plus                              _ □ ×
File  Edit  Search  Options  Help
SQL> DECLARE
  2     lv_tax_num NUMBER(2,2);
  3  BEGIN
  4    CASE :g_state
  5      WHEN 'VA' THEN lv_tax_num := .04;
  6      WHEN 'NC' THEN lv_tax_num := .02;
  7      WHEN 'NY' THEN lv_tax_num := .06;
  8    END CASE;
  9    DBMS_OUTPUT.PUT_LINE('tax rate = '||lv_tax_num);
 10    EXCEPTION
 11      WHEN CASE_NOT_FOUND THEN
 12        DBMS_OUTPUT.PUT_LINE('No tax');
 13  END;
 14  /
No tax

PL/SQL procedure successfully completed.

SQL>
```

Figure 3-41   Using the CASE_NOT_FOUND exception handler

Assignment 3-5: Work with IF Statements

Brewbean's calculates shipping cost based on the quantity of items in an order. The quantity column in the BB_BASKET table contains the total number of items in a basket. A block is needed to check the quantity provided by a host variable and determine the shipping cost. Display the determined shipping cost to the screen. Test using the basket ids of 5 and 12. Apply the shipping rates as listed in Table 3-2.

Table 3-2   Shipping charges

Quantity of Items	Shipping Cost
Up to 3	$5.00
4–6	$7.50
7–10	$10.00
Over 10	$12.00

Assignment 3-6: Perform Exception Handling of Predefined Errors

A block of code has been created to retrieve basic customer information and can be found in file assignment03-06.txt in the Chapter.03 folder. The application screen was modified so that an employee can directly enter a customer number that could potentially cause an error. An appropriate exception handler needs to be added to the block. The handler should display a message to the screen that states 'Invalid shopper id'. Use a host variable named G_SHOPPER to provide a shopper id. Test the block with a shopper id of 99.

Assignment 3-7: Perform Exception Handling of Non-Predefined Errors

Brewbean's wants to add a check constraint on the quantity column of the BB_BASKETITEM table. If a quantity value provided by a shopper on an item is greater than 20, Brewbean's wants to display a message saying "Check Quantity." The file named assignment03-07.txt can be found in the Chapter.03 folder. The first statement is an ALTER TABLE statement that needs to be executed to add the check constraint. The next item is a PL/SQL block containing an insert that tests this check constraint. Add the appropriate code to the block to trap the check constraint violation and display the desired message.

Assignment 3-8: Perform Exception Handling with User-Defined Errors

On occasion, some of Brewbean's customers mistakenly leave an item out of a basket already checked out, so they create a new basket containing the missing items. However, they request that the baskets be combined so that they are not charged extra shipping. A screen has been developed to allow an employee to modify the basket id of items in the BB_BASKETITEM table to another existing basket to combine the baskets. A block has been constructed to support this screen and can be found as file assignment03-08.txt in the Chapter.03 folder. However, an exception handler needs to be added to trap the situation in which an invalid basket id is entered for the original basket. In this case, the UPDATE affects no rows but does not raise an Oracle error. The handler should display a message stating "Invalid original basket id." Use a host variable named G_OLD with a value of 30 and a host variable named G_NEW with a value of 4 to provide the values to the block. First, verify that no item rows exist in the BB_BASKETITEM table with a basket id of 30.

CASE PROJECTS

Case 3-1: Brewbean's Application Exception Handlers

A new part-time programming employee has been reviewing some existing PL/SQL code you have previously developed. The following two blocks contain a variety of exception handlers. Explain the different types used for the new employee.

BLOCK 1:

```
DECLARE
   ex_prod_update EXCEPTION;
BEGIN
   UPDATE bb_product
     SET description = 'Mill grinder with 5 grind settings!'
     WHERE idProduct = 30;
```

```
      IF SQL%NOTFOUND THEN
        RAISE ex_prod_update;
      END IF;
    EXCEPTION
      WHEN ex_prod_update THEN
        DBMS_OUTPUT.PUT_LINE('Invalid product id entered');
    END;
```

BLOCK 2:

```
DECLARE
 TYPE type_basket IS RECORD (
   basket bb_basket.idBasket%TYPE,
   created bb_basket.dtcreated%TYPE,
   qty bb_basket.quantity%TYPE,
   sub bb_basket.subtotal%TYPE);
 rec_basket type_basket;
 lv_days_num NUMBER(3);
 lv_shopper_num NUMBER(3) := 26;
BEGIN
 SELECT idBasket, dtcreated, quantity, subtotal
  INTO rec_basket
  FROM bb_basket
  WHERE idShopper = lv_shopper_num
    AND orderplaced = 0;
 lv_days_num := SYSDATE - rec_basket.created;
EXCEPTION
 WHEN NO_DATA_FOUND THEN
   DBMS_OUTPUT.PUT_LINE('You have no saved baskets!');
 WHEN OTHERS THEN
   DBMS_OUTPUT.PUT_LINE('A problem has occurred.');
   DBMS_OUTPUT.PUT_LINE('Tech Support will be notified and
will contact you via email.');
END;
```

Case 3-2: More Movie Rentals

The More Movie Rental Company is developing an application screen that displays the total number of times a specified movie has been rented and the associated rental rating based on this count. The ratings are shown in Table 3-3.

Table 3-3 Movie rental ratings

Number of Rentals	Rental Rating
Up to 5	Dump
5–20	Low
21–35	Mid
Over 35	High

Create a block that retrieves the movie title and rental count based on a movie id provided via a host variable. The block should display the movie title, rental count, and rental rating to the screen. Add exception handlers for errors that you can and cannot anticipate. Execute the block using movie ids of 4 and 25.

4

PROCEDURES

In this chapter, you will:

♦ Use named program units
♦ Identify parameters
♦ Use the CREATE PROCEDURE statement
♦ Create a procedure in SQL*Plus
♦ Use the IN OUT parameter mode
♦ Call procedures from other blocks
♦ Use the DESCRIBE command with procedures
♦ Debug procedures using DBMS_OUTPUT
♦ Identify useful software utilities for PL/SQL
♦ Use subprograms
♦ Understand the scope of exception handling and transactions
♦ Use RAISE_APPLICATION_ERROR for error handling
♦ Remove procedures

Naming a program unit or PL/SQL block of code allows the storage and reuse of the code. This saves programmer time and computer resources. We can use SQL*Plus to create procedures. In this chapter, we first identify the need for a procedure in the Brewbean's application. Then, we obtain a fundamental understanding of procedures. Last, we develop a procedure to resolve this application need.

THE CURRENT CHALLENGE IN THE BREWBEAN'S APPLICATION

The Brewbean's company owner is anxious to get a product catalog and ordering application up and running. The application designer has already developed some screen layouts to support the coffee ordering process, as shown in Figures 4-1 and 4-2.

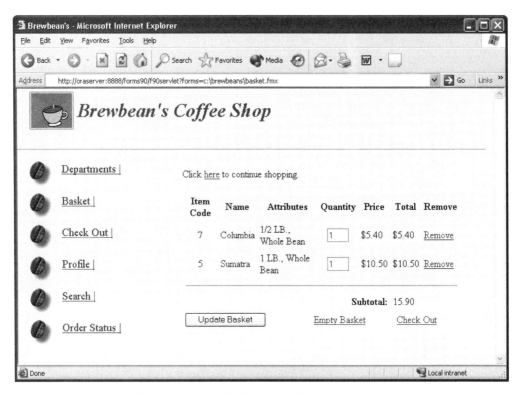

Figure 4-1   Order application screen displaying basket items

The screen in Figure 4-1 is the shopping basket, which shows all the items a shopper has selected thus far. When the shopper has selected all the desired items and is ready to complete the order, the Check Out link is used. Checking out calls the screen in Figure 4-2 to show the order total summary, including shipping and tax, and to allow the shopper to enter credit card information.

We need to create PL/SQL blocks to provide the programming logic to support tasks needed by these application screens, such as calculating the shipping cost of an order, which Brewbean's bases on the quantity of items ordered. These PL/SQL blocks need to be saved somewhere on the system to be available continuously and need to be flexible enough to process the task for any given basket of items. For example, two customers may be on the Web site ordering coffee, and one might order ten items while the other orders three items. The shipping cost calculation needs to factor in the quantity in each case and determine the appropriate shipping cost.

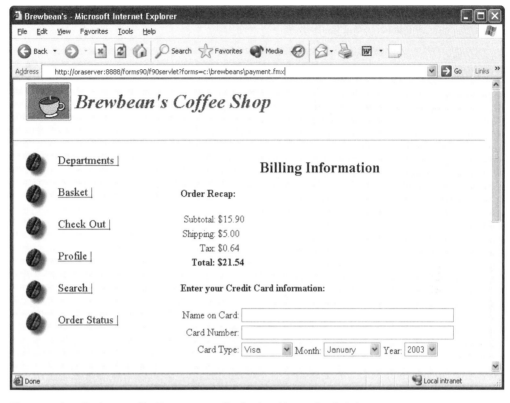

Figure 4-2   Order application screen displaying the order total summary

REBUILDING YOUR DATABASE

Follow these steps to rebuild the Brewbean's database for this chapter.

To rebuild the Brewbean's database:

1. Open SQL*Plus.

2. Create a spool file so that SQL*Plus keeps a copy of all the messages received from running the file that creates the database. On the main menu in SQL*Plus, click **File**, point to **Spool**, and then click **Spool File**. A Select File dialog box appears.

3. Browse to a folder used to contain your working files and in the File Name text box, type **DB_log4**, and then click **Save**. Now, whatever text is seen in our SQL*Plus session is saved to this file for future reference.

4. Now let's create the database. In SQL*Plus, type the code following this step. It runs all the statements contained in the c4Dbcreate.sql file. Messages verifying

the creation and data insertion steps scroll on the SQL*Plus screen. This may take a couple minutes to complete.

```
@<pathname to PL/SQL files>\Chapter.04\c4Dbcreate
```

5. Now, you can turn the spooling off so that we can review the results in the spool file created. On the main menu click **File**, point to **Spool**, and then click **Spool Off**.

6. Open a text editor.

7. Open the **DB_log4.lst** file in your working folder.

8. Review the messages for any errors.

In previous chapters, you were instructed as to when you should close windows. From this point forward in the book, that task is left to your discretion.

INTRODUCTION TO NAMED PROGRAM UNITS

So far, we have created anonymous PL/SQL blocks and executed the code in SQL*Plus. These are called **anonymous** blocks because the code is not stored to be reused and, as far as the Oracle9i server is concerned, the block of code no longer exists after it has executed. Every time we run an anonymous block of code, it is compiled and then executed.

Much of the time, we want to save the PL/SQL block so that it can be reused, which means a name needs to be assigned so that the system can find the particular block as we refer to it or call it by name. Therefore, we name our PL/SQL blocks so that they can be saved on the Oracle9i server and be continually referenced just as we would with other database objects, such as tables and sequences.

A PL/SQL block created and named is known as a **named program unit** or subprogram. The term **program unit** is used to denote that we typically create blocks of code to perform a specific task that may be needed within a number of applications. This text focuses primarily on **stored program units**, which denote that the program unit has been saved in the database. Anyone connected to the database (and with appropriate rights) can use or call a stored program unit. Therefore, stored program units can be shared or used by many applications.

As an example of the use of stored program units, let's assume we have stored our customer names in two columns called LAST and FIRST in the CUSTOMER table. If we have several application screens that display the customer name in the form of "last name, first name," instead of programming a concatenation operation for every screen displaying the name, we could create a stored program unit to accomplish this and reuse this whenever needed. This sharing or reusing of code is an important contributor to programming efficiency.

What happens when the application users decide that the name should appear differently on all the application screens? The programmer only needs to change the one stored program unit to resolve the issue rather than going to each application screen using the customer name and changing the appropriate code! This is a great time-saver.

As application programmers, one critical task to accomplish is creating code that is easy to maintain. Keep in mind that PL/SQL program units are critical in Oracle9i application development as a mechanism in providing application code modularization and reusability. In the shipping cost task that we must tackle, a program unit that is flexible enough to calculate the shipping cost for any order is needed.

Client and Server Considerations

As you might recall from your previous studies, **client-side** refers to program code that resides on the user or client machine, and **server-side** refers to program code that resides on the server, as depicted in Figure 4-3. Even though we concentrate on stored program units in this text, note that PL/SQL program unit coding is the same regardless of whether it is stored on the database server or saved to a local library on the user's computer.

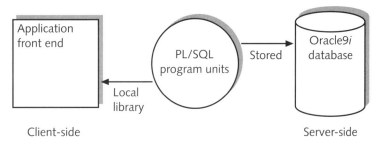

Figure 4-3   Diagram showing relationships among clients and servers

The concern in the application development arena involves the decisions as to which part of the application code should be stored on the client, or front end, and which part of the application code should be stored on the server (or the Oracle9i database, as shown in Figure 4-3). Utilizing the Oracle application developer tools allows the programmer to save program units on the client-side in local libraries that reside on the front-end computer. You can also call stored program units that are server-side or saved in the database. Typically, any PL/SQL block that has the potential of being used by a number of applications and that is intensive in database activity can be saved to the database to improve efficiency and allow sharing.

Types of Named Program Units

Named PL/SQL program units take on a variety of constructs based on where the code is stored, what type of task it performs, and how it is invoked. Table 4-1 presents a list of these various program unit types and a brief description of each. Regardless of the type of program unit being created, all of these program units consist of PL/SQL blocks.

Procedures and functions are quite similar in construct in that they can accept input values and return values. However, a procedure is used to accomplish one or more tasks, return none or many values, and can be used only in PL/SQL statements. On the other hand, a function contains a RETURN clause and is used to manipulate data and return a single resulting value. The functions we construct are similar to the Oracle9i-supplied functions, such as ROUND, that we have already used.

Table 4-1   Named program unit types

Program Unit Type	Description
Stored procedures and functions	Performs a task such as calculation of shipping cost. Can receive input values and return values to the calling program. Called explicitly from a program. Stored in the Oracle9i database.
Application procedures and functions	Same as stored procedures and functions except these are saved in an Oracle9i application or library on the client-side.
Package	Groups together related procedures and functions. Called explicitly from a program. Stored on the server-side.
Database trigger	Performs a task automatically when a DML action occurs on the table with which it is associated. Stored in the Oracle9i database.
Application trigger	Performs a task automatically when a particular application event occurs such as the user clicks a button on the screen. Stored in an Oracle9i application.

MAKING PROCEDURES REUSABLE: PARAMETERS

One important item that needs to be considered when creating stored procedures is to make them flexible so that they can be reused. Keep in mind that these procedures provide the programming logic in an application and each user may enter different values to process into the application.

We make program units flexible by using **parameters**, which are mechanisms to send values in and out of the program unit. Procedures can accept and return none, one, or many parameters. In our name formatting for Brewbean's, a procedure could be written that accepts a customer number as a parameter and the SELECT statement in the procedure could use this parameter coming in to form the WHERE clause and return the appropriate customer name in the concatenated format. Therefore, every time this procedure runs, it could return the name of a different customer; however, it always accomplishes the same task of retrieving a customer name based on the customer number and concatenating the customer first and last names.

A parameter must be assigned one of three available modes. The **mode** indicates which way the value provided for the parameter flows: into the procedure, out of the procedure, or both. When creating the procedure, we list each parameter along with a mode and data type. The available modes are listed in Table 4-2.

Table 4-2   Parameter mode types

Mode	Description
IN	Passes a value from the application environment into the procedure. This value is considered a constant, as it cannot be changed within the procedure. Is the default if no mode is indicated.
OUT	Passes a value out of the procedure to the application environment. If values are calculated or retrieved from the database within the procedure, OUT parameters are used to return these values to the calling environment.
IN OUT	Allows a value to be passed in and out using the same parameter. The value sent out can be different than the value sent in.

CREATE PROCEDURE STATEMENT

Let's first take a look at the CREATE PROCEDURE command syntax listed in Figure 4-4.

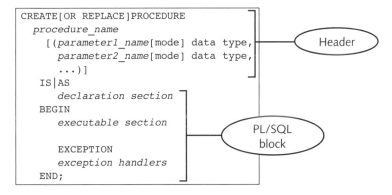

```
CREATE[OR REPLACE]PROCEDURE
    procedure_name
        [(parameter1_name[mode] data type,
         parameter2_name[mode] data type,
         ...)]
    IS|AS
        declaration section
    BEGIN
        executable section

        EXCEPTION
        exception handlers
    END;
```

Header

PL/SQL
block

Figure 4-4   CREATE PROCEDURE command syntax

In reviewing the syntax layout in Figure 4-4, notice that the top portion of the syntax is called the header section. This section indicates what type of program unit is being created, an object name, parameter information, and the keyword IS to indicate a PL/SQL block follows. Below the IS keyword, the PL/SQL block looks the same as the anonymous blocks we have created, with the exception that the DECLARE keyword is no longer needed. Why? The header section now marks the beginning of the PL/SQL block.

Three pieces of information for each parameter must be provided in the header: name, mode, and data type.

The Name

The name should be useful to the programmer for identifying what type of data it represents and must conform to Oracle9*i* naming standards, such as with table column names. Recall that names must begin with an alphabetic character and can contain up to 30 characters.

A naming convention strategy is an important survival tool to have in any programming shop. Some examples of naming conventions include using a prefix of "p_" in each parameter name or a suffix of "_num" in a numeric variable name. In both cases, the variable name itself lets us know something about it, such as it is a parameter or it holds numeric data. As we are coding or reviewing existing code to make changes, the naming conventions make for much easier reading.

The Mode

Values are passed between the application environment or user screens and program units via parameters. Three choices exist for parameter mode: IN, OUT, and IN OUT. Keep in mind that it is not required for parameters to be used in a procedure, but it is quite typical. For example, you may create a procedure that inserts logon information, such as user ID and date, into a table that does not require any input values from the application or return any values. In this scenario, no parameters are needed in the procedure.

If no mode is indicated, the default value of IN is applied. However, explicitly indicating the mode is the preferred style to make reading the code easier. If an IN or IN OUT parameter is used, the IN denotes that when the procedure is invoked or run, the procedure expects a value to be provided for this parameter and an error if no value is provided. If an OUT or an IN OUT parameter is used, the OUT denotes that a variable must be created in the calling environment that can hold the value returned from the procedure for this parameter.

The Data Type

The data type is the last required item for each parameter. An SQL data type, such as CHAR, NUMBER, or DATE, is needed. It is important to note that no size or precision information is included in the parameter data types. When a procedure is invoked, the size properties of the arguments or values that are supplied to pass into the parameters control the size and precision of the parameters. The values from the application that pass into parameters are called **actual parameters**; the parameters in the procedure declaration are called **formal parameters**.

The %TYPE and %ROWTYPE attributes may be used to provide a data type based on a table column, which is especially appropriate for parameters that are holding data retrieved from the database. Recall that using these attributes in variable declarations can minimize program maintenance. If a table column data type is modified, no changes to associated variables are required because the %TYPE attribute always uses the column's current data type. Note that a default value assignment on parameters is also legal and accomplished in the same manner as with PL/SQL variables using the DEFAULT keyword or := symbol.

The following code is a procedure header section containing two parameter declarations. This procedure is intended to sum the total for all items in the basket for the basket number that is provided as input. The P_TOTAL parameter is using both the %TYPE and the default value options.

```
CREATE OR REPLACE PROCEDURE total_calc_sp
  (p_basket IN bb_basket.idbasket%TYPE,
  p_total OUT bb_basket.total%TYPE := 0)
IS
  - - PL/SQL Block coded here - -
```

CREATE A PROCEDURE IN SQL*PLUS

The Brewbean's application needs a procedure to calculate the shipping cost of an order, so let's jump into producing our first procedure using parameters to resolve this processing need. We need to create a procedure that calculates the shipping cost for an order. The company currently calculates shipping based on the number of items in the order. Figure 4-5 displays the data from the SHIPPING table that is in the company database.

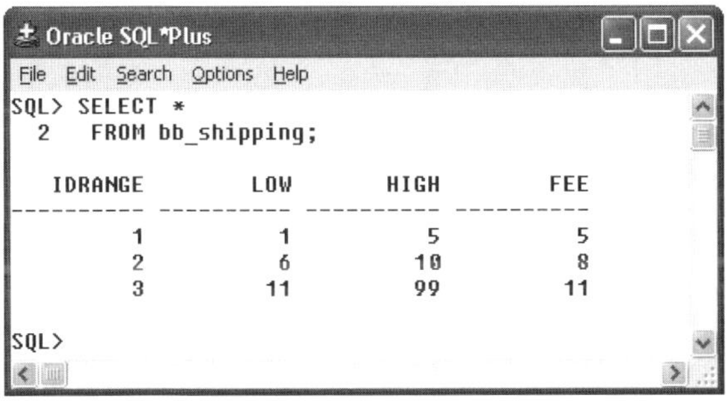

Figure 4-5   SHIPPING table data listing

Notice that the SHIPPING table depicts quantity ranges for determining shipping costs. For example, if the order consists of a total of seven items being ordered, the shipping cost is $8.00 because the quantity falls between a lower bound of six and an upper bound of ten. Assume that the quantity is an amount that is provided to the procedure.

Now, in Figure 4-6, the SHIP_COST_SP procedure has been added and executed.

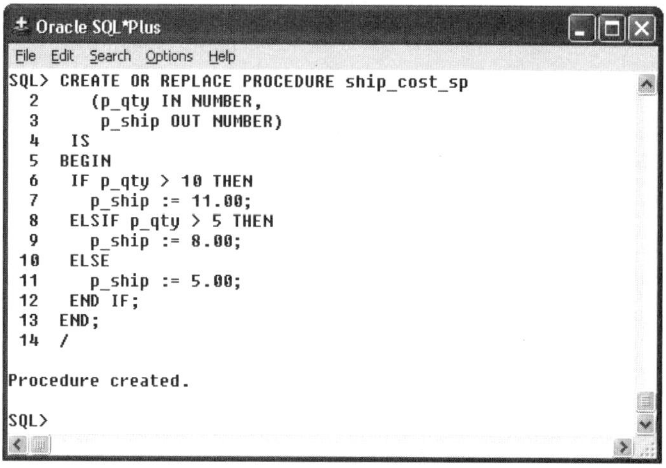

```
± Oracle SQL*Plus                                    _ □ X
File  Edit  Search  Options  Help
SQL> CREATE OR REPLACE PROCEDURE ship_cost_sp
  2      (p_qty IN NUMBER,
  3        p_ship OUT NUMBER)
  4    IS
  5  BEGIN
  6    IF p_qty > 10 THEN
  7        p_ship := 11.00;
  8    ELSIF p_qty > 5 THEN
  9        p_ship := 8.00;
 10    ELSE
 11        p_ship := 5.00;
 12    END IF;
 13  END;
 14  /

Procedure created.

SQL>
```

Figure 4-6 SHIP_COST_SP procedure is added and executed

The two parameters listed in the SHIP_COST_SP procedure are P_QTY and P_SHIP. P_QTY is an IN parameter, so it expects a value to be sent in when the procedure is run. P_SHIP is an OUT parameter that provides the mechanism to return the shipping cost to the calling environment. The parameters are used just like variables declared in the block. Notice that a forward slash needs to be on the last line so that SQL*Plus knows to run the CREATE statement. If the procedure compiles and runs without any errors, the appropriate message is displayed and the SQL> prompt returns.

When a CREATE PROCEDURE Statement Produces Errors

What if a CREATE PROCEDURE statement produces errors? Let's modify a procedure to see how to handle errors.

To work with procedure errors in SQL*Plus:

1. Copy the CREATE PROCEDURE statement from Figure 4-6 into your favorite text editor.

2. In the text editor, modify the data type for the P_QTY parameter to be **NUMBER(3)**.

3. Open or return to SQL* Plus. Copy and paste the modified statement into SQL*Plus.

4. Press the **Enter** key so that the statement runs.

5. The message "Warning: Procedure created with compilation errors" is displayed. Type **SHOW ERRORS**, and then press **Enter** to display the error messages shown in Figure 4-7. The error message indicates a problem on Line 2 at the point at which the size information is entered in the parameter declaration.

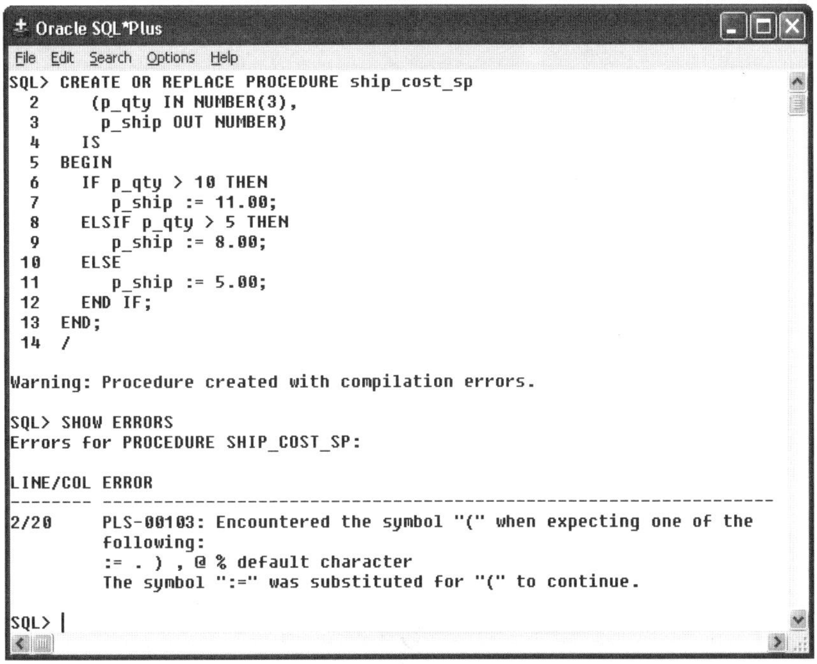

```
± Oracle SQL*Plus                                              [_][□][X]
File  Edit  Search  Options  Help
SQL> CREATE OR REPLACE PROCEDURE ship_cost_sp
  2      (p_qty IN NUMBER(3),
  3       p_ship OUT NUMBER)
  4    IS
  5  BEGIN
  6    IF p_qty > 10 THEN
  7       p_ship := 11.00;
  8    ELSIF p_qty > 5 THEN
  9       p_ship := 8.00;
 10    ELSE
 11       p_ship := 5.00;
 12    END IF;
 13  END;
 14  /

Warning: Procedure created with compilation errors.

SQL> SHOW ERRORS
Errors for PROCEDURE SHIP_COST_SP:

LINE/COL ERROR
-------- -----------------------------------------------------------
2/20     PLS-00103: Encountered the symbol "(" when expecting one of the
         following:
         := . ) , @ % default character
         The symbol ":=" was substituted for "(" to continue.

SQL> |
```

Figure 4-7   Using SHOW ERRORS to check compilation errors in SQL*Plus

6. Return to the text editor and correct the error by removing the size on the parameter data type.

7. Copy and paste the corrected statement into SQL*Plus.

8. Press **Enter** so that the statement runs. Be sure the "Procedure created" message is received so that you know you have a correct procedure.

In our error scenario, the error message pointed right to the line with the error. However, this is not the case with all errors. A common error is to forget a semicolon or comma at the end of a line in the procedure. In this case, the error points to the next line because an error is not recognized until reaching the next line. Always check the line immediately preceding the one referenced in the error message for this type of error.

Testing a Procedure

Now, we need to test the procedure to verify proper execution and results. The SQL*Plus EXECUTE statement can call or invoke a procedure by name. The call requires any arguments to be listed in parentheses and it must contain an argument for each parameter in the procedure being called. The arguments in the call match by position to the parameter list in the procedure. In other words, the first argument in the call is sent into the first parameter, which is P_QTY. However, the second parameter is an OUT parameter used to return the shipping cost, so we need to provide the name of a host or bind variable to hold the returned value.

Note that a **host** variable is a variable that exists in the application environment that can range from a field on an application screen to an SQL*Plus variable. In the scenario, we are testing or executing our procedure in SQL*Plus and, therefore, SQL*Plus is our host application environment. To reference a host variable in PL/SQL, a preceding colon must be used with the host variable name. The colon instructs the PL/SQL interpreter engine to look at the application environment for this variable rather than inside the PL/SQL block.

Keep in mind that in a complete application, the program units provide the logic that runs behind all the user screens. A host variable in a completed application is typically an object on the user screen, such as a text box or check box.

To test the SHIP_COST_SP procedure:

1. Open or return to SQL*Plus.

2. Type **VARIABLE g_ship_cost NUMBER** to create the needed host variable named G_SHIP_COST, as shown at the top of Figure 4-8, and then press **Enter**.

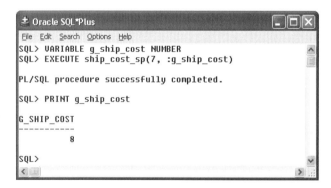

Figure 4-8   Testing the SHIP_COST_SP procedure

3. Now, we enter the SQL*Plus EXECUTE command to run the procedure. Type **EXECUTE ship_cost_sp(7, :g_ship_cost)** and then press **Enter**. The procedure is called by name and includes two arguments, one for each parameter in the procedure.

4. After execution, we need to check our results by determining what value was placed into the host variable G_SHIP_COST. Type **PRINT g_ship_cost** to display the shipping cost value. The variable contains 8, as shown in Figure 4-8, which reflects the correct shipping cost for an order containing seven items.

The arguments used in the procedure execution call were passed with a **positional method**. That is, when invoking the procedure, the first argument value is matched up with the first parameter in the procedure, the second argument value is matched up with the second parameter in the procedure, and so on. In other words, when invoking a procedure list, the values are listed in the order in which the parameters are declared in the procedure.

Another method that can be used is called the named association. In the **named association**, we associate a value to each parameter by name in the invoke statement. Let's look at the example of executing the SHIP_COST_SP procedure again but this time, we will use the named association, as shown in Figure 4-9.

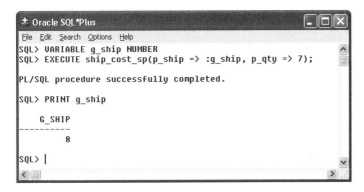

Figure 4-9   Using the named asssociation method

The special syntax of => is used to accomplish the named association. In this case, the values are passed to the parameters starting with the last parameter in the procedure and moving up through all of them. A mixture of positional and named association methods can be used.

Keep in mind when developing user screens or forms in Oracle9*i* Developer software that the application logic is coded with PL/SQL. This fact is important to you because your foundation in PL/SQL skills is applicable in completing an entire application. In addition, the application includes calls to our stored procedures to make use of them, so understanding the procedure invocation using arguments is critical.

USING THE IN OUT PARAMETER MODE

A parameter with an IN only mode can receive a value, but this value cannot be changed in the procedure. With an OUT only parameter, the parameter is empty when the procedure is invoked and a value is assigned to the parameter during the execution of the procedure and returned upon completion.

Let's now look at using the third available parameter mode: IN OUT. With this mode, the parameter can accept a value upon invocation and this value can be changed within the procedure so that a different value could be returned to the calling environment. In other words, the IN OUT combined mode can achieve the passing of a value both in and out of a procedure with a single parameter.

The Brewbean's application screens currently display customer phone numbers with no parentheses or dashes. The sales staff requested that the phone numbers be formatted to include parentheses and dashes to make them easier to read.

Let's build a procedure that formats a phone number using one IN OUT parameter to pass a phone number in and return it in a formatted form.

To use an IN OUT parameter:

1. Open or return to SQL*Plus.

2. Click **File** and then **Open** from the main menu. Navigate to and select the **phone04.sql** file from the Chapter.04 folder.

3. Click **Open** and a copy of the file contents appears in the SQL*Plus screen. Notice the code creates a procedure with one IN OUT parameter.

4. To run the code and create the procedure, click **File** and then **Run** from the main menu. You receive a "Procedure created" message.

5. Now let's test this procedure. Create a host variable by typing **VARIABLE g_phone VARCHAR2(13)** and then press **Enter**.

6. Enter the following anonymous block to initialize the host variable:
```
BEGIN
  :g_phone := '1112223333';
END;
/
```

7. Type **EXECUTE phone_fmt_sp(:g_phone);** and then press **Enter**.

8. Display the current value being held in the G_PHONE host variable by typing **PRINT :g_phone;** and pressing **Enter**. Notice a value with no parentheses and dashes went into the procedure via the G_PHONE parameter and a formatted value with parentheses and dashes was returned via the same parameter. Your screen should resemble Figure 4-10.

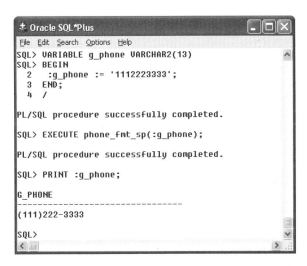

4

Figure 4-10   Testing the IN OUT parameter mode

Note that we would typically use this procedure from within another procedure; however, to test it before integrating it into another procedure, we must use an anonymous block to initialize the value of the host variable.

CALLING A PROCEDURE FROM ANOTHER PL/SQL BLOCK

Recall that the Brewbean's company application contains a screen that allows the user to select a Check Out link to calculate an order subtotal containing shipping costs (the tax cost is also displayed; however, we build up to this one step at a time, so ignore the tax part for now). A procedure is needed to accomplish this task and return the appropriate values to be displayed to the shopper.

As the shopper selects items on the Web site, a basket id number is assigned to his or her shopping cart and, therefore, a shopping basket identification number is provided as input into this procedure. The procedure needs to retrieve the basket items from the database and calculate the total product and shipping cost. The procedure needs to return the number of items being purchased, product total, shipping cost, and a total named ORDER_TOTAL_SP. We have already created a procedure to calculate the shipping cost; therefore, we will use this existing stored program unit from our new procedure to accomplish this part of the task.

Creating a Procedure That Calls Another Procedure

Use these steps to construct the ORDER_TOTAL_SP procedure, which uses the SHIP_COST_SP procedure to calculate the shipping cost.

To build the ORDER_TOTAL_SP procedure:

1. Open or return to SQL*Plus.

2. Review the code in Figure 4-11. Notice Line 13. In this line, the statement is calling the SHIP_COST_SP procedure that we previously constructed. Note that the procedure includes four OUT parameters to return all the needed values. Type the code in SQL*Plus and then execute the code.

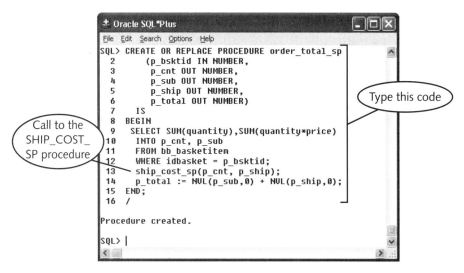

Figure 4-11   Procedure calling another procedure

3. Note that a "Procedure created" message appears.

The first task accomplished in the procedure executable section is retrieving the data from the BB_BASKETITEM table for the specific shopping basket identified with the IN parameter value. The aggregate function of SUM is used to total the number of items ordered and the cost of all items. The number of items ordered or P_CNT then becomes the IN parameter value for calling the SHIP_COST_SP procedure to calculate the shipping cost.

Recall that the shipping cost is based on the number of items ordered. Notice the line of code calling the SHIP_COST_SP procedure is an entire program statement. The execution flow is this: the SELECT statement processes, the procedure executes, and, finally, the assignment statement for P_TOTAL runs. It processes as if the program statements in the stored procedure were embedded right where the call occurs.

Testing the Procedure

As always, after we have completed a program unit, we need to verify that it executes properly.

To test the procedure:

1. Open or return to SQL*Plus.

2. The ORDER_TOTAL_SP procedure contains four OUT parameters and, therefore, we first need to create four host variables to hold the values returned. Enter the following VARIABLE commands:

```
VARIABLE g_cnt NUMBER
VARIABLE g_sub NUMBER
VARIABLE g_ship NUMBER
VARIABLE g_total NUMBER
```

3. Run the procedure for basket 12 by typing **EXECUTE order_total_sp (12, :g_cnt,:g_sub,:g_ship,:g_total);**.

4. To display the values returned, type **PRINT(host variable name)** for each host variable in the Interpreter pane, as shown in Figure 4-12.

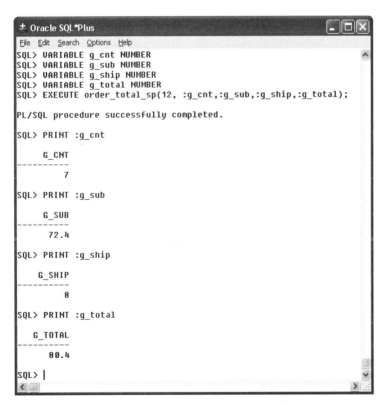

Figure 4-12   Testing the ORDER_TOTAL_SP procedure

5. To verify the correct data was calculated, run a SELECT statement querying the BB_BASKETITEM table for the quantity of items being ordered and the subtotal in the Interpreter pane, as shown in Figure 4-13. The quantity of 7 should match the value in the G_CNT host variable and the subtotal of 72.4 should match the value in the G_SUB host variable.

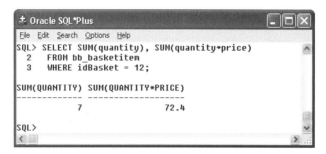

Figure 4-13   Query the BB_BASKETITEM table to verify the accuracy of the ORDER_TOTAL_SP procedure

DESCRIBE COMMAND

A stored program unit is saved to the database, just like other objects, such as a table. That being said, just as we use the DESCRIBE command to list the structure of a table, we can use the DESCRIBE command to list details about the structure of the procedure, such as information regarding the parameters. Figure 4-14 displays the DESCRIBE command on the ORDER_TOTAL_SP procedure in SQL*Plus.

Figure 4-14   Results of the DESCRIBE command on a procedure

The listing displays the parameter (or argument) name, data type, and mode. This is quite useful when we are writing blocks that call a procedure because we can quickly issue a DESCRIBE command to confirm that the procedure call provides the correct values to match the procedure parameters.

DEBUGGING IN SQL*PLUS BY DISPLAYING MESSAGES TO THE SCREEN

An important mechanism used to debug programs is displaying values to the screen to confirm logic processing or whether statements have executed. We have already used the DBMS_OUTPUT.PUT_LINE statement to confirm which, if any, exception handler executed and to display variable values. Now, we need to explore using the DBMS_OUTPUT.PUT_LINE statement within the executable section of a block to assist in debugging our code. **Debugging** is the process of identifying and removing errors from within program code.

Let's experiment with a new procedure named PROMO_TEST_SP that is used in conjunction with a new free shipping promotion. The company is experimenting with the idea of setting up some purchase incentives to encourage repeat shoppers. If a shopper spends more than $25 in a month, a free shipping offer is extended for his or her next purchase over $25. If a shopper has spent more than $50 in a month, a free shipping offer for his or her next purchase is extended regardless of the total. The SUBTOTAL column is summed in the procedure to reflect actual product purchase total. The procedure updates the BB_PROMOLIST table with the shopper information, which is used to e-mail shoppers the appropriate incentive. The PROMO_FLAG variable is used to assign the correct promotion code.

Review the procedure listed in Figure 4-15. The procedure is called PROMO_TEST_SP, and it uses a CURSOR FOR loop to review the total purchases by shopper for a given month and year. Recall a CURSOR FOR loop is a cursor that automatically sets up the necessary record and counter variable for the looping action.

To debug in SQL*Plus with DBMS_OUTPUT:

1. Open or return to SQL*Plus.

2. Click **File** and then **Open** from the main menu. Navigate to and select the **promotest04.sql** file from the Chapter.04 folder.

3. Click **Open** and a copy of the file contents appears in the SQL*Plus screen.

4. To run the code and create the procedure, click **File** and then **Run** from the main menu. You receive a "Procedure created" message.

5. Now, we need to execute the procedure so that we can test the results. In SQL*Plus, type **EXECUTE promo_test_sp('FEB', '2003');**.

6. Now, it's time to run a query to check what the procedure inserted into the BB_PROMOLIST table. Type **SELECT * FROM bb_promolist;**. Notice that five rows were inserted into the BB_PROMOLIST table. However, right off the bat, we notice that no flag value was inserted for Shopper 23, as shown in Figure 4-16.

```
± Oracle SQL*Plus                                              _ □ ×
File  Edit  Search  Options  Help
SQL> CREATE OR REPLACE PROCEDURE promo_test_sp
  2    (p_mth IN CHAR,
  3     p_year IN CHAR)
  4    IS
  5    CURSOR cur_purch IS
  6      SELECT idshopper, SUM(subtotal) sub
  7      FROM bb_basket
  8      WHERE TO_CHAR(dtCreated,'MON') = p_mth
  9        AND TO_CHAR(dtCreated,'YYYY') = p_year
 10        AND orderplaced = 1
 11      GROUP BY idshopper;
 12    promo_flag CHAR(1);
 13  BEGIN
 14   FOR rec_purch IN cur_purch LOOP
 15    If rec_purch.sub > 50 THEN
 16        promo_flag := 'A';
 17    ELSIF rec_purch.sub > 25 THEN
 18        promo_flag := 'B';
 19    END IF;
 20    IF promo_flag IS NOT NULL THEN
 21      INSERT INTO bb_promolist
 22        VALUES (rec_purch.idshopper, p_mth, p_year, promo_flag, NULL);
 23    END IF;
 24    promo_flag := '';
 25   END LOOP;
 26   COMMIT;
 27  END;
 28  /

Procedure created.

SQL> |
```

Figure 4-15   PROMO_TEST_SP procedure code

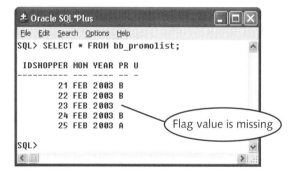

Figure 4-16   Listing of BB_PROMOLIST table data

7. To verify the results, let's do a query on the BB_BASKET table to verify the order totals of each shopper and who is eligible for the promotion. Type the **SELECT** command, as shown in Figure 4-17. Notice that Shopper 23 has a total of only 21.60 and is not eligible for the promotion.

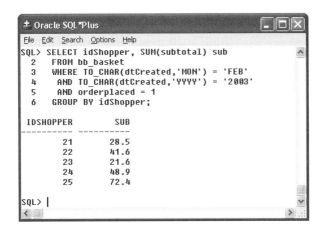

Figure 4-17   Query the BB_BASKET table to verify shopper order totals

Now we have a problem. Our procedure has included Shopper 23 in the BB_PROMOLIST table; however, this shopper is not eligible for the promotion and, therefore, should not be inserted into the table. How can we determine the problem with this procedure? Let's explore adding DBMS_OUTPUT statements to assist.

To investigate the Shopper 23 issue using DBMS_OUTPUT:

1. Open or return to SQL*Plus.

2. Type **SET SERVEROUTPUT ON**, and then press **Enter** to enable message display.

3. Type **DELETE FROM bb_promolist; COMMIT;** to delete the rows we previously inserted.

4. In a text editor, open the **promotest04.sql** file from the Chapter.04 folder.

5. Add DBMS_OUTPUT statements, as shown in Figure 4-18. Notice these are placed to confirm the values for each shopper and to confirm that the insert is processed for each row.

6. Copy the entire CREATE PROCEDURE statement in the text file.

7. Return to SQL* Plus. Click **Edit** and then **Paste** from the main menu.

8. Press **Enter**, if necessary, to execute the statement.

9. Type **EXECUTE promo_test_sp('FEB', '2003');** to run the procedure. The DBMS_OUTPUT messages should be displayed, as shown in Figure 4-19.

```
Oracle SQL*Plus
File  Edit  Search  Options  Help
SQL>
  1   CREATE OR REPLACE PROCEDURE Promo_test_sp
  2     (p_mth IN CHAR,
  3      p_year IN CHAR)
  4    IS
  5    CURSOR cur_purch IS
  6      SELECT idshopper, SUM(Subtotal) sub
  7      FROM bb_basket
  8      WHERE TO_CHAR(dtCreated,'MON') = p_mth
  9         AND TO_CHAR(dtCreated,'YYYY') = p_year
 10         AND orderplaced = 1
 11      GROUP BY idshopper;
 12    promo_flag CHAR(1);
 13  BEGIN
 14    FOR rec_purch IN cur_purch LOOP
 15      If rec_purch.sub > 50 THEN
 16                 promo_flag := 'A';
 17      ELSIF rec_purch.sub > 25 THEN
 18                 promo_flag := 'B';
 19      END IF;
 20          DBMS_OUTPUT.PUT_LINE(rec_purch.idShopper||' has sub = '||
 21                          rec_purch.sub||' and flag = '||
 22                          promo_flag);
 23      IF promo_flag IS NOT NULL THEN
 24           DBMS_OUTPUT.PUT_LINE('Insert processed for shopper '||
 25                          rec_purch.idShopper);
 26        INSERT INTO bb_promolist
 27        VALUES (rec_purch.idshopper, p_mth, p_year, promo_flag, NULL);
 28      END IF;
 29      promo_flag := '';
 30    END LOOP;
 31    COMMIT;
 32* END;

Procedure created.

SQL>
```

Figure 4-18   Add DBMS_OUTPUT statements to confirm values and execution

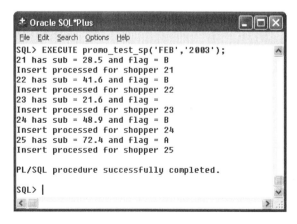

```
Oracle SQL*Plus
File  Edit  Search  Options  Help
SQL> EXECUTE promo_test_sp('FEB','2003');
21 has sub = 28.5 and flag = B
Insert processed for shopper 21
22 has sub = 41.6 and flag = B
Insert processed for shopper 22
23 has sub = 21.6 and flag =
Insert processed for shopper 23
24 has sub = 48.9 and flag = B
Insert processed for shopper 24
25 has sub = 72.4 and flag = A
Insert processed for shopper 25

PL/SQL procedure successfully completed.

SQL> |
```

Figure 4-19   DBMS_OUTPUT messages confirming promotion processing

A review of the messages confirms that the subtotal value for Shopper 23 is correct and that the flag is blank because no promotion should be received. However, the messages also confirm that the INSERT statement is being processed for Shopper 23, which indicates the flag is not NULL for this shopper. This prompts us to review all the flag assignment statements within the procedure. Notice that the last line in the loop is an assignment statement of `promo_flag :=   ";`. Is setting to a blank the same as a NULL value? Let's confirm by modifying this statement.

To correct the PROMO_TEST_SP procedure:

1. Open or return to SQL*Plus.

2. Type **DELETE FROM bb_promolist; COMMIT;** to delete the rows we previously inserted.

3. In a text editor, open the **promotest04.sql** file from the Chapter.04 folder.

4. Modify the PROMO_FLAG assignment statement at the end of the loop to be **promo_flag := NULL;**. Save the file.

5. In SQL*Plus, click **File** and then **Open** from the main menu. Navigate to and select the **promotest04.sql** file from the Chapter.04 folder.

6. Click **Open** and a copy of the file contents appears in the SQL*Plus screen.

7. To run the code and create the procedure, click **File** and then **Run** from the main menu. You receive a "Procedure created" message.

8. Type **EXECUTE promo_test_sp('FEB', '2003');** to run the procedure.

9. Run a query to check what the procedure inserted into the BB_PROMOLIST table by typing **SELECT * FROM bb_promolist;**. Note that a row for Shopper 23 is no longer inserted into the BB_PROMOLIST table, as shown in Figure 4-20.

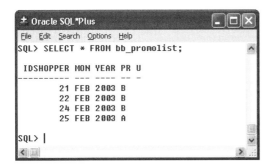

Figure 4-20   Inserts completed with modification

SOFTWARE UTILITIES AVAILABLE TO ASSIST IN PROGRAM UNIT DEVELOPMENT

SQL*Plus is available as part of the Oracle9i database and development tool suites and is used throughout this text. SQL*Plus is a great tool for learning PL/SQL code and is always available in an Oracle9i environment. However, there are a variety of software utilities available to assist in PL/SQL code development. These utilities provide additional features, including the ability to view information graphically and debug breakpoints to make code development easier.

Procedure Builder is one of these tools provided by Oracle and is included within the Oracle Developer 6i suite. Procedure Builder is not included in Oracle Developer 9i suite; however, many third-party software companies have similar products available. TOAD (or Tool for Oracle Application Development) is one of the popular third-party utilities. An introduction to using Procedure Builder is presented in Appendix B and an introduction to using TOAD is presented in Appendix C.

SUBPROGRAMS

A program unit can be defined within another program unit. This is called a **subprogram**. If a block of code is created to be used by only one specific procedure, it can be defined inside that procedure. For example, Figure 4-21 displays the ORDER_TOTAL_SP2 procedure with the SHIP_COST subprogram in the DECLARE section. Note in the earlier example, we saved the SHIP_COST_SP as a stand-alone procedure and called it from within the ORDER_TOTAL procedure.

Only the program unit in which the subprogram is declared can use the subprogram. In this scenario, the subprogram SHIP_COST can be used only by ORDER_TOTAL_SP2. Therefore, we would not want to save a block of code that could be used by a number of procedures as a subprogram. Instead, we would want to save it as a stand-alone procedure so that it can be shared by a number of program units. A subprogram is created in the DECLARE section of the procedure block. If any variables are also declared in the procedure, they must be listed before the subprogram in the DECLARE section. A subprogram is typical in a procedure to address blocks of code that could be called numerous times throughout the procedure.

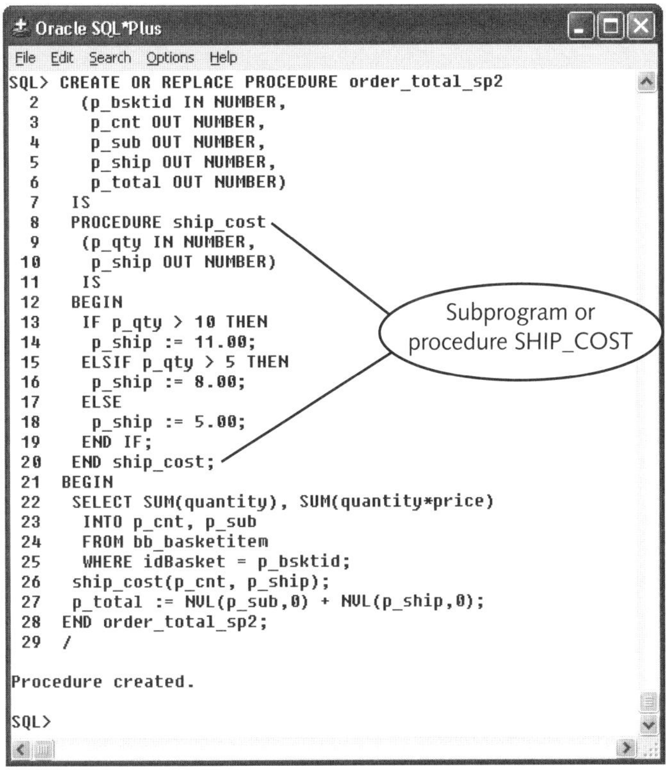


```
SQL> CREATE OR REPLACE PROCEDURE order_total_sp2
  (p_bsktid IN NUMBER,
   p_cnt OUT NUMBER,
   p_sub OUT NUMBER,
   p_ship OUT NUMBER,
   p_total OUT NUMBER)
 IS
 PROCEDURE ship_cost
  (p_qty IN NUMBER,
   p_ship OUT NUMBER)
  IS
 BEGIN
  IF p_qty > 10 THEN
   p_ship := 11.00;
  ELSIF p_qty > 5 THEN
   p_ship := 8.00;
  ELSE
   p_ship := 5.00;
  END IF;
 END ship_cost;
BEGIN
 SELECT SUM(quantity), SUM(quantity*price)
  INTO p_cnt, p_sub
  FROM bb_basketitem
  WHERE idBasket = p_bsktid;
 ship_cost(p_cnt, p_ship);
 p_total := NVL(p_sub,0) + NVL(p_ship,0);
END order_total_sp2;
/

Procedure created.

SQL>
```

Figure 4-21 A subprogram within a procedure

EXCEPTION HANDLING AND TRANSACTION SCOPE

From your studies, you probably recall that the **exception handling** area of a block determines what happens if an error occurs, and that **transaction scope** refers to the logical group of DML actions that is affected by a transaction control statement. We must recognize if we call a procedure from within another procedure, the flow of exception handling and transactions can cross over multiple program units. If an exception is raised in procedure B that has been called from procedure A, the control initially moves to the exception handler section of procedure B. If it is handled, the control then returns to procedure A to the next statement after the call to procedure B. If the exception is not handled within procedure B, the control moves to the EXCEPTION section of procedure A and looks for an appropriate handler. Figure 4-22 depicts the error handler flow across multiple program units.

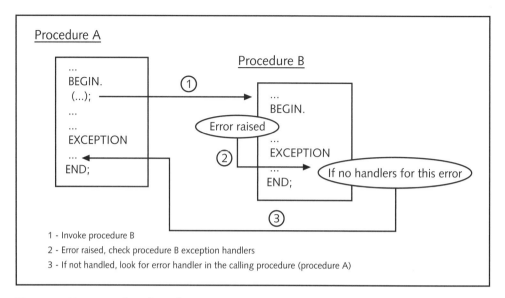

Figure 4-22   Error handling flow

In regards to transaction control, the transaction of a procedure includes not only DML actions within that procedure, but also any DML actions in any procedures that are called from that procedure. If any DML action occurs in the procedure that has been called from another procedure, these statements become part of a continuous transaction started in the main procedure. Therefore, any transaction control statements affect any DML actions that have executed thus far, regardless of which procedure executed the DML actions and regardless of which procedure executes the transaction control statement. Let's look at an example using two new procedures to illustrate this point.

To test transaction control across multiple program units:

1. Open or return to SQL*Plus.

2. In SQL*Plus, click **File** and then **Open** from the main menu, and then navigate to and select the **tctest04b.sql** file from the Chapter.04 folder.

3. Click **Open** and a copy of the file contents appears in the SQL*Plus screen.

4. To run the code and create the procedure, click **File** and then **Run** from the main menu. You receive a "Procedure created" message. Notice that this procedure contains only an INSERT statement.

5. In SQL*Plus, click **File** and then **Open** from the main menu, and then navigate to and select the **tctest04a.sql** file from the Chapter.04 folder.

6. Click **Open** and a copy of the file contents appears in the SQL*Plus screen.

7. To run the code and create the procedure, click **File** and then **Run** from the main menu. You receive a "Procedure created" message. Notice this procedure contains an INSERT statement, a call to the other procedure, and a COMMIT.

8. Now let's test the procedures. To do so, type **EXECUTE tc_test_sp1();** and then press **Enter** to invoke the procedure.

9. To determine what data has actually been committed to the database, we use a second session. Start another SQL*Plus session and log on.

10. In the second SQL*Plus session, list all the data in the BB_TEST1 table by typing **SELECT * FROM bb_test1;** and then press **Enter**. Two rows should now be in the table: a value of 1 from procedure TC_TEST_SP1 and a value of 2 from procedure TC_TEST_SP2. Notice that the COMMIT was the last statement to execute and the inserts from both procedures were, therefore, committed.

11. Now let's change the scenario. While in the first session of SQL*Plus, empty the BB_TEST1 table by typing **DELETE FROM bb_test1;** and **COMMIT;**.

12. In a text editor, open the **tctest04a.sql** file from the Chapter.04 folder.

13. Move the COMMIT statement above the TC_TEST_SP2 procedure invocation, as shown in Figure 4-23. Save the changes to the file.

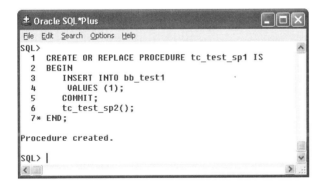

Figure 4-23   Move the COMMIT statement

14. In the first session of SQL*Plus, click **File** and then **Open** from the main menu. Navigate to and select the **tctest04a.sql** file from the Chapter.04 folder.

15. Click **Open** and a copy of the file contents appears in the SQL*Plus screen.

16. To run the code and create the procedure, click **File** and then **Run** from the main menu. You receive a "Procedure created" message.

17. Let's test the procedures again. Type **EXECUTE tc_test_sp1();** and then press **Enter** to invoke the procedure.

18. In SQL*Plus, list all the data in the BB_TEST1 table by typing
SELECT * FROM bb_test1; and then press **Enter**. Note that two rows
are inserted as expected by the execution.

19. Now move to the second session of SQL*Plus and type
SELECT * FROM bb_test1; and then press **Enter**. Only one row is in the
table: a value of 1 from procedure TC_TEST_SP1. Notice that the INSERT
statement in the TC_TEST_SP2 procedure is executed after the COMMIT
statement and, therefore, is still in the transaction queue awaiting a COMMIT or
rollback action.

At times, it is more desirable to be able to treat some SQL statements in a transaction
separately from the rest. For example, what if we want to commit the INSERT in the
TC_TEST_SP2 procedure but we do not want the COMMIT operation to affect the
INSERT in the TC_TEST_SP1 procedure? We can accomplish this by using
autonomous transactions, which are transactions created within another transaction
called a parent transaction. Note that they can be treated independently of the parent
transaction.

To accomplish this task, we use a **pragma**, which is a compiler directive or, in other
words, additional instructions for the PL/SQL compiler to use during program unit
compilation. Let's look at an example of an autonomous transaction.

To create an autonomous transaction:

1. First empty the BB_TEST1 table. In the first session of SQL*Plus, type
DELETE FROM bb_test1; COMMIT;.

2. In a text editor, open the **tctest04b.sql** file from the Chapter.04 folder.

3. In the DECLARE section, type the statement **PRAGMA AUTONOMOUS_
TRANSACTION;**, as shown in Figure 4-24. Also type a **COMMIT;** statement after
the insert action. The pragma instructs the compiler to treat the transaction of
this program unit separately from the transaction in the calling block. Save the
changes to the file.

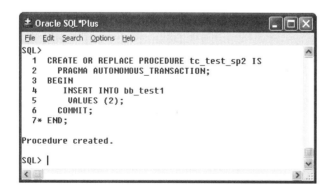

Figure 4-24   Making a procedure an autonomous transaction

4. In a text editor, open the **tctest04a.sql** file from the Chapter.04 folder.

5. Remove the **COMMIT;** statement. Save the file.

6. In the first session of SQL*Plus, click **File** and then **Open** from the main menu. Navigate to and select the **tctest04a.sql** file from the Chapter.04 folder.

7. Click **Open** and a copy of the file contents appears in the SQL*Plus screen.

8. To run the code and create the procedure, click **File** and then **Run** from the main menu. You receive a "Procedure created" message.

9. In the first session of SQL*Plus, click **File** and then **Open** from the main menu. Navigate to and select the **tctest04b.sql** file from the Chapter.04 folder.

10. Click **Open** and a copy of the file contents appears in the SQL*Plus screen.

11. To run the code and create the procedure, click **File** and then **Run** from the main menu. You receive a "Procedure created" message.

12. Type **EXECUTE tc_test_sp1();**, and then press **ENTER** to execute the procedures.

13. Move to the second SQL*Plus session to check the table data by typing **SELECT * FROM bb_test1 ;**. Only one row is committed to the table with a value of 2. This is from the INSERT in TC_TEST_SP2. Notice that even though the INSERT from the TC_TEST_SP1 procedure executes before the COMMIT statement, that INSERT action was not committed. The COMMIT affected only the TC_TEST_SP2 INSERT because it was declared as autonomous.

14. What happens with the INSERT from the TC_TEST_SP1 procedure? It is still in the transaction queue waiting for a transaction control statement. Return to the first session of SQL*Plus. Type **COMMIT;** and then press **Enter**.

15. Return to the second session of SQL*Plus and type **SELECT * FROM bb_test1;**. Now, the INSERT from the TC_TEST_SP1 procedure has also been committed and both rows are displayed.

ERROR HANDLING USING RAISE_APPLICATION_ERROR

Error handling using exception handlers was covered in Chapter 3. In most cases, a DBMS_OUTPUT statement was used to display a message to the screen. Another option to display error messages exists when we are working with stored program units. The RAISE_APPLICATION_ERROR built-in function is provided by Oracle to create our own error messages. This function accepts two arguments: an error number and an error message. The error number needs to be assigned from the range −20,000 to −20,999, which are error numbers reserved for assignment by developers. The error message is a text string up to 512 characters.

Figure 4-25 displays a procedure that checks if the stock level of an item is sufficient to fill the order. Notice that the IF statement checks if the quantity requested is greater than the stock level.

```
Oracle SQL*Plus
File  Edit  Search  Options  Help
SQL> CREATE OR REPLACE PROCEDURE stock_ck_sp
  2    (p_qty IN NUMBER,
  3     p_prod IN NUMBER)
  4    IS
  5     lv_stock_num NUMBER(4);
  6  BEGIN
  7     SELECT stock
  8       INTO lv_stock_num
  9       FROM bb_product
 10      WHERE idProduct = p_prod;
 11     IF p_qty > lv_stock_num THEN
 12      RAISE_APPLICATION_ERROR(-20000,
 13         'Not enough in stock.  Request = '||p_qty||
 14         ' and stock level = '||lv_stock_num);
 15     END IF;
 16  EXCEPTION
 17     WHEN NO_DATA_FOUND THEN
 18       DBMS_OUTPUT.PUT_LINE('No stock found');
 19  END;
 20  /

Procedure created.

SQL>
```

Figure 4-25   Using the RAISE_APPLICATION_ERROR function

If the IF statement is TRUE, then the processing halts and the error message −20,000 is displayed. Let's look at an execution that raises the error to see what this function returns. Figure 4-26 displays the execution of the procedure in which the quantity requested is greater than the stock level.

```
Oracle SQL*Plus
File  Edit  Search  Options  Help
SQL> EXECUTE stock_ck_sp(20,2);
BEGIN stock_ck_sp(20,2); END;

*
ERROR at line 1:
ORA-20000: Not enough in stock.  Request = 20 and stock level = 15
ORA-06512: at "SCOTT2.STOCK_CK_SP", line 12
ORA-06512: at line 1

SQL> |
```

Figure 4-26   Message display of RAISE_APPLICATION_ERROR function

REMOVING PROCEDURES

As with all database objects, we can use the DROP command to remove procedures, as shown in the following code:

```
DROP PROCEDURE procedure_name;
```

CHAPTER SUMMARY

- ❐ Named program units are PL/SQL blocks that have been named and saved so that they can be reused. A stored program unit is a named program unit that is saved in the database.

- ❐ A procedure is one type of named program unit that can perform actions and accept parameters.

- ❐ Parameters allow values to be passed into and out of program units. Parameters have three modes: IN, OUT, and IN OUT.

- ❐ The header refers to the section of a named program unit that contains the name and parameters.

- ❐ Procedures can be created using SQL*Plus or other software utilities. Procedure Builder is a tool that is part of the Oracle Developer 6i suite and is used to create program units.

- ❐ You can test procedures by invoking them with appropriate arguments to match the parameters in the procedure.

- ❐ Any OUT mode parameters must have a host or bind variable argument to receive the value returned.

- ❐ The DBMS_OUTPUT package can be used for debugging by checking values and confirming which statements are executing.

- ❐ Two methods can be used to pass values into parameters: positional and named association.

- ❐ Subprograms are procedures that are declared within another procedure. The subprogram can be used only by the containing procedure.

- ❐ Exceptions raised within procedures called from another procedure can propagate to the calling procedure.

- ❐ Separate transactions are not created for procedures called from within other procedures unless a procedure is declared as autonomous.

- ❐ The RAISE_APPLICATION_ERROR function can be used in stored program units to return unique error messages.

- ❐ The DROP PROCEDURE statement is used to remove a procedure from the system.

REVIEW QUESTIONS

1. The difference between an anonymous PL/SQL block and a named program unit is that _____.

   a. an anonymous block cannot issue transaction control

   b. a named block cannot issue transaction control

   c. an anonymous block has a header

   d. a named block has a header

2. Which of the following are valid modes for parameters in procedures? (Choose all that apply.)

   a. IN

   b. IN OUT

   c. OUT

   d. OUT IN

3. Which parameter mode must be used to have one value sent into a parameter and a different value returned by the same parameter?

   a. IN

   b. OUT

   c. IN OUT

   d. OUT IN

4. A host variable is _____.

   a. a PL/SQL variable declared in a PL/SQL block

   b. a variable that contains a calculation

   c. a variable created in the calling environment of program units

   d. any variable created in anonymous blocks

5. Which of the following methods can be used to pass values to parameters? (Choose all that apply.)

   a. positional

   b. ordered

   c. name association

   d. list

6. The top section of a named program unit containing the program unit name and parameters is called the _____.

   a. named unit

   b. cap

   c. header

   d. title

4

7. If procedure c calls procedure d and an error is raised in procedure d and no exception handler exists for the error in procedure d, _____.

   a. an error message is displayed to the user

   b. an error is raised in procedure c

   c. program control moves to the EXCEPTION section of procedure c

   d. the results are unpredictable

8. How many values can a procedure return to a calling environment?

   a. none

   b. the same as the number of parameters

   c. the same as the number of parameters that include an OUT mode

   d. at least one

9. Which built-in function can be used to raise errors numbered –20,000 to –20,999?

   a. RAISE_ERROR

   b. RAISE_APPLICATION_ERROR

   c. WHEN EXCEPTION

   d. RAISE_EXCEPTION

10. Which statement successfully deletes the SHIP_SP procedure from the system?

    a. `DELETE ship_sp;`

    b. `DELETE PROCEDURE ship_sp;`

    c. `DROP PROCEDURE ship_sp;`

    d. `REMOVE PROCEDURE ship_sp;`

11. What is meant by the term "named program units"?

12. Describe how the DBMS_OUTPUT package can be used for debugging.

13. Describe the role of parameters in procedures.

14. Describe transaction handling across multiple procedures, such as when procedure A calls procedure B. What can be used to override the default transaction scope?

15. What advantages are gained by storing a program unit in the database?

ADVANCED REVIEW QUESTIONS

1. Review the following procedure header. Which of the following procedure calls would be valid to include in another procedure?

```
PROCEDURE  order_change_sp
    (p_prodid IN NUMBER,
     p_prodqty IN OUT  NUMBER )
IS
```

   a. `order_change_sp(100, 362);`

   b. `EXECUTE order_change_sp(100, 362);`

   c. `order_change_sp(100, :g_qty);`

   d. `order_change_sp()`

2. Regarding the following code, which of the subsequent statements is correct?

```
CREATE OR REPLACE PROCEDURE test_sp
    (p_num IN NUMBER)
  IS
BEGIN
    UPDATE test_table
      SET range = v_range
      WHERE id = p_num;
END;
```

   a. The procedure compiles successfully.

   b. The procedure does not compile successfully due to not including a COMMIT statement in a block containing a DML action.

   c. The procedure does not compile successfully due to not declaring V_RANGE.

   d. The procedure does not compile successfully due to not declaring P_NUM.

3. A procedure named RESET_SP contains two parameters: the first listed is named P_NUM1 and the second listed is named P_NUM2. Both parameters have a mode of IN. P_NUM1 has a data type of NUMBER and P_NUM2 has a data type of CHAR. Which of the following are valid invocations of this procedure?

   a. `reset_sp(101, '33');`

   b. `reset_sp(p_num1 => :g_start, '33');`

   c. `reset_sp(p_num2 => '33', p_num1 => 101);`

   d. all of the above

4. Which of the following procedure parameter declarations is not valid in a procedure?

   a. `P_test VARCHAR2;`

   b. `P_test IN OUT NUMBER :=1;`

   c. `P_test IN NUMBER(5);`

   d. `P_test OUT CHAR;`

5. If the following procedure is called from the NEW_BASK_SP procedure, what is the effect of the COMMIT statement in this procedure?

```
CREATE OR REPLACE PROCEDURE new_shop_sp
       (p_last IN VARCHAR2(15),
        p_first IN VARCHAR2(10),
        p_id OUT NUMBER   )
  IS
    PRAGMA_AUTONOMOUS_TRANSACTION;
  BEGIN
    p_id := id_seq.NEXTVAL;
    INSERT INTO shopper (id, last, first)
       VALUES (p_id, p_last, p_first);
    COMMIT;
  END;
  /
```

a. INSERTS contained in both the NEW_BASK_SP and NEW_SHOP_SP procedures are committed.

b. Any DML statements in both the NEW_BASK_SP and NEW_SHOP_SP procedures are committed.

c. Only the INSERT statement in NEW_SHOP_SP is committed.

d. Only DML statements in NEW_BASK_SP are committed.

HANDS-ON ASSIGNMENTS

Assignment 4-1: Create a Procedure

Complete the following steps to create a procedure that allows a company employee to make corrections to the product name assigned to a product. Review the BB_PRODUCT table and identify the PRODUCT NAME column and the PRIMARY KEY column. The procedure needs two IN parameters to identify the product id and provide the new description. This procedure need only accomplish a DML action, so no OUT parameters are necessary.

1. Open or return to SQL*Plus.

2. Type the following code to create the procedure:

```
CREATE OR REPLACE PROCEDURE prod_name_sp
 (p_prodid IN bb_product.idproduct%TYPE,
  p_descrip IN bb_product.description%TYPE)
  IS
BEGIN
  UPDATE bb_product
    SET description = p_descrip
    WHERE idproduct = p_prodid;
  COMMIT;
END;
/
```

3. Run the code.

4. Invoke the procedure by typing **EXECUTE prod_name_sp(1,' CapressoBar Model #388');**.

5. Check if the update was executed by querying the table by typing **SELECT * FROM bb_product;**.

Assignment 4-2: Use a Procedure with IN Parameters

Complete the following steps to create a procedure that allows a company employee to add a new product to the database. This procedure needs only IN parameters.

1. Open or return to SQL*Plus.

2. Create a procedure named PROD_ADD_SP that adds a row for a new product into the BB_PRODUCT table. Keep in mind that the user provides values for the product name, description, image file name, price, and active status. Address the input values or parameters in the order listed in the previous sentence.

3. Run the code to create the procedure.

4. Invoke the procedure by typing **EXECUTE prod_add_sp('Roasted Blend',' Well-balanced mix of roasted beans, a medium body', 'roasted.jpg',9.50,1);**.

5. Check if the update was executed by querying the table.

Assignment 4-3: Calculate the Tax on an Order

Complete the following steps to create a procedure to calculate the tax on an order. The BB_TAX table contains the states that require taxes to be submitted for Internet sales. If the state is not listed in the table, then no tax should be assessed on the order. The shopper's state and basket subtotal are the inputs into the procedure while the tax amount should be returned.

1. Open or return to SQL*Plus.

2. Create a procedure named TAX_COST_SP to accomplish the tax calculation task. Keep in mind that the state and subtotal values are inputs into the procedure and the procedure is to return the tax amount. Review the contents of the BB_TAX table, which contains the tax rate for each state that needs to be taxed.

3. Create a host variable named G_TAX to hold the value returned by the procedure.

4. Invoke the procedure using the values of "VA" for the state and $100 for the subtotal.

5. Display the tax amount returned by the procedure (it should be $4.50).

Assignment 4-4: Update Columns in a Table

After a shopper completes the ordering process, a procedure is called to update the following columns in the BASKET table: ORDERPLACED, SUBTOTAL, SHIPPING, TAX, and TOTAL. A value of 1 is to be entered into the ORDERPLACED column to indicate the shopper has completed the order. Inputs to the procedure are the basket id and the amounts for the subtotal, shipping, tax, and total.

1. Open or return to SQL*Plus.
2. Create a procedure named BASKET_CONFIRM_SP that accepts the input values outlined in the introduction to this assignment. Keep in mind that you are modifying an existing row of the BB_BASKET table in this procedure.
3. Enter the following inserts in SQL*Plus to test this procedure:

```
INSERT INTO BB_BASKET (IDBASKET, QUANTITY, IDSHOPPER,
                       ORDERPLACED, SUBTOTAL, TOTAL,
                       SHIPPING, TAX, DTCREATED, PROMO)
        VALUES (15, 2, 22, 0, 0, 0, 0, 0, '28-FEB-03', 0);
INSERT INTO BB_BASKETITEM (IDBASKETITEM, IDPRODUCT,
                           PRICE, QUANTITY, IDBASKET,
                           OPTION1, OPTION2)
        VALUES (39, 7, 10.8, 3, 15, 2, 3);
INSERT INTO BB_BASKETITEM (IDBASKETITEM, IDPRODUCT,
                           PRICE, QUANTITY, IDBASKET,
                           OPTION1, OPTION2)
        VALUES (40, 8, 10.8, 3, 15, 2, 3);
```

4. Type **COMMIT;** to save the data from the inserts.
5. Invoke the procedure by typing **EXECUTE basket_confirm_sp(15, 64.80, 8.00, 1.94, 74.74);**. Note these values represent the basket id and the amounts for the subtotal, shipping, tax, and total, respectively.
6. Type **SELECT subtotal, shipping, tax, total, orderplaced FROM bb_basket WHERE idbasket = 15;**, and then press **Enter** to check that the insert works correctly.

Assignment 4-5: Update the Status of an Order

Create a procedure named STATUS_SHIP_SP that allows a company employee in the Shipping Department to update the status of an order to add shipping information. The BB_BASKETSTATUS table maintains a list of events for each order so that a shopper can see the current status, date, and comments as each stage of the order process is completed. The IDSTAGE column of the BB_BASKETSTATUS table identifies each stage and an IDSTAGE of 3 indicates the order has been shipped.

The procedure should allow the addition of a row indicating an IDSTAGE of 3, date shipped, tracking number, and shipper. The sequence BB_STATUS_SEQ is used to provide a value for the primary key column. Test the procedure with the following information:

```
Basket # = 3
Date shipped = 20-FEB-03
Shipper = UPS
Tracking # = ZW2384YXK4957
```

Assignment 4-6: Return Order Status Information

Create a procedure that returns the most recent order status information for a particular basket. This procedure should determine the most recent stage entry from the BB_BASKETSTATUS table and return the data. An IF or CASE clause needs to be used to return the stage description rather than the IDSTAGE number, which means little to the shopper. The IDSTAGE column of the BB_BASKETSTATUS table identifies each stage as follows:

◻ 1 = Submitted and received

◻ 2 = Confirmed, processed, sent to shipping

◻ 3 = Shipped

◻ 4 = Cancelled

◻ 5 = Backordered

The procedure needs to accept a basket id number and return the most recent status description and the date that status was recorded. If no status is available for the basket id queried, return a message stating that no status is available. Use a VARCHAR2 host variable to hold the date value returned. Name the procedure STATUS_SP. Test the procedure twice using the basket id of 4 and then 6.

Assignment 4-7: Identify Customers

The company wants to offer an incentive of free shipping to those customers who have not returned for two months. Create a procedure named PROMO_SHIP_SP that determines who these customers are and then updates the BB_PROMOLIST table accordingly. The procedure uses the following information:

1. Date cutoff = Any customers who have not shopped on the site since this date should be included as incentive participants. Use the basket creation date to reflect shopper activity dates.

2. Month = Three-character month (such as APR) that should be added to the promotion table to indicate which month the free shipping is available.

3. Year = Four-digit year indicating the year the promotion is effective.

4. PROMO_FLAG = 1 (representing free shipping).

The BB_PROMOLIST table also has a USED column, which contains a default value of 'N' and is updated to a 'Y' when the shopper uses the promotion. Test the procedure with a cutoff date of 15–FEB–03. Assign the free shipping for the month of APR and the year 2003.

Assignment 4-8: Add Items to a Cart

As a shopper selects products on the Web site, a procedure is needed to add the newly selected item to the current shopper's basket. Create a procedure named BASKET_ADD_SP that accepts a product id, basket id, price, quantity, size code (1 or 2), and form code (3 or 4) and that uses this information to add a new item to the BB_BASKETITEM table. Note the PRIMARY KEY column of the table is generated by BB_IDBASKETITEM_SEQ. Execute the procedure using the following values:

- Basket id = 14
- Product id = 8
- Price = 10.80
- Quantity = 1
- Size code = 2
- Form code = 4

Assignment 4-9: Create a Logon Procedure

On the home page of the Brewbean's Web site, an option is available for existing members to log on with their IDs and passwords. Develop a procedure named MEMBER_CK_SP that accepts the ID and password as inputs, checks if it is a valid logon, and returns the member name and cookie value. The name should be returned as a single text string containing the first and last name.

The head developer asked that the number of parameters should be minimized so that the same parameter is used to accept the password and return the name value. Also, if a valid user name and password are not provided by the user, return the value of INVALID in a parameter named P_CHECK. Execute a valid member first using the username "rat55" and password of "kile." Then execute an invalid logon changing the previous user name to "rat." Be sure to set all the host variables to NULL after running the first execution so that they contain no values upon the second execution.

CASE PROJECTS

Case 4-1: Reporting and Analysis Summary Tables

The Reporting and Analysis Department has a database containing a number of tables that hold summarized data to make report generation simpler and more efficient. They want to schedule some jobs to run nightly to update some of the summary tables. Create two procedures to update the following two tables (assume that existing data in the tables is deleted before these procedures execute):

❑ Procedure PROD_SALES_SUM_SP to update table BB_PROD_SALES. The table holds total sales dollars and total sales quantity by product id, month, and year. The date ordered should be used for the month and year information.

❑ Procedure SHOP_SALES_SUM_SP to update table BB_SHOP_SALES. The table holds total dollar sales by shopper id. The total should include only product amounts, excluding shipping and tax.

The BB_SHOP_SALES and BB_PROD_SALES tables have already been created. Use the DESCRIBE command to review the table structure of each.

Case 4-2: The More Movie Rentals Company Rental Process

The More Movie Rentals company is experimenting with a new concept to make rentals more convenient. The company allows members to request movies via the Internet or they can check out at the store location. In either case, a small barcode sticker of the member id and movie id is printed at the time of rental. This is affixed to a paper slipcase for the movie, which can serve as an envelope. The member can return the movie by dropping it in the U.S. Mail or by dropping it off at the store location. In either case, the clerk scans the member id and movie id from the slipcase barcodes.

The first two procedures requested are to record rentals and associated returns to the database. Create a procedure named MOVIE_RENT_SP that adds a new record to the MM_RENTAL table and updates the movie inventory, which is the MOVIE_QTY column of the MM_MOVIE table. This procedure needs to accept a member id, movie id, and a payment method. The member id is scanned in from a barcode on the membership card and the movie id is scanned in from a barcode on the movie case. The cashier selects the payment type. Test using member id = 13, movie id = 12, and payment method = 4. Verify that the rental has been added and that the movie inventory has been updated.

The second procedure needs to record the return of the rental movie. Create a procedure named MOVIE_RETURN_SP that records the current date in the CHECKIN_DATE column and updates the movie inventory. Inputs are the member id and movie id from the barcodes on the slipcase; therefore, the procedure first needs to determine the rental id. Test for the rental we recorded above: member id = 13 and movie id = 12.

5

FUNCTIONS

In this chapter, you will:

♦ Learn about functions

♦ Create a stored function in SQL*Plus

♦ Use OUT parameters in functions

♦ Include multiple RETURN statements in a function

♦ Use a RETURN statement in a procedure

♦ Use constraints of actual and formal parameters

♦ Understand and control the passing of parameter values

♦ Work with function purity levels

♦ Reference the data dictionary for program units

♦ Delete program units

This chapter explores creating functions, using functions, differences between functions and procedures, parameter constraints, actual versus formal parameters, parameter value passing techniques, and function restrictions within SQL statements. In the process of working through this chapter, you'll learn that functions are similar to procedures but have several distinctive qualities, such as the use of a RETURN statement and the ability to be used in both PL/SQL and SQL statements.

THE CURRENT CHALLENGE IN THE BREWBEAN'S APPLICATION

The development of the Brewbean's company catalog and ordering application is underway. Currently, two application screens are being reviewed in an effort to determine the programming logic required. First, repeat shoppers are encouraged to set up a member account consisting of a user name and password. This simplifies the ordering process by saving information such as the shopper's address so that it does not have to be entered for each shopping session. Figure 5-1 shows the screen that is displayed after a successful member login.

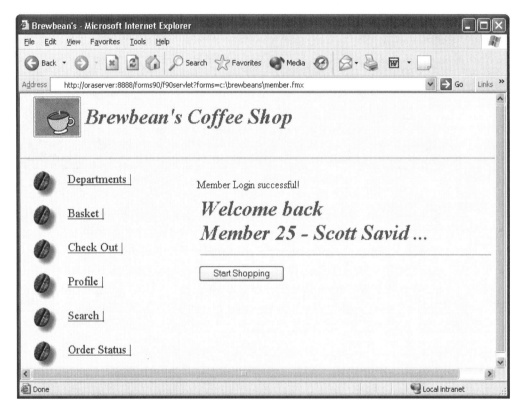

Figure 5-1   Member login confirmation screen

A number of application screens, such as the one in Figure 5-1, identify the shopper by displaying the string including the member id number and name. A program unit that retrieves the member id, name, and format as shown is needed to fulfill this task for all applicable screens.

Another screen being reviewed is the initial check out summary screen. This screen displays the order amounts, including the subtotal, shipping, tax, and total, as shown in Figure 5-2.

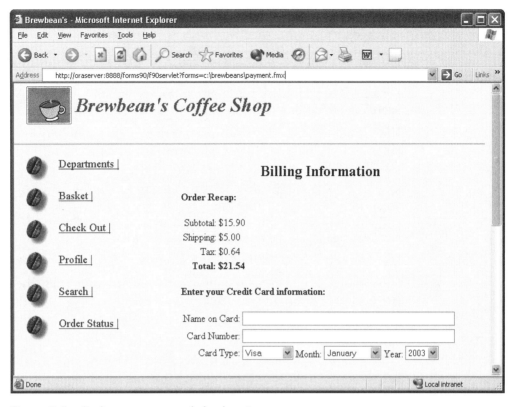

Figure 5-2   Order summary and check out screen

The current calculation of concern on this screen is the shipping cost. Brewbean's calculates shipping cost based on the number of items in the shopping basket. A program unit that counts the total number of items in the order and determines the appropriate shipping cost is needed for this screen. In addition, it would be ideal if this program unit could be used in SQL queries. Some shoppers receive free shipping specials and the Brewbean's manager wants to calculate the total shipping that would be charged if discounts were not used. This calculation allows the manager to track the total cost of the free shipping offers.

REBUILDING YOUR DATABASE

Follow these steps to rebuild the Brewbean's database for this chapter.

To rebuild the Brewbean's database:

1. Open SQL*Plus.

2. You need to create a spool file so that SQL*Plus keeps a copy of all the messages received from running the file that creates the database. On the main menu in SQL*Plus, click **File**, point to **Spool**, and then click **Spool File**. A Select File dialog box appears.

3. Browse to a folder used to contain your working files. In the File name text box, type **DB_log5**, and then click **Save**. Now, whatever text is seen in our SQL*Plus session is saved to this file for future reference.

4. Now let's create the database. In SQL*Plus, enter the following command, which runs all the statements contained in the c5Dbcreate.sql file. Messages verifying the creation and data insertion steps scroll on the SQL*Plus screen. This may take a couple minutes to complete.

   `@<pathname to PL/SQL files>\Chapter.05\c5Dbcreate`

5. Now, you need to turn the spooling off so that you can review the results in the spool file created. On the main menu, click **File**, point to **Spool**, and then click **Spool Off**.

6. Open a text editor.

7. Open the file named **DB_log5.lst** in your working folder.

8. Review the messages for any errors.

AN INTRODUCTION TO FUNCTIONS

At this point, we know that a named program unit is a PL/SQL block saved with a name so that it can be reused. We also know that a procedure is one type of program unit that is used to accomplish one or more tasks, return none or many values, and is used only in PL/SQL statements. Now, let's explore another type of program unit—a function.

A **function** is quite similar to a procedure in that it is a program unit that achieves a task, can receive input values, and returns values to the calling environment. The main difference between functions and procedures is that a function is part of an expression. It cannot serve as an entire statement. A procedure, on the other hand, can serve as an entire statement. This difference is important because it makes functions available for use in both SQL commands and PL/SQL statements (procedures cannot be used in SQL statements).

Consider the Oracle9*i* built-in functions that you have already used in your SQL programming. The following listing displays an SQL query and its results. The query contains the Oracle9*i* built-in ROUND function that rounds a numeric value to a given place.

```
SELECT idProduct, price, ROUND(price, 0)
 FROM bb_product
 WHERE idProduct < 4;
IDPRODUCT       PRICE  ROUND(PRICE,0)
---------- ---------- --------------
        1      99.99             100
        2     129.99             130
        3       32.5              33
```

Notice that the ROUND function is used as part of the entire SQL expression. Its task is to use the two arguments provided, a numeric value (the price column) and the degree of rounding (0 or whole numbers), and return the resulting value. In this case, it looks at the value of price in every row of the BB_PRODUCT table, accomplishes the rounding action, and returns the appropriate result. This is true of all the single-row functions with which we are familiar thus far.

The ROUND function can also be used in a PL/SQL statement, as shown in the following listing:

```
DECLARE
  v_amt1 number(5,2);
  v_amt2 number(3,0);
BEGIN
  v_amt1 := 32.50;
  v_amt2 := ROUND(v_amt1,0);
  DBMS_OUTPUT.PUT_LINE(v_amt2);
END;
/
33
PL/SQL procedure successfully completed.
```

Again, note that the function is only part of the entire PL/SQL assignment statement for variable V_AMT2. The function's job is to accept the values provided (the value to be rounded, or V_AMT1, and the rounding level or whole dollars), manipulate them according to rounding rules, and return the desired value. Notice that each time a function is used, a single value is returned. In this example, the function returned a single value of 33, which is displayed using DBMS_OUTPUT.

Many Oracle-supplied functions are available (for instance, the ROUND function in the preceding examples is Oracle-supplied), but there are always instances in which you need a function that is not available as an Oracle-supplied function to accomplish a particular task. In such a case, you can create your own function. We discuss creating functions next.

5

CREATING A STORED FUNCTION IN SQL*PLUS

The syntax to create a function is exactly like the syntax to create a procedure with one critical exception—a RETURN statement must be used to handle the value to be returned by the function. In the header, a line is added to indicate the data type of the value to be returned. In the function PL/SQL block (also called the body), a RETURN statement in the executable section indicates which value is returned. Figure 5-3 reviews the syntax layout for creating a function.

```
         ┌─CREATE [OR REPLACE] FUNCTION function_name
         │   [ (parameter1_name [mode] datatype,
  Header │      parameter2_name [mode] datatype,
         │       . . .) ]
         │    RETURN return_value_datatype
         └─IS|AS
         ┌─     declaration section
         │  BEGIN
PL/SQL   │      executable section
 block   │      RETURN variable_name;
         │  EXCEPTION
         │      exception handlers
         └─ END;


   --------------------------------------------------
 Notes on syntax layout:
      [ ] - indicates optional portions of the statement
      Key commands - in all upper case
      User provided - in lower case
      | - offers an OR option
      . . . - indicates continuation possible
```

Figure 5-3   CREATE FUNCTION syntax

After the parameters are listed in the header, a RETURN statement must be included to indicate the data type of the value that is returned. This data type does not include any sizing information, as used in declaring PL/SQL block variables. At least one RETURN statement must be included in a function body to instruct which value to return.

Let's look at an example to visualize the creation of a function to meet the Brewbean's shipping cost calculation need. Figure 5-4 displays a CREATE FUNCTION statement in SQL*Plus. The cost of shipping is calculated based on the quantity of items being purchased. This task is appropriately addressed with a function because we are concerned with returning a single value with the operation.

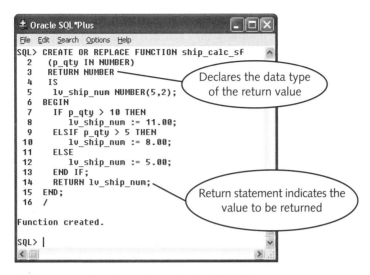

Figure 5-4   Creating a function in SQL*Plus

Notice that the function declares the RETURN value data type in the header and then uses the RETURN statement in the function body to identify which value is to be returned. In this case, the RETURN statement instructs the system to return the value contained in the LV_SHIP_NUM variable.

 The RETURN statement can also return constant values, such as a text string, or use expressions to accomplish calculations within the statement.

Invoking and Testing a Created Function

Now that we have a function, how do we invoke (execute) and test it? Figure 5-5 demonstrates invoking or calling a function within a PL/SQL statement.

Figure 5-5   Invoking a function from a PL/SQL block

Figure 5-5 uses the function as part of an assignment statement within an anonymous PL/SQL block. The SHIP_CALC_SF function accepts 12 as the input argument, which feeds this value to the IN parameter of the function. The function checks which range the 12 falls within and then assigns the correct shipping cost of $11 to the LV_SHIP_NUM variable. The last statement or RETURN statement in the function returns the value in the LV_SHIP_NUM variable back to the calling statement. After the value is returned to the calling block, the LV_COST_NUM variable now holds a value of 11.

Note that after the assignment statement in Figure 5-5 completes, the DBMS_OUTPUT statement is then used in the anonymous block to simplify testing by displaying the value to the screen. Remember to issue the SET SERVEROUTPUT ON command in SQL*Plus to see the results of the DBMS_OUTPUT.PUT_LINE operation. The value of 12 was hard-coded as the input to the function; however, the argument would more likely be a variable in the block.

After working with procedures, a common error in invoking functions is attempting to invoke them by name only as a stand-alone call instead of as part of a statement. Figure 5-6 displays an invocation of the SHIP_CALC_SF function by name only and the error that is received.

 Keep in mind that a function is a part of an expression because the value returned needs to be held in a variable and, therefore, it cannot be invoked in a stand-alone fashion, as is the case with a procedure.

```
Oracle SQL*Plus
File  Edit  Search  Options  Help
SQL> EXECUTE ship_calc_sf(12);
BEGIN ship_calc_sf(12); END;

      *
ERROR at line 1:
ORA-06550: line 1, column 7:
PLS-00221: 'SHIP_CALC_SF' is not a procedure or is undefined
ORA-06550: line 1, column 7:
PL/SQL: Statement ignored

SQL>
```

Figure 5-6   Invoking a function in SQL*Plus

Notice that the error indicates that this type of program unit invocation is used for procedures, not functions.

Using a Function in an SQL Statement

Unlike a procedure, a function you create can be used in SQL statements, just as you use Oracle9*i* built-in functions. This is a great advantage in creating stored functions, because the functionality can be available to both PL/SQL and SQL coding.

Recall that the Brewbean's manager wants to be able to determine the cost of free shipping promotions by comparing the actual shipping costs charged versus the charges that would have been charged if no discounts were used. Figure 5-7 demonstrates the use of the SHIP_CALC_SF function in two SQL statements that assist in this analysis.

5

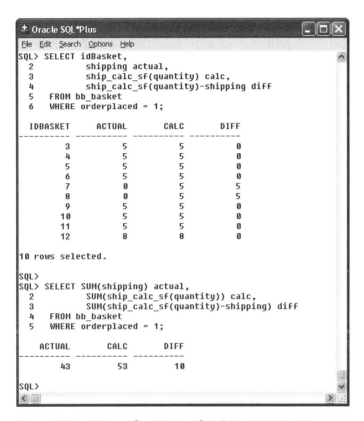

Figure 5-7   Using a function within SQL statements

The first SQL query displays details regarding each shopper basket in the database. The output shows the basket id, total item quantity, actual shipping charges applied, and the applicable shipping cost. The SHIP_CALC_SF function uses a table column, QUANTITY, as the IN argument. The function is applied or processed for each row of the BB_BASKET table.

In Figure 5-7, the second SQL query uses an aggregate function to determine the overall SUM of the shipping amounts and the difference. In this case, we use nested functions. First, our SHIP_CALC_SF function is used on each row of the table, and then the SUM function computes the total of all the SHIP_CALC_SF results. Because an aggregate function is being used with no GROUP BY clause, a single value results for each SUM calculation.

Building and Testing a Function for the Brewbean's Member Name Display

At the beginning of the chapter, you learned that the Brewbean's application needed formatting for member ids and names. Let's create and test a function to fulfill this need.

To create the function:

1. Open or return to SQL*Plus.

2. Type **SET SERVEROUTPUT ON**, and then press **Enter** to enable the display of output with DBMS_OUTPUT.

3. Enter the following code to create MEMFMT1_SF, which accepts three input values and returns the formatted id and name string:

```
CREATE OR REPLACE FUNCTION memfmt1_sf
 (p_id IN NUMBER,
  p_first IN VARCHAR2,
  p_last IN VARCHAR2)
 RETURN VARCHAR2
 IS
  lv_mem_txt VARCHAR2(35);
BEGIN
lv_mem_txt := 'Member '||p_id||' - '||p_first||' '||p_last;
RETURN lv_mem_txt;
END;
/
```

4. If you receive an error on the CREATE FUNCTION statement, use the SHOW ERRORS command to show the error messages. Also, compare your typing closely with the code listed in Step 3.

5. Now, let's test the function to ensure it is working properly. Enter the following anonymous block to test the function:

```
DECLARE
  lv_name_txt VARCHAR2(35);
  lv_id_num NUMBER(4) := 25;
  lv_first_txt VARCHAR2(15) := 'Scott';
  lv_last_txt VARCHAR2(20) := 'Savid';
BEGIN
  lv_name_txt := memfmt1_sf(lv_id_num, lv_first_txt,
                                        lv_last_txt);
  DBMS_OUTPUT.PUT_LINE(lv_name_txt);
END;
/
```

6. Compare your results to Figure 5-8.

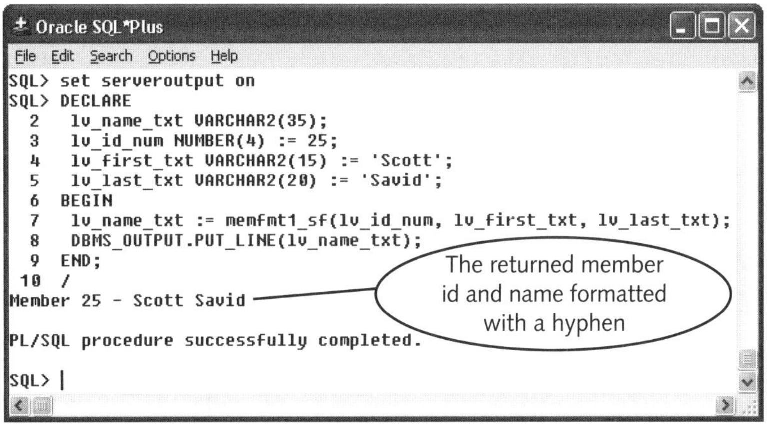

```
± Oracle SQL*Plus                                            _ □ ✕
File  Edit  Search  Options  Help
SQL> set serveroutput on
SQL> DECLARE
  2    lv_name_txt VARCHAR2(35);
  3    lv_id_num NUMBER(4) := 25;
  4    lv_first_txt VARCHAR2(15) := 'Scott';
  5    lv_last_txt VARCHAR2(20) := 'Savid';
  6  BEGIN
  7    lv_name_txt := memfmt1_sf(lv_id_num, lv_first_txt, lv_last_txt);
  8    DBMS_OUTPUT.PUT_LINE(lv_name_txt);
  9  END;
 10  /
Member 25 - Scott Savid

PL/SQL procedure successfully completed.

SQL> |
```

The returned member id and name formatted with a hyphen

Figure 5-8   Test the MEMFMT1_SF function

Now that we have resolved the member id and name formatting need, let's continue by using this function within a procedure to complete the member logon process. This gives us an appreciation for how procedures are used in conjunction. That is, the logon procedure needs to verify the user name and password entered and if a match is found, use the MEMFMT1_SF function to format the member information. The procedure returns a flag (to indicate if logon succeeded), the member id, and the formatted member information.

To create a member logon procedure using the MEMFMT1_SF function:

1. Open or return to SQL*Plus.

2. Click **File**, and then click **Open** from the main menu. Navigate to and select the **login05.sql** file from the Chapter.05 folder.

3. Click **Open** and a copy of the file contents appears in the SQL*Plus screen. Notice the procedure contains two IN parameters, three OUT parameters, and a call to the MEMFMT1_SF function.

4. To run the code and create the procedure, click **File** and then **Run** from the main menu. You receive a "Procedure created" message.

5. Now let's test this procedure. Create and initialize the host variables by typing the following commands:

```
VARIABLE g_user VARCHAR2(8)
VARIABLE g_pass VARCHAR2(8)
VARIABLE g_id NUMBER
VARIABLE g_flag CHAR(1)
VARIABLE g_mem VARCHAR2(35)
BEGIN
```

```
    :g_user := 'fdwell';
    :g_pass := 'tweak';
  END;
  /
```

6. Invoke the procedure by typing **EXECUTE login_sp(:g_user,:g_pass,
   :g_id,:g_flag,:g_mem);**, and then press **Enter**.

7. Finally, check the values returned by using the PRINT command, as shown
   in Figure 5-9.

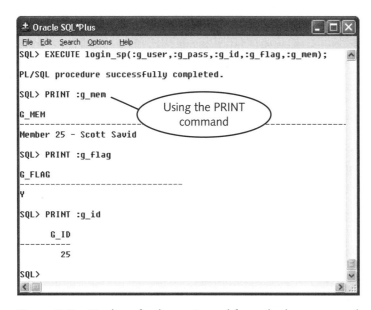

Figure 5-9   Display of values returned from the logon procedure

Now that we have a basic understanding of creating and invoking a function, the next
sections explore various elements within a function in more detail.

USING THE OUT PARAMETER MODE IN A FUNCTION

Did you notice that the functions we have worked with thus far have contained parameters only with an IN mode? Recall that we worked with procedures that contained
parameters with an OUT mode that enable the return of a value from the procedure to
the calling environment. Well, can a parameter with an OUT mode be used in a
function? Yes. However, in practice, OUT parameters are rarely used in functions for
two primary reasons. First, developers typically include OUT parameters only in
procedures because mixing the RETURN value with OUT parameters in functions
tends to be confusing. Second, after OUT parameters are included in functions, the
function can no longer be used in SQL statements.

Using and coding functions following a guideline of returning a single value with the RETURN statement simplifies the use of functions. Otherwise, the developer must review the code closely to determine what data is being returned via the RETURN statement and what is being returned via parameters. In addition, if a development shop does include OUT parameters in functions, then the developer must confirm that no OUT parameters are included in a particular function prior to attempting to use the function in an SQL statement. It is considered good form to return only one value from a function and to do so using the RETURN statement.

 Don't forget: One of the primary reasons OUT parameters are typically not used in functions is that the function does not work if used in an SQL statement.

Let's create a function including a parameter with an OUT mode and experiment with its execution.

To create and run a function with an OUT parameter:

1. Open or return to SQL*Plus.

2. Click **File**, and then click **Open** from the main menu. Navigate to and select the **testout05.sql** file from the Chapter.05 folder.

3. Click **Open** and a copy of the file contents appears in the SQL*Plus screen. Notice the function contains a parameter with an IN OUT mode.

4. To run the code and create the function, click **File** and then **Run** from the main menu. You receive a "Function created" message.

5. Now, let's test this procedure within a PL/SQL block. Create the host variables and run an anonymous block by typing the following commands:

```
VARIABLE g_var1 NUMBER
VARIABLE g_var2 VARCHAR2(5)
BEGIN
  :g_var1 := 1;
  :g_var2 := test_out_sf(:g_var1);
END;
/
```

6. Check the values returned with PRINT commands, as shown in the following code listing. This confirms that we have a function that returns two values.

```
PRINT :g_var1
PRINT :g_var2
```

7. Now, what happens when we attempt to use this function in an SQL statement? Let's try. Enter the SQL query shown in Figure 5-10 and review the error message.

5

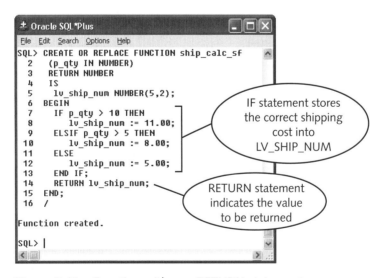

Figure 5-10   SQL query using a function containing an OUT parameter

Notice the error message from the SQL statement execution indicates that OUT parameters cannot be handled. In the PL/SQL block, note that the RETURN statement in the function returns a value to the :G_VAR2 variable and the IN OUT parameter returns a new value into the :G_VAR1 variable.

Multiple RETURN Statements

All the function examples we have reviewed thus far contain only a single RETURN statement in the body. This was done for simplicity. Of course, you can include multiple RETURN statements in the body. Why might we choose to do this? Let's look at the SHIP_CALC_SF function we created earlier in this chapter (shown in Figure 5-11). Note that the appropriate shipping cost value is saved to a variable named LV_SHIP_NUM in the IF statement and then a single RETURN statement references the LV_SHIP_NUM variable at the end.

```
SQL> CREATE OR REPLACE FUNCTION ship_calc_sf
  2    (p_qty IN NUMBER)
  3    RETURN NUMBER
  4    IS
  5      lv_ship_num NUMBER(5,2);
  6  BEGIN
  7    IF p_qty > 10 THEN
  8      lv_ship_num := 11.00;
  9    ELSIF p_qty > 5 THEN
 10      lv_ship_num := 8.00;
 11    ELSE
 12      lv_ship_num := 5.00;
 13    END IF;
 14    RETURN lv_ship_num;
 15  END;
 16  /

Function created.

SQL>
```

IF statement stores the correct shipping cost into LV_SHIP_NUM

RETURN statement indicates the value to be returned

Figure 5-11   Function with one RETURN statement

This same function could be written to include RETURN statements within the IF statement, as listed in Figure 5-12. You might want to include RETURN statements within the IF statement to avoid having to create a variable to hold the value to be returned. After an IF clause evaluates to true, the corresponding RETURN statement executes and returns the explicit shipping dollar amount to the calling statement.

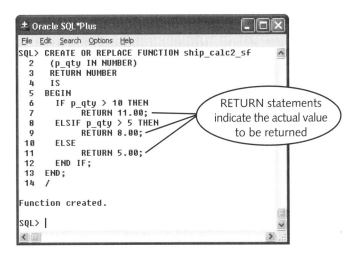

Figure 5-12   Function with multiple RETURN statements

In practice, it is considered good form to use a single RETURN statement in a function to simplify the review of code.

USING A **RETURN** STATEMENT IN A PROCEDURE

A RETURN statement can be used in a procedure as well as in a function; however, the statement serves a different purpose in each case. The RETURN statement in a procedure is used to control the flow of execution. In a procedure, the RETURN statement includes no arguments and is followed with a semicolon. When the RETURN is executed, the procedure immediately returns the execution to the next statement following the procedure call. The values of any OUT parameters at this time are passed back to the arguments in the procedure call.

Figure 5-13 displays an example of including a RETURN statement in the logon procedure we created earlier in this chapter. Notice if a match to the logon user name and password is not found, the execution of the procedure ends because the RETURN statement is issued.

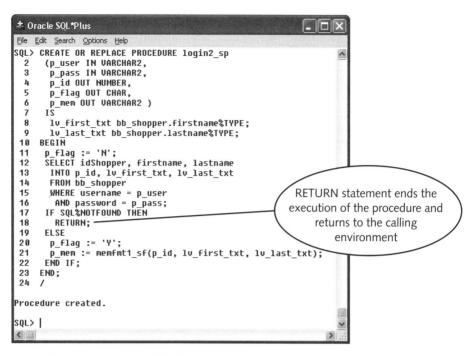

Figure 5-13 Using a RETURN statement in a procedure

ACTUAL AND FORMAL PARAMETER CONSTRAINTS

You might encounter the terms "formal parameters" and "actual parameters" in regards to program units. **Formal parameters** refer to the parameters that are listed in the program unit. **Actual parameters** refer to the arguments that are used when calling or invoking the program unit. If an OUT or IN OUT parameter is used in a program unit, the calling statement must provide variables for the arguments or actual parameters. A variable is necessary because a formal parameter with an OUT mode returns a value that needs to be held by the caller.

Recall that the formal parameters including the RETURN data type cannot include any size information. So, are there any restrictions on the length of character or numeric parameters? Yes, the actual parameters determine the size restrictions. Actual parameters are either program unit or host variables. In either case, these variables are declared with a size. For example, if a PL/SQL variable is declared as a VARCHAR(10) and used as an actual parameter to call a procedure or function, the value of the corresponding formal parameter can hold a string only up to ten characters in length.

Tip

You can't include size information when declaring a formal parameter. Including size information causes an error in the program unit creation.

For demonstration, let's return to the MEMFMT1_SF function created earlier in this chapter to return a string containing the member id and name. We use an anonymous PL/SQL block to call this function and demonstrate the size set on the actual parameter and to determine the size constraint on the formal parameter. Figure 5-14 shows two executions of the anonymous block. The actual parameter LV_NAME_TXT is set to a length of 35 in the first run and set to 15 in the second run.

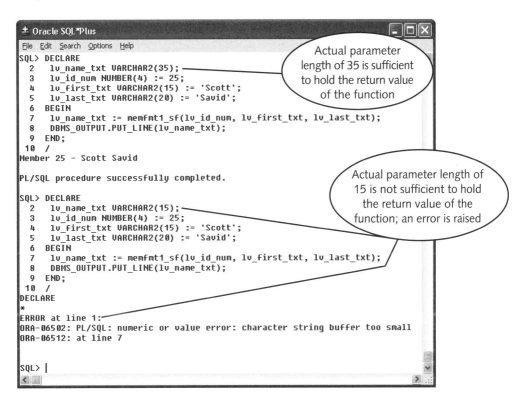

Figure 5-14 Parameter size constraint testing

The block invokes MEMFMT1_SF and uses the actual parameter named LV_NAME_TXT to hold the return value. The first execution succeeds as the actual parameter or LV_NAME_TXT variable is declared with a length of 35. However, in the second execution, LV_NAME_TXT is declared with a length of 15. Notice the error references the line in which the function is invoked and states that the character string buffer is too small. This occurs because we attempted to retrieve a character value with a length of 23 into a formal parameter that has been restricted to 15 characters based on the actual parameter construct.

Keep in mind that a good method to handle the size issue when declaring variables that hold values from a database table is to use the %TYPE attribute to use the size of the database column. Along these lines, if actual parameters are dealing with database column values returned by a program unit, then the %TYPE declaration should still be used to ensure sufficient parameter size.

TECHNIQUES OF PASSING PARAMETER VALUES

PL/SQL uses two techniques for passing values between actual parameters and formal parameters. In the first method, IN parameter values are **passed by reference**, which means a pointer to the value in the actual parameter is created instead of copying the value from the actual parameter to the formal parameter. In the second, OUT and IN OUT parameters are **passed by value** in which the value is copied from the actual to the formal parameter.

Using the first method (the reference method) is generally faster especially if you are dealing with collections such as tables and varrays, which can hold a great deal of data. This method is faster because the copy action does not have to take place.

CONTROLLING WHICH VALUE PASSING TECHNIQUE IS USED

Beginning with Oracle8i, the default behaviors of value passing can be overridden by using a compiler hint named NOCOPY. A **compiler hint** is a request a programmer includes within his or her code that asks Oracle to modify the default processing in some manner. Oracle acts on the hint as long as any restrictions on this operation are met.

Figure 5-15 displays the TEST_NOCOPY_SP procedure, which uses the NOCOPY hint on the P_OUT parameter. By including this hint immediately after the mode for the parameter, we have asked the system to pass the value of the actual parameter used to call this procedure by reference rather than by value (i.e., NOCOPY). Notice that the procedure assigns the value of 5 to the OUT parameter and raises an error if the IN parameter is equal to 1.

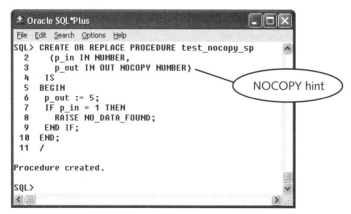

Figure 5-15   NOCOPY compiler hint for a parameter

The no_data_found error raised in the TEST_NOCOPY_SP procedure has no particular significance in this example. Any error could have been used.

Now, we can construct another procedure that calls the TEST_NOCOPY_SP procedure so that we can see the effect of passing parameter values between actual and formal parameters by reference. Figure 5-16 displays the procedure named RUN_NOCOPY_SP. This procedure contains one variable named LV_TEST_NUM that is used as the actual parameter for the P_OUT formal parameter when calling the TEST_NOCOPY_SP procedure. It also contains the general exception handler of WHEN OTHERS.

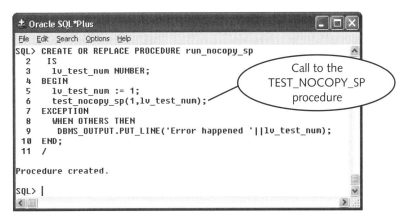

Figure 5-16   Procedure to test the NOCOPY hint

To demonstrate how the NOCOPY hint works, we can invoke the RUN_NOCOPY_SP procedure, as shown in Figure 5-17.

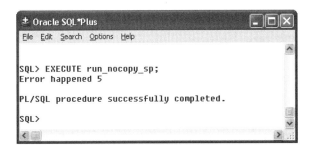

Figure 5-17   Invoking the RUN_NOCOPY_SP procedure

The LV_TEST_NUM variable is set to 1 and then the TEST_NOCOPY_SP procedure is called. The TEST_NOCOPY_SP procedure is called with a 1 in the IN parameter to force an error to be raised. The TEST_NOCOPY_SP procedure assigns a value of 5 to the P_OUT parameter and then the error is raised. Execution is returned to the calling procedure or RUN_NOCOPY_SP, and the exception handler prints a message to the screen showing the current value of the LV_TEST_NUM variable. The LV_TEST_NUM has a new value of 5 even though the TEST_NOCOPY_SP procedure raised an error. Because the LV_TEST_NUM actual parameter was passed by reference when the formal parameter P_OUT is altered, the actual parameter (LV_TEST_NUM) is altered because only one value really exists. The formal parameter maintains a pointer to the value only in the actual parameter because the NOCOPY hint was included for this parameter.

If the NOCOPY hint were not present, the value in LV_TEST_NUM would not have changed because the TEST_NOCOPY_SP procedure raised an error. After an error is raised, values are not passed or copied between parameters. Let's remove the NOCOPY hint and run the test again to confirm this operation. Figure 5-18 shows the modified procedure and execution result. Notice LV_TEST_NUM remains at a value of 1.

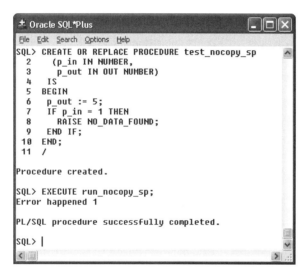

Figure 5-18   Removing the NOCOPY hint

FUNCTION PURITY LEVELS

To use functions in SQL, you must follow certain restrictions and requirements. Restrictions are limits on how the function can be used in an SQL statement, whereas requirements are facets that must be present to use the function in an SQL statement. The types of structures and actions that are affected by a particular function are described in terms of its purity level. The **purity level** is identified with a set of acronyms that indicate the restrictions on using the function.

Before reviewing the purity level acronyms, let's identify the restrictions on functions used in SQL by reviewing the following list:

- Functions cannot modify any tables in Oracle8 and prior versions. Beginning with Oracle8*i*, the function cannot modify a table used in the SQL statement that calls the function; however, it may alter other tables if called from a non-SELECT statement.

- If used in a remote or parallel operation, no reading or writing of packaged variables is allowed.

- If used in a SELECT, VALUES, or SET clause, the function can write values to packaged variables; otherwise, it is not allowed.

- Functions cannot be used in a CHECK constraint or as a default value of a table column.

- If the function calls other subprograms, the subprograms cannot break these rules.

In addition to the restrictions for specific function uses, any function you use in an SQL statement—regardless if it is created by you or supplied by Oracle9*i*—*must* meet the following list of requirements:

- Must be a stored database object (or in a stored package)

- Can use only IN parameters

- Formal parameter data types must use database data types (no PL/SQL data types such as BOOLEAN are permitted)

- Return data types must be a database data type

- Must not issue transaction control statements to end the current transaction prior to execution

- Cannot issue ALTER SESSION or ALTER SYSTEM commands

Now, let's return to the purity level acronyms. Each purity level defines what objects the function, when used in SQL, can legally modify or read. You know which purity level applies to your programming situation by identifying which of the restrictions apply to the function. Table 5-1 lists the four purity levels.

Table 5-1   Function purity levels

Level Acronym	Level Name	Level Description
WNDS	Writes No Database State	Function does not modify any database tables (no DML)
RNDS	Reads No Database State	Function does not read any tables (no SELECT)
WNPS	Writes No Package State	Function does not modify any packaged variables (packaged variables are variables declared in a package specification)
RNPS	Reads No Package State	Function does not read any packaged variables

We can see how the purity level acronyms are included in our coding when creating functions within packages later in this book. In the meantime, let's do an example to visualize how the restrictions impact the use of our functions within SQL statements.

To test function restriction within SQL statements:

1. Open or return to SQL*Plus.

2. Enter the following code to create a function named FCT_TEST1_SF that performs an update on the BB_TEST1 table:

```
CREATE OR REPLACE FUNCTION fct_test1_sf
  (p_num IN NUMBER)
    RETURN NUMBER
 IS
BEGIN
 UPDATE bb_test1
  SET col1 = p_num;
 RETURN p_num;
END;
/
```

3. Enter the following code to create a function named FCT_TEST2_SF that performs an update on the BB_TEST2 table:

```
CREATE OR REPLACE FUNCTION fct_test2_sf
  (p_num IN NUMBER)
    RETURN NUMBER
 IS
BEGIN
 UPDATE bb_test2
  SET col1 = p_num;
 RETURN p_num;
END;
/
```

4. Enter the following UPDATE statement and compare your error to Figure 5-19. The error indicates that a function used in an SQL statement cannot update the same table that is being modified by the SQL statement that calls the function.

```
UPDATE bb_test1
  SET col1 = fct_test1_sf(2);
```

5. Enter the following UPDATE statement and note that this UPDATE executes with no errors. The function is updating a different table (BB_TEST2) than the SQL statement calling the function (BB_TEST1) and, therefore, it executes just fine.

```
UPDATE bb_test1
  SET col1 = fct_test2_sf(2);
```

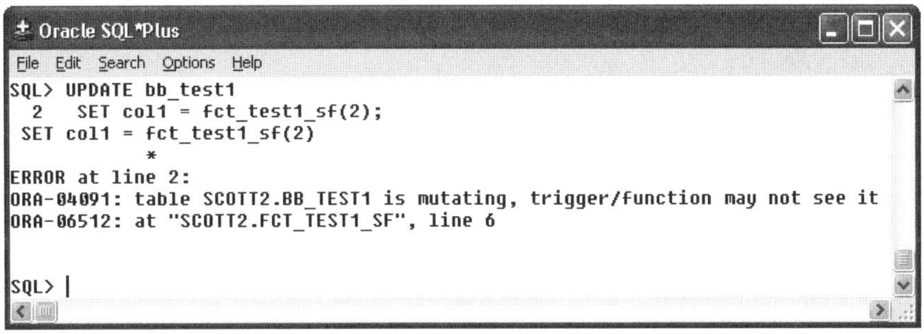

Figure 5-19   Error in function called by an UPDATE statement

6. Enter the following SELECT statement and compare your error to Figure 5-20. The error indicates that a function called from a SELECT statement cannot modify table data. Recall in the restrictions list that functions in SQL statements can alter other tables if called from a non-SELECT statement.

```
SELECT fct_test2_sf(col1)
  FROM bb_test1;
```

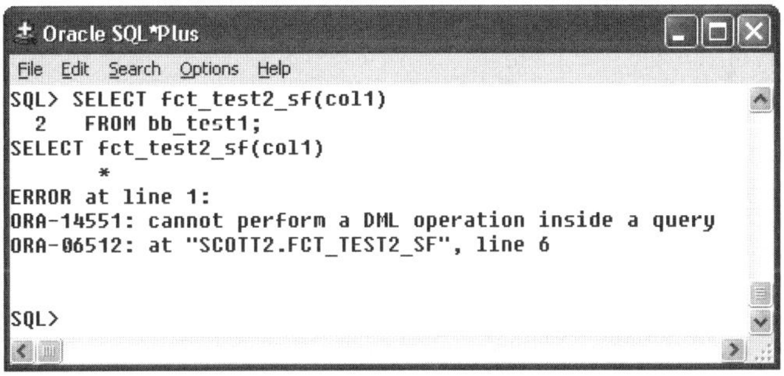

Figure 5-20   Error in function called by a SELECT statement

DATA DICTIONARY INFORMATION ON PROGRAM UNITS

We can quickly view parameter information by issuing a DESC command on the program unit of interest, as shown in Figure 5-21.

In addition, we can reference the data dictionary to determine what code is contained in a program unit. This can be important when walking into a shop that has little manual documentation to reference. Use the text column of the USER_SOURCE data dictionary view to display the code contained in the program unit, as shown in Figure 5-22.

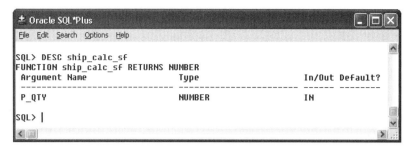

Figure 5-21   Using the DESC command

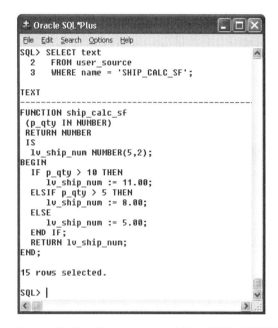

Figure 5-22   Text column of the USER_SOURCE data dictionary view

Listing the code to display on your screen also assists if you need to change a program unit but do not have a file with the original script at hand. Just copy and paste the program unit code from the display to a text editor to edit.

DELETING PROGRAM UNITS

At times, we want to eliminate particular procedures and functions. This is typically the case when an application or particular functionality becomes outdated and the program unit is no longer needed.

Deleting a program unit can be accomplished by issuing a DROP command. The DROP command must include the object type and object name. The following are sample DROP commands:

```
DROP PROCEDURE procedure_name;
DROP FUNCTION function_name;
```

CHAPTER SUMMARY

- A function is another type of program unit like a procedure; however, a function can be used in SQL as well as PL/SQL statements.

- A function is part of an expression, not an entire expression like a procedure.

- Functions use parameters and must return a value. A RETURN statement is included in the header to indicate the return value data type and at least one RETURN statement must be included in the function body.

- Even though an OUT parameter is legal in a function, it is rarely used. A function usually returns only a single value via the RETURN statement so that it can be used in SQL statements.

- Multiple RETURN statements can be included in a function; however, only one is executed. In addition, for simplicity, it is considered good form to use only a single RETURN statement.

- A RETURN statement in a procedure is different in that it simply controls the flow of statement execution.

- A formal parameter refers to the parameters included in the program unit. Actual parameters are the variables used to pass values to the program unit when invoking it. The actual parameters are also referred to as arguments in the call statement.

- Parameter values can be passed by value or by reference. Passing by value copies the value to and from the actual and formal parameters. Passing by reference creates a pointer from the formal parameter to the actual parameter and only one copy of the value exists. The NOCOPY compiler hint is used to override the default actions of parameter passing methods.

- Function purity refers to the rules that a function must comply with to be used in SQL statements.

- The USER_SOURCE data dictionary view can be referenced to view the code contained in a program unit.

- The DROP statement is used to delete a program unit.

REVIEW QUESTIONS

1. A RETURN statement in a function header accomplishes what task?

   a. declares the return value data type

   b. declares the return value size and data type

   c. moves execution to the body

   d. A RETURN statement cannot be used in a function header.

2. A function body can include many RETURN statements but _____.

   a. then the function cannot be used in SQL statements

   b. an OUT parameter needs to exist for each RETURN

   c. only one RETURN statement can execute

   d. RETURN statements are used only in function headers

3. How do the header of a function and a procedure differ?

   a. A function cannot use IN OUT parameters.

   b. No parameters are included in a function.

   c. The function header contains a RETURN statement.

   d. The procedure header contains a RETURN statement.

4. A formal parameter is _____.

   a. a parameter in a program unit

   b. a parameter used to call a program unit

   c. a parameter with a constant value

   d. the same as an actual parameter

5. What does a RETURN statement in a function body accomplish?

   a. It stops execution.

   b. It returns a value to the calling statement.

   c. It changes the flow of execution.

   d. RETURN statements cannot be used in a function body.

6. Are any size restrictions placed on formal parameters?

   a. No.

   b. No, except if a length is included in the formal parameter declaration.

   c. Yes, size constraints are assumed from actual parameters.

   d. Yes, the size must be included in the formal parameter declaration.

7. Passing a parameter value by reference means the value is _____.

   a. used in an assignment statement

   b. copied to the formal parameter

   c. referenced with a pointer and not copied

   d. used in several parameters

8. Passing a parameter value by value means the value is _____.

   a. used in an assignment statement

   b. copied to the formal parameter

   c. referenced with a pointer and not copied

   d. used in several parameters

9. A compiler hint named _____ can be used to override the default parameter passing methods.

   a. ByReference

   b. NOCOPY

   c. ByValue

   d. COPY

10. A function purity level of WNDS indicates that the function _____.

    a. reads from database tables

    b. writes to database tables

    c. does not read from database tables

    d. does not write to database tables

11. Describe the main differences between a function and a procedure.

12. What is the difference between using a RETURN statement in a function versus a procedure?

13. Describe what is meant by the terms "actual parameters" and "formal parameters."

14. What is the difference between passing parameter values by reference versus by value?

15. Name two requirements that must be met for a function to be used in an SQL statement.

Advanced Review Questions

1. Based on the following anonymous PL/SQL block, could the program unit named CHECK_SUM be a function?

```
DECLARE
 v_test NUMBER;
BEGIN
 IF v_test > 100 THEN
  check_sum(v_test);
 END IF;
END;
/
```

   a. Yes.

   b. Yes, but only if the function includes no SQL statements.

   c. Yes, if the V_TEST variable is initialized.

   d. No.

2. If the function named CALC_IT contains an UPDATE to the JUNK1 table, can it be legally used in the following PL/SQL block?

```
DECLARE
 v_num NUMBER(1) := 5;
BEGIN
 UPDATE junk2
  SET col1 = calc_it(v_num);
 COMMIT;
END;
/
```

   a. Yes.

   b. No, a mutating table error is raised.

   c. No, functions cannot include DML actions.

   d. No, functions cannot be called from SQL statements.

3. If the function named CALC_IT contains an UPDATE to the JUNK1 table, can it be legally used as shown in the following SQL statement?

```
SELECT id, calc_it(cost)
  FROM junk2;
```

   a. Yes.

   b. No, a mutating table error is raised.

   c. No, functions called from a query cannot include DML actions.

   d. No, functions cannot be called from SQL statements.

4. Given the following function, which statement would execute successfully?

```
CREATE FUNCTION calc_it
  (p_cost NUMBER)
  RETURN NUMBER
 IS
 v_num NUMBER(8);
BEGIN
  v_num := p_cost*100;
  RETURN v_num;
END;
/
```

a. `calc_it(55);`

b. `EXECUTE calc_it(55);`

c. `SELECT calc_it(cost) FROM orders;`

d. `SELECT calc_it(p_cost => 55) FROM orders;`

5. Which of the following CREATE FUNCTION statements produces a compile error?

a.
```
CREATE OR REPLACE FUNCTION calc_it
    (p_num NUMBER)
     RETURN NUMBER;
   IS
     v_num NUMBER(8);
   BEGIN
     v_num := p_num * 100;
     RETURN v_num;
   END;
```

b.
```
CREATE OR REPLACE FUNCTION calc_it
    (p_num NUMBER(8))
     RETURN NUMBER;
   IS
     v_num NUMBER(8);
 BEGIN
     v_num := p_num * 100;
     RETURN v_num;
   END;
```

c.
```
CREATE OR REPLACE FUNCTION calc_it
     RETURN NUMBER;
   IS
     v_num NUMBER(8);
   BEGIN
     v_num := :p_num * 100;
     RETURN v_num;
   END;
```

```
d. CREATE OR REPLACE FUNCTION calc_it
     (p_num NUMBER)
      RETURN VARCHAR2;
     IS
      v_num VARCHAR2(8);
     BEGIN
      v_num := TO_CHAR(p_num * 100);
     RETURN v_num;
     END;
```

HANDS-ON ASSIGNMENTS

Assignment 5-1: Format Numbers as Currency

Many of the application screens and reports that are generated from the database display dollar amounts. Complete the following steps to create a function that formats the number provided as an argument with a dollar sign, commas, and two decimal places.

1. Open or return to SQL*Plus

2. Create a function using the following code:

```
CREATE OR REPLACE FUNCTION dollar_fmt_sf
  (p_num NUMBER)
   RETURN VARCHAR2
 IS
   lv_amt_txt VARCHAR2(20);
BEGIN
   lv_amt_txt := TO_CHAR(p_num,'$99,999.99');
   RETURN lv_amt_txt;
END;
/
```

3. Type **SET SERVEROUTPUT ON** to enable display from DBMS_OUTPUT.

4. Test the function by executing the following anonymous PL/SQL block. Your results should match Figure 5-23.

```
DECLARE
   lv_amt_num NUMBER(8,2) := 9999.55;
BEGIN
   DBMS_OUTPUT.PUT_LINE(dollar_fmt_sf(lv_amt_num));
END;
/
```

5. Test the function using the following SQL statement. Your results should match Figure 5-23.

```
SELECT dollar_fmt_sf(shipping), dollar_fmt_sf(total)
  FROM bb_basket
  WHERE idBasket = 3;
```

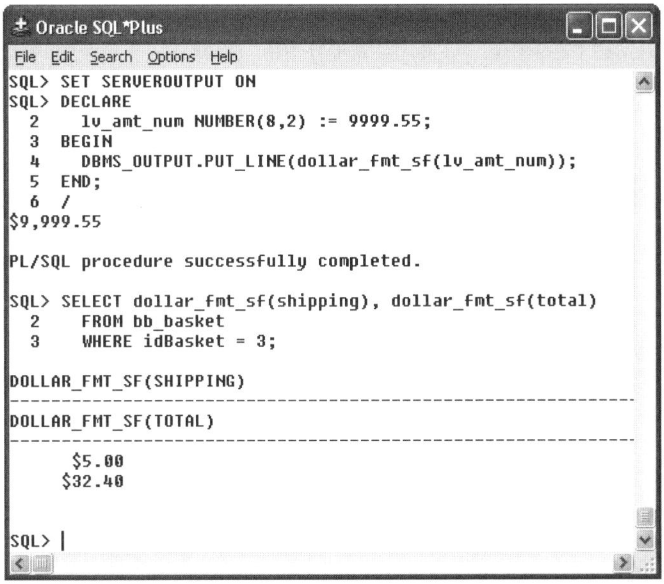

Figure 5-23   Testing the dollar format function

Assignment 5-2: Calculate Total Shopper Spending

Many of the reports generated from the system calculate the total dollars in purchases for a shopper. Complete the following steps to create a function named TOT_PURCH_SF that accepts a shopper id as input and returns the total dollars that the shopper has spent with the company. Use the function in a SELECT statement that shows the shopper id and total purchases for every shopper in the database.

1. Open or return to SQL*Plus.

2. Develop and run a CREATE FUNCTION statement to create the TOT_PURCH_SF function. The function code needs a formal parameter for the shopper id and to sum the total column from the BB_BASKET table.

3. Develop a SELECT statement using the BB_SHOPPER table to produce a list of each shopper in the database and his or her respective total.

Assignment 5-3: Calculate the Count of Orders by a Shopper

Another commonly used statistic in reports is the total number of orders placed by a shopper. Complete the following steps to create a function named NUM_PURCH_SF that accepts a shopper id and returns the total number or orders the shopper has placed. Use the function in a SELECT statement to display the number of orders for shopper 23.

1. Open or return to SQL*Plus.

2. Develop and run a CREATE FUNCTION statement to create the NUM_PURCH_SF function. The function code needs to tally the number of orders

(think of the appropriate Oracle9*i* function to use) by shopper. Keep in mind that the ORDERPLACED column contains a 1 if the order has been placed.

3. Create a SELECT query using the function on the IDSHOPPER column of the BB_SHOPPER table. Be sure to select only shopper 23.

Assignment 5-4: Identify the Day of the Week for the Order Date

The day of the week that baskets are created is quite often analyzed to determine consumer-shopping patterns. Create a function named DAY_ORD_SF that accepts an order date and returns the day of the week. Use the function in a SELECT statement to display each basket id and the day of the week the order was created. Do a second SELECT statement using this function to display the total number of orders for each day of the week that has orders. (*Hint:* Call the TO_CHAR function to retrieve the day of week from a date.)

1. Open or return to SQL*Plus

2. Develop and run a CREATE FUNCTION statement to create the DAY_ORD_SF function. Use the DTCREATED column of the BB_BASKET table as the date the basket was created. Call the TO_CHAR function using the DAY option to retrieve the day of week for a date value.

3. Create a SELECT statement that lists the basket id and the day of the week placed for every basket.

4. Create a SELECT statement using a GROUP BY clause to list the total number of baskets per day of the week. Which is the most popular shopping day? (You should discover it is Sunday.)

Assignment 5-5: Calculate Days Between Ordering and Shipping

An analyst in the quality assurance office reviews the time lapse between receiving an order and shipping the order. Any orders that have not been shipped within a day of the order being placed are investigated. Create a function named ORD_SHIP_SF that calculates the number of days between the date the basket was created and the shipped date. The function should return a character string that indicates 'OK' if the order was shipped within a day or 'CHECK' if it was not. The IDSTAGE column of the BB_BASKETSTATUS table indicates the item is shipped with a value of 5 and the DTSTAGE column is the shipping date. The DTORDERED column of the BB_BASKET table is the order date. Use the function in an anonymous block that uses a host variable to receive the basket id to check basket 3.

Assignment 5-6: Identify the Description of an Order Status Code

When a shopper returns to the Web site to check the status of an order, the information from the BB_BASKETSTATUS table is displayed. However, only the status code is available in the BB_BASKETSTATUS table, not the status description. Create a function named STATUS_DESC_SF that accepts a stage id and returns the status description. The descriptions by stage id are listed in Table 5-2. Test the function in a SELECT statement that retrieves all rows in the BB_BASKETSTATUS table for basket 4 and displays the date of the stage item and stage description.

Table 5-2   Basket stage code descriptions

idStage	Description
1	Order submitted
2	Accepted, sent to shipping
3	Backordered
4	Cancelled
5	Shipped

Assignment 5-7: Calculate the Tax Amount for an Order

Create a function named TAX_CALC_SF that accepts a basket id, calculates the tax amount using the basket subtotal, and returns the appropriate tax amount for the order. The tax is determined by the shipping state, which is stored in the BB_BASKET table. The table named BB_TAX contains the tax rate applicable to states that require tax on Internet purchases. If the state is not listed in the tax table, then a tax amount of zero should be applied to the order (set a default value of zero). Use the function in a SELECT statement that displays the shipping cost for basket 3.

Assignment 5-8: Identify Products That Are on Sale

When a product is placed on sale, Brewbean's records the sale start and end dates in columns of the BB_PRODUCT table. A function is needed to provide sales information when a shopper selects an item. If the product is on sale, then the function should return the value of ON SALE! However, if it is not on sale, then the function should return the value of Great Deal! These values are used in the product display page. Create a function named CK_SALE_SF that accepts a date and product id as arguments, checks if the date falls within the sale period for the product, and returns the appropriate string value. Test using product id 6 and two dates: 10-MAY-03 and 19-MAY-03. Verify results by reviewing the product sales information.

CASE PROJECTS

Case 5-1: Update Basket Data Upon Order Completion

A number of functions created in this chapter assumed that the basket amounts including shipping, tax, and total were already posted to the BB_BASKET table. However, we have not yet developed the program units needed to update these columns when a shopper checks out. A procedure is needed to update the following columns in the BB_BASKET table when an order is completed: ORDERPLACED, SUBTOTAL, SHIPPING, TAX, and TOTAL.

Three functions should be constructed to accomplish each of the following tasks: Calculate the subtotal using the BB_BASKETITEM table based on a basket id as input, calculate shipping costs based on a basket id as input, and calculate the tax based on basket id and the subtotal as input. Use these functions in a procedure to complete this task.

A value of 1 is to be entered into the ORDERPLACED column to indicate the shopper has completed the order. The subtotal is determined by totaling the item lines of the BB_BASKETITEM table for the applicable basket number. The shipping is based on the number of items in the basket: 1-4 = $5, 5-9 = $8, and 10+ = $11.

The tax is based on the rate applied by referencing the SHIPSTATE column of the BB_BASKET table to the state column of the BB_TAX table. This rate should be multiplied by the basket subtotal, which should be an INPUT parameter to the tax calculation because the subtotal is being calculated in this same procedure. The total tallies all these amounts.

The only INPUT parameter for the procedure is a basket id. The procedure needs to update the appropriate row in the basket table with all these amounts. To test, first set all the column values to NULL for basket 3 with the following UPDATE statement. Then invoke the procedure for basket 3 and check the INSERT results.

```
UPDATE bb_basket
   SET orderplaced = NULL,
           Subtotal = NULL,
           Tax = NULL,
           Shipping = NULL,
           Total = NULL
      WHERE idBasket = 3;
COMMIT;
```

Case 5-2: More Movies Rentals

More Movies receives numerous requests checking if movies are in stock. The company needs a function that retrieves movie stock information and formats a friendly message to display for user requests. The display should resemble the following: Star Wars is Available: 11 on the shelf.

Use movie id as the input value for this function. Assume that the MOVIE_QTY column in the MM_MOVIES table indicates the number of movies currently available for check out.

6

PL/SQL PACKAGES

In this chapter, you will:

- ◆ Create package specifications
- ◆ Create package bodies
- ◆ Invoke packaged program units
- ◆ Identify public versus private construct scope
- ◆ Test global construct value persistence
- ◆ Forward declare program units
- ◆ Create one time only procedures
- ◆ Overload packaged program units
- ◆ Manage restrictions on packaged functions used in SQL
- ◆ Determine execution privileges
- ◆ Identify data dictionary information regarding packages
- ◆ Delete or remove packages

A **package** is another type of PL/SQL construct, and it is a container that can hold multiple program units, such as procedures and functions. A package is similar to an ice tray in that the tray represents the package and the individual ice cubes represent the program units.

Using packages not only allows programmers to organize program units into related groups, but also enables the establishment of private program units, the sharing of variable values across program units, the overloading of program units, easier user privilege granting, and improved performance. All of these are discussed in this chapter. Packages also lead to improved handling of dependencies, which is addressed in Chapter 7.

THE CURRENT CHALLENGE IN THE BREWBEAN'S APPLICATION

The Brewbean's development group has some new issues to be resolved in their application effort. First, the group is starting to accumulate quite a few program units and needs to organize them in a more manageable manner. Second, the group needs to resolve how values can be stored throughout a user session. For example, the manager wants to be able to run daily percentage discount specials and the value needs to be stored during a shopper session and applied to the basket subtotal when appropriate. Third, a product search screen is being developed and the goal is to allow the user to enter a product id or a product name to use in the search.

Note that the program unit invoked to handle the search needs to handle either the number or character value provided and appropriately search based on id or name. However, the group is confused as to how to accomplish this because a program unit must have the data type defined for the parameters. Lastly, the database administrator is concerned with the time it takes to grant all the users privileges to all the objects and operations needed to run the application and wants to identify a mechanism to simplify this process.

REBUILDING YOUR DATABASE

Complete the following steps to rebuild the Brewbean's database for this chapter.

To rebuild the Brewbean's database:

1. Locate the **c6Dbcreate.sql** file in the Chapter.06 directory to ensure it exists. This file contains the script to create the database.

2. Open SQL*Plus.

3. We need to create a spool file so that SQL*Plus keeps a copy of all the messages received from running the file that creates the database. On the main menu in SQL*Plus, click **File**, point to **Spool**, and then click **Spool File**. A Select File dialog box appears.

4. Browse to a folder used to contain your working files and in the File name text box, type **DB_log6**, and then click **Save**. Now, whatever text is seen in our SQL*Plus session is saved to this file for future reference.

5. Now let's create the database. In SQL*Plus, enter the following command, which runs all the statements contained in the c6Dbcreate.sql file. Messages verifying the creation and data insertion steps scroll on the SQL*Plus screen. This may take a couple minutes to complete.

   ```
   @<pathname to PL/SQL files>\Chapter.06\c6Dbcreate
   ```

6. Now, we turn the spooling off so that we can review the results in the spool file created. On the main menu, click **File**, point to **Spool**, and then click **Spool Off**.

7. Open a text editor.

8. Open the file named **DB_log6.lst** in your working folder.

9. Review the messages for any errors.

PACKAGE SPECIFICATION

The **package specification** declares all the contents of the package and is referred to as the package header. The specification is required and must be created before the body.

Declarations in a Package Specification

A package specification can contain declarations for procedures, functions, variables, exceptions, cursors, and types. As you'll recall, a declaration is a statement used to define a data construct, such as a variable, name, and data type. Declarations of procedures and functions contain the program unit header information (name, parameters, and return data type if applicable) but not the PL/SQL block portion of the program unit. The full procedure or function code is contained in the package body. In regards to program units, the package specification serves as a map to what program units are available in the package and the parameters of the program unit.

Figure 6-1 depicts the syntax layout for creating a package specification.

```
CREATE [OR REPLACE] PACKAGE package_name
  IS|AS
    declaration section...
  END;


-----------------------------------------------
Notes on syntax layout:
      [ ] - indicate optional portions of the statement
      Key commands - in all upper case
      User provided - in lower case
      | - offers an OR option
      . . . - indicate continuation possible
```

Figure 6-1   CREATE PACKAGE specification syntax

Let's use our knowledge of the syntax layout to create a package specification to include two procedures, one function, and a variable, as shown in Figure 6-2.

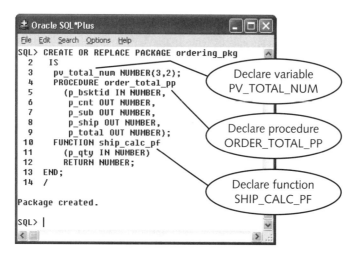

Figure 6-2   CREATE PACKAGE specification named ORDERING_PKG

These package elements are all related to the ordering process within the Brewbean's application. The ORDER_TOTAL_PP procedure and the SHIP_CALC_PF function are both used to calculate the order summary amounts upon shopper check out. The PV_TOTAL_NUM variable holds the total.

To create the package specification:

1. Open or return to SQL*Plus.

2. Open the **ordpkg06.sql** file from the Chapter.06 folder.

3. To run the code and create the package specification, click **File** and then **Run** from the main menu. You receive a "Package created" message.

Packages can be created only as stored program units. A package cannot be created in local libraries, as can procedures and functions.

Ordering of Items Within a Specification

The order of the items (such as procedures, functions, and variables) declared within a specification is not important—with the exception that if one declaration item is referenced by another declaration item, then the referenced item must be declared first within the package specification. For example, if we declared a cursor in our package specification that referenced the PV_TOTAL_NUM package variable, the variable must be declared before the cursor.

Note that any mixture of declarations may be used and no one type of declaration is required in a package specification. For example, a specification may contain only one variable and one cursor. It is not a requirement for a package specification to contain a procedure or function even though this is quite common. In the next section, we explore completing the creation of our package by creating the package body.

PACKAGE BODY

A **package body** is the program unit that contains the code for any procedures and functions declared in the specification. The package body must be created using the same name of an existing specification. The same name ties the specification and the body code together. In addition, all code in the procedure and function header sections in the package body must match exactly to the declarations in the corresponding specification. A diagram of the CREATE PACKAGE specification and body statements for the ORDERING_PKG package we are developing for Brewbean's is displayed in Figure 6-3.

Did you notice that the END statements that close the procedure and function statements in the package body reference the program unit name? Adding the procedure or function name in the END statement is optional; however, it is considered a good coding practice because it eases the reading of the code in the body by specifically marking which program unit code ends at this point.

The package body can also include declarations of variables, cursors, types, and program units not found in the corresponding specification. Any items declared in the body and not in the specification can be used only by other procedures and functions within this same package body. These are considered private items which are discussed in a later section in this chapter.

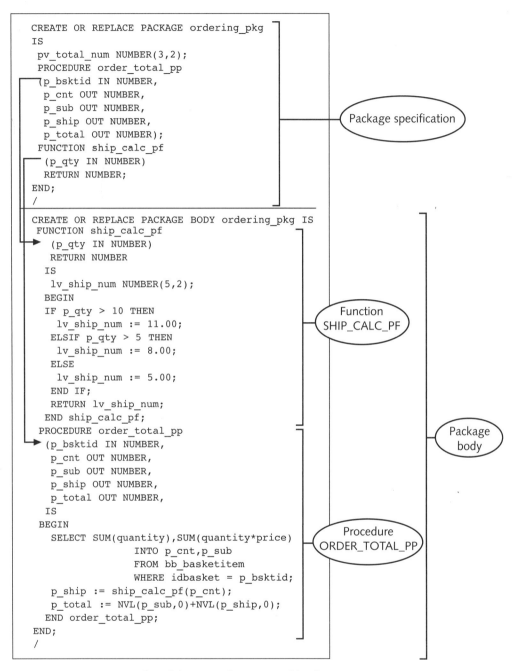

```
CREATE OR REPLACE PACKAGE ordering_pkg
IS
 pv_total_num NUMBER(3,2);
 PROCEDURE order_total_pp
(p_bsktid IN NUMBER,
 p_cnt OUT NUMBER,
 p_sub OUT NUMBER,
 p_ship OUT NUMBER,
 p_total OUT NUMBER);
 FUNCTION ship_calc_pf
 (p_qty IN NUMBER)
  RETURN NUMBER;
END;
/
```

Package specification

```
CREATE OR REPLACE PACKAGE BODY ordering_pkg IS
 FUNCTION ship_calc_pf
   (p_qty IN NUMBER)
   RETURN NUMBER
   IS
   lv_ship_num NUMBER(5,2);
   BEGIN
   IF p_qty > 10 THEN
     lv_ship_num := 11.00;
    ELSIF p_qty > 5 THEN
     lv_ship_num := 8.00;
    ELSE
     lv_ship_num := 5.00;
    END IF;
    RETURN lv_ship_num;
   END ship_calc_pf;
 PROCEDURE order_total_pp
  (p_bsktid IN NUMBER,
   p_cnt OUT NUMBER,
   p_sub OUT NUMBER,
   p_ship OUT NUMBER,
   p_total OUT NUMBER,
   IS
   BEGIN
    SELECT SUM(quantity),SUM(quantity*price)
              INTO p_cnt,p_sub
              FROM bb_basketitem
              WHERE idbasket = p_bsktid;
   p_ship := ship_calc_pf(p_cnt);
   p_total := NVL(p_sub,0)+NVL(p_ship,0);
  END order_total_pp;
END;
/
```

Function
SHIP_CALC_PF

Package
body

Procedure
ORDER_TOTAL_PP

Figure 6-3   Diagram of package specification and body

Let's familiarize ourselves with the syntax for creating the package body so that we can complete the ORDERING_PKG package. Figure 6-4 displays the syntax layout for creating a package body. Notice the statement specifically denotes package body in the CREATE clause at the beginning.

```
CREATE [OR REPLACE] PACKAGE BODY package_name
  IS|AS
    declaration section
    program units
END;


------------------------------------------------------------
Notes on syntax layout:
      [ ] - indicate optional portions of the statement
      Key commands - in all upper case
      User provided - in lower case
      | - offers an OR option
      . . . - indicate continuation possible
```

Figure 6-4   CREATE PACKAGE body syntax layout

Let's create the package body so we can explore using the package.

To create the package body:

1. Open or return to SQL*Plus.

2. Open the **ordbod06.sql** file from the Chapter.06 folder.

3. To run the code and create the package body, click **File** and then **Run** from the main menu. You receive a "Package body created" message, as shown in Figure 6-5.

To simplify package creation, it is quite typical for a developer to first create stand-alone procedures and functions. Then, he or she includes all the needed program unit code into the package. With this method, the program units are already tested before being included in a package. If this practice is used, the stand-alone procedure or function is normally deleted so that it is not accidentally referenced. In other words, we should only have a program unit in one place in the database.

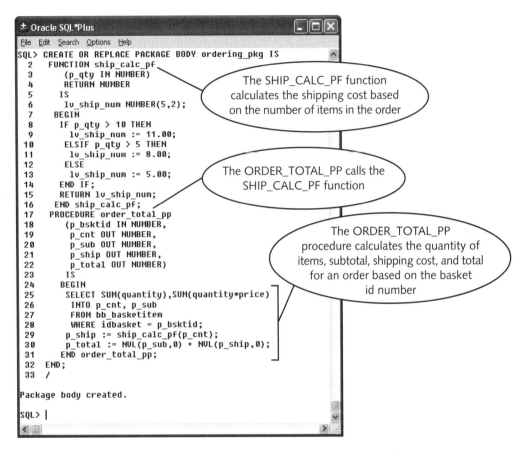

Figure 6-5   CREATE PACKAGE body execution

INVOKING PACKAGE CONSTRUCTS

Now, we have a completed package with a specification and body. Let's test the package by invoking one of the procedures. First, let's test the ORDER_TOTAL_PP procedure, which is one of the more heavily used procedures in the Brewbean's application.

To test the ORDER_TOTAL_PP procedure:

1. Open or return to SQL*Plus.

2. Enter the host variable commands to create CNT, SUB, SHIP, and TOTAL_P, as shown in the following code. These hold the values returned from the ORDER_ TOTAL_PP procedure for each of the parameters containing OUT modes.

```
VARIABLE cnt NUMBER
VARIABLE sub NUMBER
VARIABLE ship NUMBER
VARIABLE total NUMBER
```

3. Invoke ORDER_TOTAL_PP with the following statement, which uses a basket id of 12. Notice the procedure is invoked by using the notation of *package_name.procedure_name*.

```
EXECUTE ordering_pkg.order_total_pp
                   (12,:cnt,:sub,:ship,:total);
```

4. Issue the PRINT commands and the query shown in Figure 6-6 to check if the packaged procedure is operating properly. The query displays the count and sum of all the items in basket 12. These results need to match the values of the count (CNT) and subtotal (SUB) variables.

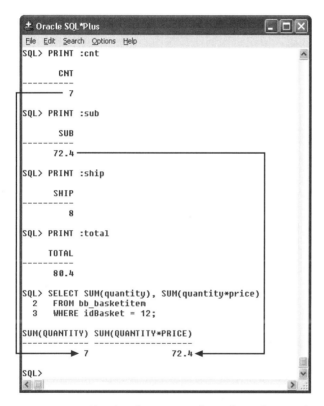

Figure 6-6 Check values to verify packaged procedure results

In addition, in reviewing the IF statement in the SHIP_CALC_PF, which is called by the ORDER_TOTAL_PP procedure, we can confirm that the correct shipping cost of $8 for 7 items was also calculated. What if we had other procedures outside of the ORDERING_PKG package that could use the SHIP_CALC_PF function? Any procedure or function that is declared in the package specification can be invoked from any program unit. Therefore, we could also call the SHIP_CALC_PF function directly from a PL/SQL block that exists outside of the package. Let's test a call to the packaged function named SHIP_CALC_PF from outside the package via an anonymous block.

To test a call:

1. Open or return to SQL*Plus.

2. Issue the following statement to create a variable named CALC to hold the return value of the function:

   ```
   VARIABLE calc NUMBER
   ```

3. Execute the following anonymous PL/SQL block to invoke the function with the value of 7 and place the return value into the CALC variable:

   ```
   BEGIN
    :calc := ordering_pkg.ship_calc_pf(7);
   END;
   /
   ```

4. Use the PRINT command as shown at the bottom of Figure 6-7 to check what value was returned to the CALC variable. This Brewbean's function should return a shipping cost of $8 for seven items ordered.

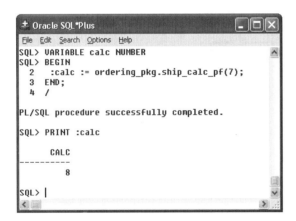

Figure 6-7   Test the packaged shipping calculation function

In each of the cases in this section, we invoked a packaged procedure and function. Note that the invoke command was exactly like the invoke command used with stand-alone program units, except the program unit name must be qualified or prefixed by the package name and a period notation. This instructs the system to find the package named ORDERING_PKG and then, within this package, find the specific program unit by name.

PACKAGE CONSTRUCT SCOPE

Package scope is the range of visibility for a particular element or construct contained in a package. In this context, "range of visibility" is a shorthand way of describing whether a packaged item, such as the SHIP_CALC_PF function, can be accessed from outside the package or only from other program units within the same package. Any elements declared in a package specification are considered **public**, which means they can be referenced from outside of the package. The elements are considered **private** if they can be called only from other program units within the same package.

Let's bring this concept to the package we have been developing for the Brewbean's application. First, consider that we were able to invoke the packaged function named SHIP_CALC_PF from an anonymous block. We could do this because we declared this function in the package specification, thus making it public. However, what would happen if this function declaration were not included in the package specification? Let's modify the ORDERING_PKG package to test this concept by removing the SHIP_CALC_PF function declaration from the specification and then testing.

To test public versus private package constructs:

1. Open or return to SQL*Plus.

2. Open the **ordpkgb06.sql** file from the Chapter.06 folder.

3. To run the code and re-create the package specification, click **File** and then **Run** from the main menu. You receive a "Package created" message, as shown in Figure 6-8.

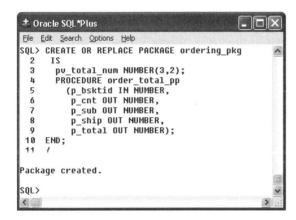

Figure 6-8   Create the ORDERING_PKG package without the SHIP_CALC_PF declaration

4. Now attempt to invoke the packaged SHIP_CALC_PF function with the following variable command and anonymous block (you should receive the error shown in Figure 6-9):

```
VARIABLE calc NUMBER
BEGIN
 :calc := ordering_pkg.ship_calc_pf(7);
END;
/
```

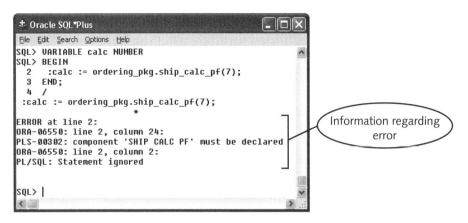

Figure 6-9 Error when calling the SHIP_CALC_PF function

The function is now private because it is not declared in the package specification. Therefore, it can only be called by other procedures within the same package. The call from an anonymous block in our example receives an error that indicates the function must be declared. This call causes the system to go to the ORDERING_PKG package and check the specification for a program unit named SHIP_CALC_PF declaration. As the function is not declared, the error is returned.

> When a program unit is called from within the same package body, the program unit name does not have to be preceded by the package name. The ORDER_TOTAL_PP procedure within the package called the SHIP_CALC_PF function with this statement: `p_ship := ship_calc_pf(p_cnt);`. However, any calls to a packaged program unit from outside the package must use the notation of *package_name.program_unit_name*.

PACKAGE GLOBAL CONSTRUCTS

One advantage that packages provide is that constructs, such as variables, cursors, types, and exceptions, declared in the specification are **global**. In other words, the values of the element persist throughout a user session and, therefore, can be referenced (used) in code within various parts of the application during a user session. This capability is valuable because it allows values to persist or be stored and used throughout an application session, not just in a single program unit. In addition, each user session maintains a separate copy of these elements that applies to their session only.

For example, recall the ordering package specification created earlier in this chapter contained a variable named PV_TOTAL_NUM. Let's suppose user A logs onto the system at 8:00 a.m. and calls a procedure that sets the value of this packaged variable to zero. After this point, whenever user A runs a program unit that references this variable, it returns a value of zero. However, let's suppose that at 10:00 a.m. user A runs a procedure that sets the value of the variable to five. Now, any references to this variable return a value of five from this point forward. What happens if user A logs off at lunchtime? When user A returns from lunch and logs back onto the system, a new session is started and the value of the variable is now returned to its initial state. In this case, the variable is not initialized in the variable declaration so it contains no value or is NULL until user A runs a command that sets a value into the variable. This concept is visually depicted on the left side of Figure 6-10.

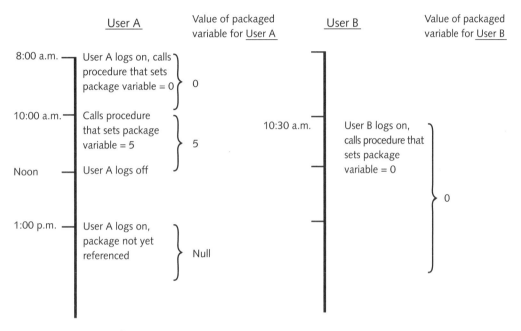

Figure 6-10   Initialization issues

What if user B logs onto the same system at 10:30 a.m.? Will user B see a value of five for the packaged variable? No, the values of these packaged elements are global to each individual user session. In other words, a set of global variables is maintained for each user or session. Therefore, in this scenario at 10:30 a.m., user A sees a value of 5 in the LV_TOTAL_NUM packaged variable and user B sees a value of 0 in the LV_TOTAL_NUM packaged variable.

Testing the Persistence of Packaged Variables

Let's try testing packaged variable persistence with the PV_TOTAL_NUM variable in the ORDERING_PKG package.

To test how the values of package variables persist throughout a session:

    1. Open or return to SQL*Plus.

2. First, you need to create the package specification again including an initial value for the PV_TOTAL_NUM variable. Open the **pkgscope06.sql** file from the Chapter.06 folder. Notice the PV_TOTAL_NUM variable is initialized to zero using := 0 in the declaration.

3. To run the code and re-create the package specification, click **File** and then **Run** from the main menu. You receive a "Package created" message.

4. Next, we need to create procedures that set and test the value held by the packaged variable. Execute the **pkgscopeb06.sql** script using the following command:

```
@<pathname to PL/SQL files>\Chapter.06\pkgscopeb06
```

Review the code in the procedures.

> Any scripts that contain multiple statements such as this one must be executed using the @ command. You will notice the command SET ECHO ON at the beginning of each of these script files to enable the display of the statements executed for your review.

5. Test the initial value of the packaged variable with the following statements. An initial value of 0 will be displayed.

```
VARIABLE g_test NUMBER
EXECUTE global_test_sp(:g_test);
PRINT :g_test
```

6. Change the packaged variable value to 5 and test with the following statements.

```
EXECUTE global_set_sp(5);
EXECUTE global_test_sp(:g_test);
PRINT :g_test
```

See Figure 6-11, which shows the packaged variable persistence.

7. Now open a second SQL*Plus session (using the same logon used to open the first session).

8. Test that the initial value of the packaged variable is zero with the following statements. Note that the user in the first session of SQL*Plus currently sees a value of 5 in the packaged variable, whereas the second session user sees a value of 0.

```
VARIABLE g_test NUMBER
EXECUTE global_test_sp(:g_test);
PRINT :g_test
```

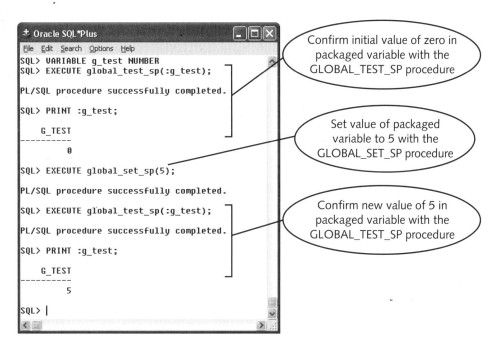

Figure 6-11   Set and test a packaged variable value

9. Change the packaged variable value to 7 with the following statements. Your screen should match Figure 6-12.

```
EXECUTE global_set_sp(7);
EXECUTE global_test_sp(:g_test);
PRINT :g_test
```

10. Move back to session 1 and test the current value of the packaged variable with the following statements. Note that the value remains at 5.

```
EXECUTE global_test_sp(:g_test);
PRINT :g_test
```

11. Close the second session of SQL*Plus.

As this example demonstrates, packaged variable values are persistent throughout a user session and each user of the package has his or her own copy of the packaged variable. This also holds true for the other types of constructs that can be declared in a package specification: cursors, types, and exceptions.

Package Specifications with No Body

Because we can store and share values via package variables, this is quite often used as a mechanism to hold static data values that are regularly referenced in programs. For example, let's say Brewbean's is developing some application screens to support the roasting operation and metric conversions that are commonly needed. As shown in the following code, we can create a package specification to hold just these conversion factors to reference whenever needed.

```
CREATE OR REPLACE PACKAGE metric_pkg IS
  cup_to_liter CONSTANT NUMBER := .24;
  pint_to_liter CONSTANT NUMBER := .47;
  qrt_to_liter CONSTANT NUMBER := .95;
END;
/
```

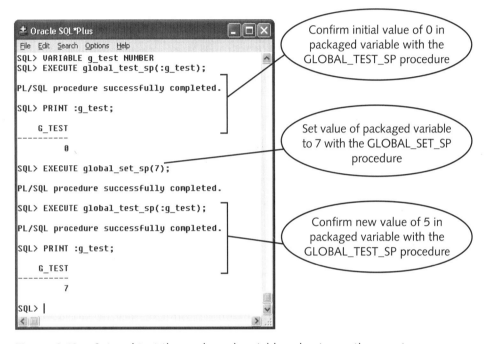

Figure 6-12   Set and test the packaged variable value in another session

There should not be a body created for this package because no program units are included in the package. Also, notice the option CONSTANT is used for each packaged variable declared. This option prohibits the values of these variables being modified.

Improving Processing Efficiency

As we have seen, packages allow the sharing of values across a user session. This not only allows the sharing of data across various procedures and functions during a session, but also provides an avenue of improving processing efficiency in an application. How? Let's consider the value of this efficiency in our own application.

For example, the Brewbean's manager wants an application that projects increased profits by product type and is based on various price increases. In this case, we can use a packaged cursor to retrieve the needed current product pricing, the costs, and the sales volumes that are needed in the analysis. If this data is put into a packaged cursor, it can be reused for each price increase scenario the user generates in a session with only a single SQL query run. In other words, no matter how many different pricing schemes the manager tests, only one SQL query is processed to initially retrieve the data into a cursor. The query results are cached in memory and can be reused for each pricing scenario processed.

Let's follow an example using a package with a cursor to demonstrate this concept. The cursor holds the product data needed to accomplish the described pricing analysis and the procedure calculates a profit increase based on a given percent price increase by each of the two types of products, equipment and coffee. First, review the package creation code in the following code example. The package specification includes a cursor and procedure declaration. The package body contains the procedure code that calculates the total sales dollar increase based on given percentage increases for both equipment and coffee products.

```
CREATE OR REPLACE PACKAGE budget_pkg
 IS
 CURSOR pcur_sales IS
   SELECT p.idProduct, p.price, p.type, SUM(bi.quantity) qty
   FROM bb_product p, bb_basketitem bi, bb_basket b
   WHERE p.idProduct = bi.idProduct
     AND b.idBasket = bi.idBasket
     AND b.orderplaced = 1
   GROUP BY p.idProduct, p.price, p.type;
 PROCEDURE project_sales_pp
   (p_pcte IN OUT NUMBER,
    p_pctc IN OUT NUMBER,
    p_incr OUT NUMBER);
END;
/
CREATE OR REPLACE PACKAGE BODY budget_pkg
 IS
 PROCEDURE project_sales_pp
   (p_pcte IN OUT NUMBER,
    p_pctc IN OUT NUMBER,
    p_incr OUT NUMBER)
  IS
   equip NUMBER := 0;
   coff NUMBER := 0;
  BEGIN
   FOR rec_sales IN pcur_sales LOOP
    IF rec_sales.type = 'E' THEN
     equip := equip + ((rec_sales.price*p_pcte)*rec_sales.qty);
    ELSIF rec_sales.type = 'C' THEN
     coff := coff + ((rec_sales.price*p_pctc)*rec_sales.qty);
    END IF;
   END LOOP;
    p_incr := equip + coff;
  END;
END;
/
```

All three parameters in the procedure have an OUT mode, so we need a host variable for each. Figure 6-13 displays the host variable creation and initialization. The PL/SQL block initializes two of the host variables, assigning 3% for the equipment price increase and 7% for the coffee price increase.

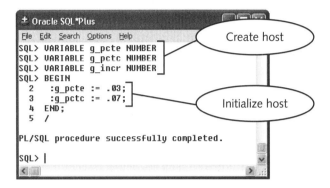

Figure 6-13   Host variable creation and initialization

To be able to compare the execution times of running the sales price analysis several times to determine if the packaged cursor improves performance after the initial execution, the TIMING feature in SQL*Plus needs to be turned on. Figure 6-14 displays setting the TIMING to on and executing the packaged procedure. The PRINT command is used to check the calculated result returned by the procedure.

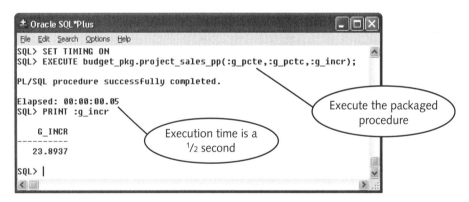

Figure 6-14   Initial execution of the packaged procedure

The second scenario for projected sales uses a 5% price increase for equipment and a 10% increase for coffee. Figure 6-15 displays the resetting of host variable values and the second execution of the packaged procedure. Notice that the execution time has reduced to less than 1/10 of a second.

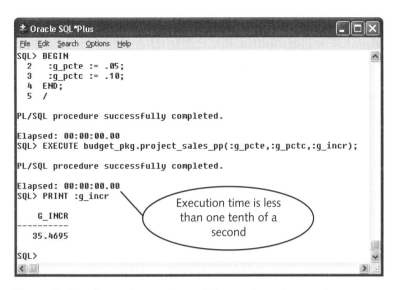

Figure 6-15 Second execution of the packaged procedure

This example demonstrates that the cursor data is being stored in cache or memory when the package is first invoked. After the initial invoke, all other invocations process much faster because the SQL query does not have to be processed again. This is extremely important in considering efficiency of application code because SQL statements or database retrieval tends to be some of the most costly processing statements.

Package code is cached, as is package data such as variables and cursors; therefore, the timing reduction shown from the first run of procedure PROJECT_SALES and the second run is not 100% due to the caching of the cursor data. A slight savings in time would also be generated from the procedure being cached as well. However, the SQL processing generates the main portion of processing time in this example, so the savings due to the procedure code caching is minimal. Also, keep in mind that execution times vary based on software and hardware configuration.

FORWARD DECLARATION IN PACKAGES

In the first package we created in this chapter, the ORDERING_PKG package, one procedure called or referenced a function from within the same package. Specifically, the ORDER_TOTAL_PP procedure in the ORDERING_PKG package called the SHIP_CALC_PF function. We made the SHIP_CALC_PF function private by not declaring it in the package specification. In the case of private packaged program units, does the order of the program units in the package body matter? Let's discover the answer by modifying the ordering package body by moving the SHIP_CALC_PF function to be the last program unit in the body.

To test the packaged program unit order:

1. Open or return to SQL*Plus.

2. First, you need to re-create the package body listing the SHIP_CALC_PF function as the last program unit. Open the **fwddec06.sql** file from the Chapter.06 folder.

3. To run the code and re-create the package specification, click **File** and then **Run** from the main menu. You receive a "Warning: Package Body created with compilation errors" message.

4. To view the error information, type **SHOW ERRORS** and then press **Enter**. The error information appears, as shown in Figure 6-16.

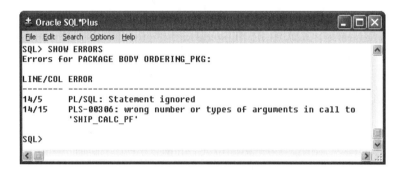

Figure 6-16 Compilation error for package body

So, what happened? Well, the private package function of SHIP_CALC_PF has no declaration in the specification and it is was the last item declared in the body. Thus, an error was raised in the first procedure in the package, ORDER_TOTAL_PP, because it attempted to call the SHIP_CALC_PF function. As far as the package body is concerned, the function does not exist.

Note that we could move the function back to the top of the package body and quit bothering with this compilation error. However, what if we have two program units that are mutually recursive, or they call each other? Or, once we get a lot of program units in a package, what if we want to organize the program units in some order, such as alphabetical or logical groupings, so that they are easier to locate? In such a case, we may have program units called prior to being declared.

You can work around this issue by using forward declarations within the package body. A **forward declaration** is a declaration of a program unit in a package body by placing the header code at the top of the package body code. Let's return to our current error in the ordering package body and add a forward declaration for the SHIP_ CALC_PF function.

To add a forward declaration:

1. Open or return to SQL*Plus.

2. First, you need to re-create the package body adding the forward declaration at the top of the body code. Open the **addfwd06.sql** file from the Chapter.06 folder.

3. A copy of the file contents appears in the SQL*Plus screen. Review the code, as shown in Figure 6-17.

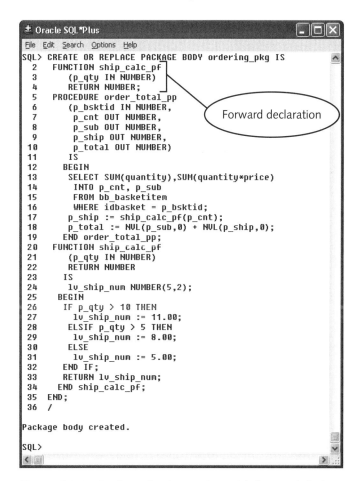

Figure 6-17   Package body creation with forward declaration

4. To run the code and re-create the package specification, click **File** and then **Run** from the main menu. You receive a "Package body created" message. There are no longer any compilation errors.

The package body containing the forward declaration for the SHIP_CALC_PF function compiles successfully because the ORDER_TOTAL_PP procedure now recognizes the function exists in the package.

ONE TIME ONLY PROCEDURE

Sometimes, it is desirable to assign a value to a global package variable dynamically from the database. For example, the Brewbean's company wants to allow the Marketing Department to run periodic daily percent specials. Each order subtotal would be reduced by this percent (hereafter called a "bonus" because it is a bonus for the customer).

To allow marketing direct access to assigning and making these bonuses available, we could develop an application that enables marketing to access and update the bonus percent in a database table named BB_PROMO. Then, when the ordering package is called during the ordering process, we can set a package variable to use this bonus percent in the order total calculation. This can be accomplished with a **one time only procedure**, which is a procedure in a package that runs only once—when the package is initially invoked. Regardless of how many times the package is referenced during a user session, the one time only procedure is executed only once upon the initial call to the package.

A one time only procedure is included in a package body as an anonymous PL/SQL block at the end of the body code. For example, the CREATE PACKAGE BODY statement shown in Figure 6-18 displays the needed change to the ORDERING_PKG package body to support the Brewbean's bonus application process.

```
CREATE OR REPLACE PACKAGE ordering_pkg
IS
 pv_bonus_num NUMBER(3,2);
 pv_total_num NUMBER(3,2) :=0;
 PROCEDURE order_total_pp
  (p_bsktid IN NUMBER,
   p_cnt OUT NUMBER,
   p_sub OUT NUMBER,
   p_ship OUT NUMBER,
   p_total OUT NUMBER);
END;
/

CREATE OR REPLACE PACKAGE BODY ordering_pkg IS
FUNCTION ship_calc_pf
 (p_qty IN NUMBER)
 RETURN NUMBER;
PROCEDURE order_total_pp
```

Add the packaged variable PV_ BONUS_NUM to the package specification

Figure 6-18   Include a one time only procedure in a package

```
(p_bsktid IN NUMBER,
 p_cnt OUT NUMBER,
 p_sub OUT NUMBER,
 p_ship OUT NUMBER,
 p_total OUT NUMBER)
IS
BEGIN
  SELECT SUM(quantity),SUM(quantity*price)
           INTO p_cnt,p_sub
           FROM bb_basketitem
           WHERE idbasket = p_bsktid;
  p_sub := p_sub - (p_sub*pv_bonus_num);
  p_ship := ship_calc_pf(p_cnt);
  p_total := NVL(p_sub,0) + NVL(p_ship,0);
 END order_total_pp;
FUNCTION ship_calc_pf
  (p_qty IN NUMBER)
  RETURN NUMBER
 IS
  lv_ship_num NUMBER(5,2);
 BEGIN
  IF p_qty > 10 THEN
   lv_ship_num := 11.00;
   ELSIF p_qty > 5 THEN
   lv_ship_num := 8.00;
   ELSE
   lv_ship_num := 5.00;
  END IF;
  RETURN lv_ship_num;
 END ship_calc_pf;
BEGIN
 SELECT amount
 INTO pv_bonus_num
 FROM bb_promo
 WHERE idPromo='B';
END;
/
```

Add calculation to reduce the subtotal by the bonus percentage amount

One time only procedure to retrieve the bonus amount from the BB_PROMO table and place it into the PV_BONUS_NUM variable

Figure 6-18   Include a one time only procedure in a package (continued)

Notice the one time only procedure starts with the key word BEGIN at the end of the package body. Let's modify and test the ordering package to include the one time only bonus amount procedure.

To modify and test a one time only procedure:

1. Open or return to SQL*Plus.

2. First, you need to re-create the package specification and body by adding the bonus variable and the one time only procedure. Execute the **onetime06.sql** script using the following command: **@<*pathname to PL/SQL files*>\ Chapter.06\onetime06**. Review the package code.

3. Enter the following query to check the values that currently exist in the BB_PROMO table. Note that this example applies a 5% bonus.

```
SELECT *
FROM bb_promo;
```

4. Use the following statements to create host variables, execute the packaged procedure, and check the results. Your results should match those displayed in Figure 6-19.

```
VARIABLE cnt NUMBER
VARIABLE sub NUMBER
VARIABLE ship NUMBER
VARIABLE total NUMBER
EXECUTE
    ordering_pkg.order_total_pp(12,:cnt,:sub,:ship,:total);
PRINT :cnt
PRINT :sub
PRINT :ship
PRINT :total
```

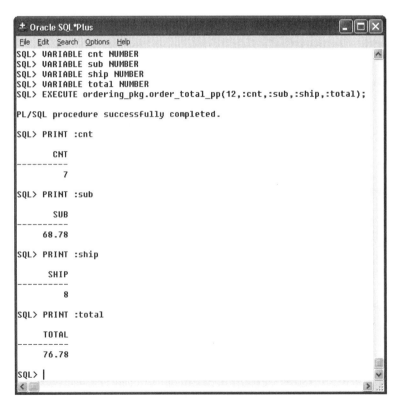

Figure 6-19   Testing the one time only procedure

5. As always, we can verify our results by querying the database, as shown in Figure 6-20.

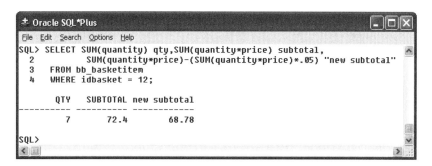

Figure 6-20   Verify results of the one time only procedure

When the ORDER_TOTAL_PP packaged procedure is invoked, the sequence of events is as follows: the ORDERING_PKG package is loaded, the one time only procedure executes and sets the bonus percent value into the PV_BONUS_NUM packaged variable, and then the ORDER_TOTAL_PP procedure executes and calculates the basket amounts.

OVERLOADING PROGRAM UNITS IN PACKAGES

The feature of **overloading** in packages is the ability to use the same name on multiple program units within the same package. Why would we want to have two procedures or functions with the same name? For example, one part of the Brewbean's application allows a product search to look up a particular product's regular price and sale price. The application screen shows a text box in which the user enters either a product id or name. In this scenario, we want to accomplish the same action (return the product's regular and sale price), but in one case, a number or id is provided as input, and in the other a string or product name is provided as input. When the user clicks the Search button, we want to invoke a procedure that can accept either a numeric or character argument as provided by the user.

How can this be accomplished when a data type must be assigned to a parameter when a procedure is created? The solution is overloading or creating two procedures in a package with the same name, and have the first contain a numeric parameter while the second contains a character parameter. When the procedure is invoked, the system identifies which one of the procedures has parameters that match the input data based on number of parameters and data types. In the Brewbean's example, the system can identify the correct procedure because one of the procedures has a numeric IN parameter while the other has a character IN parameter.

Let's test the concept of overloading by creating a new package that functions as described in the preceding scenario.

To create an overloaded procedure in a package:

1. Open or return to SQL*Plus.

2. First, you need to re-create the package specification and body to contain an overloaded procedure for the product search action. Review the package code, as shown in Figure 6-21.

```
CREATE OR REPLACE PACKAGE product_info_pkg IS
 PROCEDURE prod_search_pp
  (p_id IN bb_product.idproduct%TYPE,
   p_sale OUT bb_product.saleprice%TYPE,
   p_price OUT bb_product.price%TYPE);
   PROCEDURE prod_search_pp
  (p_id IN bb_product.productname%TYPE,
   p_sale OUT bb_product.saleprice%TYPE,
   p_price OUT bb_product.price%TYPE);
END;
/
CREATE OR REPLACE PACKAGE BODY product_info_pkg
IS
 PROCEDURE prod_search_pp
  (p_id IN bb_product.idproduct%TYPE,
   p_sale OUT bb_product.saleprice%TYPE,
   p_price OUT bb_product.price%TYPE)
  IS
 BEGIN
    SELECT saleprice, price
    INTO p_sale, p_price
    FROM bb_product
    WHERE idProduct = p_id;
 END;
 PROCEDURE prod_search_pp
  (p_id IN bb_product.productname%TYPE,
   p_sale OUT bb_product.saleprice%TYPE,
   p_price OUT bb_product.price%TYPE)
  IS
 BEGIN
    SELECT saleprice, price
    INTO p_sale, p_price
    FROM bb_product
    WHERE productname = p_id;
 END;
END;
/
```

Two procedures declared with the same name: PROD_SEARCH_PP

Package specification

Procedure code in the package body is the same in each, with the exception of the P_ID data type

Package body

Figure 6-21   Overloaded procedure in a package

3. Execute the **overload06.sql** script using the following command:

`@<pathname to PL/SQL files>\Chapter.06\overload06.`

4. Let's test by invoking with a product id. Use the following statements to execute the product search based on an id, and compare your results to Figure 6-22.

```
VARIABLE g_sale NUMBER
VARIABLE g_price NUMBER
EXECUTE
product_info_pkg.prod_search_pp(8,:g_sale,:g_price);
PRINT :g_sale
PRINT :g_price
```

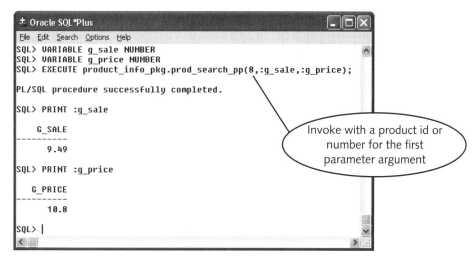

Figure 6-22 Invoking the overloaded procedure using a product id

5. Run the following anonymous block to set the host variables to 0 before the next invocation:

```
BEGIN
  :g_sale := 0;
  :g_price := 0;
END;
/
```

6. Next, you will test by invoking with a product name. Use the following statements to execute the product search based on a name, and compare your results to Figure 6-23:

```
EXECUTE
product_info_pkg.prod_search_pp('Brazil',:g_sale,:g_price);
PRINT :g_sale
PRINT :g_price
```

For overloading to be successful, the formal parameters in the procedures or functions must differ in at least one of the following categories: total number, data type family, or listed order. Note that the data types cannot just differ but have to be from different data type families. For example, let's say we build two procedures with the same name, and one has the IN parameter as CHAR and the other has the IN parameter as VARCHAR. Will this overloading work? No. CHAR and VARCHAR are in the same data type family and, therefore, do not qualify to be overloaded.

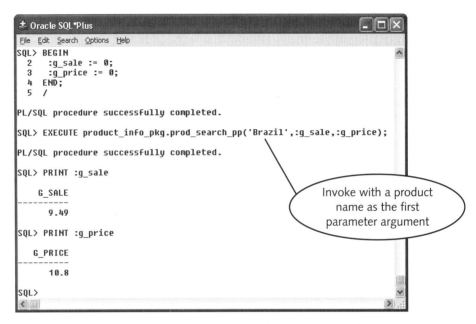

Figure 6-23   Invoking the overloaded procedure using a product name

Let's take a moment and consider what is meant by listed order. An example of differing listed orders would be if procedure A and B both contained two parameters, one character and one numeric, but A listed the character one first whereas B listed the numeric one first. This difference qualifies the procedures to be overloaded successfully. In addition, different data types set on a return value in a function do not qualify functions to be overloaded.

In any case, the invocation of the overloaded program unit is the driving force as to which of the packaged program units should be used. The number, data types, and order of arguments in the invocation are matched to the formal parameter lists of the overloaded program units to determine the appropriate program unit to execute.

MANAGING PACKAGED FUNCTION SQL RESTRICTIONS

Scan the following list of restrictions on functions used in SQL statements.

- Functions cannot modify any tables in Oracle8*i* and prior versions. Beginning with Oracle8*i*, the function cannot modify a table used in the SQL statement that calls the function; however, it may alter other tables if called from a nonselect statement.

- If used in a remote or parallel operation, no reading or writing of packaged variables is allowed.

- If used in a SELECT, VALUES, or SET clause, the function can write values to packaged variables; otherwise, it is not allowed.

- Functions cannot be used in a CHECK constraint or as a default value of a table column.

- If the function calls other subprograms, the subprograms cannot break these rules.

Considering these restrictions in regards to packaged functions versus stand-alone functions introduces additional connotations. What is different with packaged functions? Keep in mind that when we create a program unit, it is compiled or, in other words, the syntax is checked, the objects referenced are verified for existence, and a match of invocation arguments and parameters is completed. If we create a procedure that uses a function in an SQL statement, during compilation of the procedure, the function is checked to verify that the appropriate restrictions are abided by. However, compilation works differently when referencing packaged program units.

When compiling objects that call packaged program units, only the package specification or program unit header is used for verification. For example, if we create a procedure that uses a packaged function in an SQL statement, during compilation of the procedure, only the function header information is reviewed for existence and for appropriate parameters. However, because only the header information is used, the PL/SQL compiler does not see the entire block of code for the function and cannot determine if the restrictions on functions within SQL statements are being respected. We would not discover restriction violation issues until runtime, when an error is produced if the proper restrictions have not been followed in the function.

On the other hand, the fact that only the header is needed to compile an object referencing a packaged program unit does present an opportunity for developers. This provides an advantage in development shops in the sense that after a package specification exists, other program units can be created calling any of the program units within the package even though the package body may not yet exist. The PL/SQL compiler checks the package specification only to determine if the parameters of the call match the parameters in the specification for the function.

Why Developers Indicate Purity Levels

There is a mechanism that can be used to enable the compiler to verify that proper restrictions are being respected by the packaged function being called. Developers can use the PRAGMA RESTRICT_REFERENCES compiler instruction to indicate the purity level of the function in the package specification.

The **purity level** defines what type of data structures the function reads or modifies. If the purity level is declared in the package specification, the PL/SQL compiler can use this to determine whether the function meets all the restrictions required by the calling program unit. In addition, when the package body is created, the PL/SQL compiler can determine whether the function code adheres to the purity level the function was declared with in the specification. With this, we discover any errors at compile time rather than at runtime.

The purity levels are presented in Table 6-1. The PRAGMA RESTRICT_REFERENCES directive specifies each of the purity levels that apply to the packaged function.

Table 6-1   Packaged function purity levels

Level Acronym	Level Name	Level Description
WNDS	Writes No Database State	Function does not modify any database tables (No DML)
RNDS	Reads No Database State	Function does not read any tables (No select)
WNPS	Writes No Package State	Function does not modify any packaged variables (packaged variables are variables declared in a package specification)
RNPS	Reads No Package State	Function does not read any packaged variables

Not only is it helpful to discover issues at compile time but, in addition, using PRAGMA RESTRICT_REFERENCES leads to some performance advantages because the purity level is verified at compile time rather than at runtime. When the object using the packaged function is compiled, the restriction compliance is verified. In addition, when the completed package body including the entire function code is compiled, this code is checked to verify it matched the purity levels declared in the specification. All the verification is completed during compilation and, therefore, does not have to be accomplished during application execution.

PRAGMA RESTRICT_REFERENCES in Action

The PRAGMA RESTRICT_REFERENCES directive needs to be included in the package specification to clearly state the purity levels applicable to the function. The following code displays an example package specification including a function header. In the PRAGMA RESTRICT_REFERENCES statement, the first argument defines the function to which it applies; the remaining arguments state all the purity levels that apply to the function.

```
CREATE OR REPLACE PACKAGE pack_purity_pkg IS
  FUNCTION tax_calc_pf
   (p_amt IN NUMBER)
   RETURN NUMBER;
  PRAGMA RESTRICT_REFERENCES(tax_calc_pf,WNDS,WNPS);
END;
/
```

In this example, the WNDS and WNPS purity levels are indicated, which means the function does not contain any DML statements and does not modify any packaged variables. This leads us to assume the function does include reads on database tables and packaged variables given that the RNDS and RNPS purity levels are not listed in the PRAGMA RESTRICT_REFERENCES statement. Also, note that the PRAGMA statement must follow the function declaration in the package specification.

 The PRAGMA RESTRICT_REFERENCES compiler directive is required in Oracle versions prior to Oracle8*i*. Starting with Oracle8*i*, PRAGMA RESTRICT_REFERENCES is optional; however, keep in mind that you do reap the benefit of determining errors earlier at compile time and the performance gain of not checking the restrictions during runtime.

Default Purity Level for Packaged Functions

A default purity level for packaged functions can also be set using PRAGMA RESTRICT_REFERENCES for Oracle8*i* and higher. The default purity level applies to all the functions in that package specification that do not have their own PRAGMA RESTRICT_REFERENCES statement specified. In the following code, the word default (bolded for reference) is used instead of a specific function name.

```
PRAGMA RESTRICT_REFERENCES(DEFAULT,WNDS,WNPS);
```

Functions Written in External Languages

Functions written in other languages such as JAVA can also be called from Oracle programs. Even though the PL/SQL compiler cannot confirm the purity of an externally written function, we can still use PRAGMA RESTRICT_REFERENCES to assist in compilation. In this case, the TRUST option must be used with the PRAGMA RESTRICT_REFERENCES statement to address purity level declarations. Because the compiler cannot confirm the purity level of externally written functions, the TRUST option allows the calling program unit to compile successfully. It basically instructs the compiler to trust the purity levels indicated are indeed respected in the actual function code. An example of the PRAGMA RESTRICT_REFERENCES statement with the TRUST keyword listed as the last argument is in the following code:

```
CREATE OR REPLACE PACKAGE java_pkg AS
  FUNCTION phone_fmt_pf
    (p_phone IN VARCHAR2)
  RETURN VARCHAR2
    IS
  LANGUAGE JAVA
    NAME  'External.phone (char[])  return char[]';
  PRAGMA RESTRICT_REFERENCES (phone_fmt_pf, WNDS, TRUST);
END;
/
```

Program Unit and Package Execute Privileges

Another advantage of using program units and packages is that they ease the process of granting needed privileges to application users. Users need specific rights or privileges granted for each object (tables, sequences, etc) accessed and for each type of action (INSERT, DELETE, etc) to be accomplished. If you have multiple objects and multiple actions, granting these rights can become a time-consuming process.

Fortunately, you can grant the execute privilege to a user for a program unit or package and the user by default is allowed to use the program unit owner's privileges while running the program unit. The owner or developer of the package would, of course, already have all the needed privileges needed to execute the code. This is called definer-rights; the user assumes the rights of the program unit owner just during the processing of that program unit. It is important to note that the user can use these rights only during execution of that particular program unit. This provides another level of security on the system in that the user is never granted direct access to objects.

There may be situations in which this default privilege behavior of definer-rights is not desired. If we would rather have the user's own privileges in effect during the execution of a program unit, the default definer-rights behavior must be overridden. This is done using the AUTHID CURRENT_USER clause in the program unit header. This forces the user to use his or her own privileges and not the owner's privileges; it is called invoker-rights.

The following package specification demonstrates how to include the AUTHID clause to apply to the entire package (bolding is added for reference):

```
CREATE OR REPLACE PACKAGE pack_purity_pkg
   AUTHID CURRENT_USER IS
 FUNCTION tax_calc_pf
  (p_amt IN NUMBER)
  RETURN NUMBER;
  PRAGMA RESTRICT_REFERENCES(tax_calc_pf,WNDS,WNPS);
END;
/
```

This can also be accomplished with individual program units by including the AUTHID clause as the last item in the header.

Data Dictionary Information for Packages

The USER_SOURCE data dictionary view allows us to view the source code of packages, as shown in Figure 6-24.

```
± Oracle SQL*Plus                                    _ □ X
File  Edit  Search  Options  Help
SQL> SELECT text
  2    FROM user_source
  3      WHERE name = 'PRODUCT_INFO_PKG';

TEXT
-------------------------------------------------------
PACKAGE product_info_pkg IS
  PROCEDURE prod_search_pp
    (p_id IN bb_product.idproduct%TYPE,
     p_sale OUT bb_product.saleprice%TYPE,
     p_price OUT bb_product.price%TYPE);
  PROCEDURE prod_search_pp
    (p_id IN bb_product.productname%TYPE,
     p_sale OUT bb_product.saleprice%TYPE,
     p_price OUT bb_product.price%TYPE);
END;
PACKAGE BODY product_info_pkg IS
  PROCEDURE prod_search_pp
    (p_id IN bb_product.idproduct%TYPE,
     p_sale OUT bb_product.saleprice%TYPE,
     p_price OUT bb_product.price%TYPE)
    IS
  BEGIN
    SELECT saleprice, price
     INTO p_sale, p_price
     FROM bb_product
     WHERE idProduct = p_id;
  END;
  PROCEDURE prod_search_pp
    (p_id IN bb_product.productname%TYPE,
     p_sale OUT bb_product.saleprice%TYPE,
     p_price OUT bb_product.price%TYPE)
     IS
  BEGIN
    SELECT saleprice, price
     INTO p_sale, p_price
     FROM bb_product
     WHERE productname = p_id;
  END;
END;

34 rows selected.
```

Figure 6-24   Viewing package code from the data dictionary

The USER_OBJECTS data dictionary view is also useful to identify what packages exist on the system, as shown in Figure 6-25.

Recall that all the data stored in the data dictionary is in uppercase, so the WHERE clause conditions need to be in uppercase or use the UPPER function to work correctly.

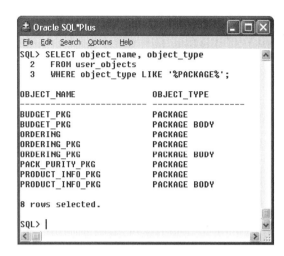

Figure 6-25   Listing package objects from USER_OBJECTS view

DELETING PACKAGES

At times, we want to eliminate procedures and functions, and a number of options exist to accomplish this task. You can issue a DROP command just as we can delete any object in the database. The DROP command must include the object type and object name.

The following code shows examples of DROP commands. The first example removes the package specification and body. The second example removes the package body only.

```
DROP PACKAGE package_name;
DROP PACKAGE BODY package_name;
```

It is typical to create individual program units, test them, and then place them into packages. It is a good practice to delete the stand-alone program units after you have placed the program unit in a package. Otherwise, it is easy to mistakenly call the stand-alone versus the packaged program unit. By keeping only one copy of any program unit on the system, you can be assured as to which one should be used and updated.

CHAPTER SUMMARY

▫ A package is a program construct that can house multiple program units and contain other constructs, such as variables and cursors. Packages provide performance benefits and provide new capabilities, such as overloading and global constructs.

▫ A package specification contains declarations for all the constructs to be used in the package and is referred to as the package header. A specification can exist without a body. Everything declared in the specification is considered public and can be referenced from outside the package.

❑ A package body contains the code for any program units and declaration of additional constructs, such as variables. Anything in the body that is not declared in the specification is considered private or cannot be referenced from outside the package.

❑ The private and public nature of items in a package is called the construct scope.

❑ Values of package constructs, such as variables and cursors, are considered global in that they persist throughout a user session.

❑ Any program unit referenced from within the same package must be declared before it is called. Forward declaration can be used if this cannot be achieved due to mutually recursive program units or the need to organize a large package in some logical order.

❑ When a package is first called, a one time only procedure can be used to accomplish such tasks as initializing global variable values from the database. The one time only procedure runs only once when the package is first referenced.

❑ Overloading is a feature available with packages and allows multiple program units to be created with the same name. This allows the same program unit to accept a variety of different parameter settings, enabling the calls to the unit to be flexible in handling different data types. The parameters of overloaded units must differ in parameter number, data type family, or order.

❑ Additional restrictions referred to as purity levels apply to functions used in SQL statements. Because package specifications insulate the PL/SQL compiler from the function code, the PRAGMA RESTRICT_REFERENCES directive is used in the function declaration to indicate to what purity levels the function code adheres.

❑ The USER_OBJECTS data dictionary view makes information regarding all the database objects you have created available. The USER_SOURCE view allows access to the source code of your program units.

6

REVIEW QUESTIONS

1. If a procedure is included in the package body but is not declared in the specification, it is considered _____.

   a. illegal

   b. private

   c. public

   d. restricted

2. If we have a function named CALC in a package named TAX_INFO and this function accepts one numeric value, which of the following is a legal invocation of this function?

   a. `calc(12);`

   b. `calc.tax_info(12)`

   c. `tax_info.calc(12,10)`

   d. `tax_info.calc(12);`

3. If we have a procedure named PROC_A in a package named PACK_A and this procedure calls a procedure named PROC_B, which is in the same package, which of the following are legal calls of the PROC_B from within the PROC_A? (Choose all that apply. Assume PROC_B accepts one value and returns two.)

   a. `proc_b(var1,var2,var3);`

   b. `proc_b(var1);`

   c. `pack_a.proc_b();`

   d. `pack_a.proc_b(var1,var2,var3);`

4. If a package specification includes the following code, which of the following declarations is a legal overloading declaration? (Choose all that apply.)

   ```
   PROCEDURE test_it (g_one IN NUMBER,
                      g_two OUT CHAR,
                      g_three OUT NUMBER);
   ```

   a.
   ```
   PROCEDURE test_it (g_one IN DECIMAL,
          g_two OUT CHAR,
          g_three OUT NUMBER);
   ```

   b.
   ```
   PROCEDURE test_it (g_one IN NUMBER,
          g_two OUT VARCHAR2,
          g_three OUT NUMBER);
   ```

   c.
   ```
   PROCEDURE test_it (g_one IN NUMBER,
          g_two OUT CHAR);
   ```

   d.
   ```
   PROCEDURE test_it (g_one IN NUMBER,
          g_two OUT NUMBER,
          g_three OUT NUMBER);
   ```

5. Which of the following conditions regarding parameters of two packaged program units with the same name would enable overloading? (Choose all that apply.)

   a. different data types

   b. different numbers

   c. different names

   d. different data type family

6. If invoker-rights are in effect when a user executes a program unit, what privileges are being used?

   a. those of the user

   b. those of the program unit owner

   c. those of the schema owner for which the program unit exists

   d. the system default privileges

7. What directive does Oracle provide to indicate the purity level of a function within a package specification?

   a. PRAGMA RESTRICT_FUNCTION

   b. PRAGMA RESTRICT_SQL

   c. PRAGMA RESTRICT_REFERENCES

   d. PRAGMA REFERENCES_RESTRICT

8. Which of the following data dictionary views allows a developer to review the code contained within a package?

   a. CODE

   b. SOURCE

   c. USER_CODE

   d. USER_SOURCE

9. Which of the following is not an advantage of packages?

   a. They allow function restrictions regarding use in SQL statements to be ignored.

   b. They allow global variables in a session.

   c. They allow logical grouping of multiple program units.

   d. They allow program units to be private.

10. Where is a one time only procedure placed within a package?

    a. at the top of the specification

    b. at the bottom of the specification

    c. at the top of the body

    d. at the bottom of the body

11. What resources could you use to identify all the packages that have been created on the system and the source code each contains?

12. How does persistence of value for global constructs contribute to processing efficiency?

13. Explain the concept of overloading and why it would be used.

14. If a package specification can exist without a body, why might you do this?

15. Explain what a private package construct is.

ADVANCED REVIEW QUESTIONS

1. Which of the following statements is true in regards to the following package specification?

```
CREATE OR REPLACE PACKAGE sales_process IS
    FUNCTION calc_tax
        (p_bask IN NUMBER)
        RETURN NUMBER;
    PRAGMA RESTRICT_REFERENCES(calc_tax, WNDS, WNPS);
END;
```

   a. The function CALC_TAX is private.

   b. A package body is not needed for this specification.

   c. The function CALC_TAX does not write to a database or package construct.

   d. An error will be raised by the PRAGMA statement.

2. Given the following package specification, how many private or local functions does the package contain?

```
CREATE OR REPLACE PACKAGE sales_process IS
    Procedure sale_sum
            (p_bask IN NUMBER,
             p_total OUT NUMBER,
             p_sub OUT NUMBER);
END;
```

   a. 1

   b. 2

   c. none

   d. cannot determine

3. To declare a set of public constants, the variable declarations in the package specification must _____.

   a. include the CONSTANT option

   b. include the CONSTANT option and initialize the variable values

   c. match the variables declared in the body

   d. retrieve the appropriate data from the database

4. Which of the following data dictionary views can be referenced to view the code contained within a package?

   a. USER_OBJECTS

   b. USER_PACKAGE_TEXT

   c. USER_SOURCE

   d. USER_TEXT

5. What items in the package coding must match between the specification and body of the same package? (Choose all that apply.)

   a. declared variables

   b. parameter lists of public program units

   c. package names

   d. parameter lists of public program units

HANDS-ON ASSIGNMENTS

Assignment 6-1: Create a Package

Follow the steps outlined to create a package containing a procedure and a function pertaining to basket information. Note that the initial compilation of the package body will fail to provide practice with error messages.

1. Locate the **Assignment06-01.txt** file in the Chapter.06 folder.

2. Open the file in a text editor and review the package code.

3. Highlight and copy all the code in the text file.

4. Open or return to SQL*Plus.

5. From the main menu, select **Edit**, **Paste** and then run the code. You will receive a "Warning: Package Body created with compilation errors" message from the package body CREATE statement.

6. To review the error, type **SHOW ERRORS** and then press **Enter**. The error indicates that the procedure named BASKET_INFO_PP is declared in the specification but not included in the body. Note the procedure name is mistakenly written as BASKET_INF_PP, missing the O in INFO.

7. Return to the text file and fix the procedure name in the package body.

8. Cut and paste the package body CREATE statement to build successfully.

Assignment 6-2: Use Packaged Program Units

In this assignment, you will execute packaged program units using a package pertaining to basket information. The package contains a function that returns the recipient's name and a procedure that retrieves the shopper id and order date for a basket.

1. Open or return to SQL*Plus.

2. First, create the ORDER_INFO_PKG package using the Assignment06-02.sql file located in the Chapter.06 folder. Review the code to become familiar with the two program units in the package.

3. Create an anonymous block that will invoke both the packaged procedure and function in this package to test them. Invoke each using basket id 12. Use the DBMS_OUTPUT command to display values returned from the program units to verify the data.

4. Also, test the packaged function in an SQL statement. Use the function in a SELECT clause on the BB_BASKET table. Use a WHERE clause to select only the basket 12 row.

Assignment 6-3: Create a Package with Private Program Units

In this assignment, you will modify a package to make program units private. The programming group decided that the SHIP_NAME_PF function in the ORDER_INFO_PKG package should be used only from within this package. Follow the steps to accomplish the needed package modification.

1. Open the Assignment06-03.txt file with a text editor and review the package code. You will modify the package in this assignment.

2. Modify the package to add to the BASKET_INFO_PP procedure so that it also returns the name to which the order is shipped by using the SHIP_NAME_PF function. In doing so, make the necessary changes to make the SHIP_NAME_PF function private to the package.

3. Create and execute an anonymous block that will invoke the BASKET_INFO_PP procedure and display the shopper id, order date, and shipped to name to check the values returned. Use DBMS_OUTPUT to display the values.

Assignment 6-4: Use Packaged Variables

In this assignment, you will create a package that will use packaged variables to assist in the user logon process. When a returning shopper logs on, the user name and password entered need to be verified against the database. In addition, two values need to be stored in packaged variables for reference during the user session. First, the shopper id needs to be stored. Second, the first three digits of the shopper's zip code need to be stored to keep them available for regional advertisements displayed on the site.

1. Create a function that will accept a user name and password as arguments and verify these values against the database for a match. If a match is found, return a value of 'Y.' Set the initial value of the variable holding the return value to 'N.' Include a NO_DATA_FOUND exception handler to display a message that the logon values are invalid.

2. Create a host variable named G_CK to receive the return value of the function.

3. Use an anonymous block to execute the procedure using a user name of "gma1" and password of "goofy".

4. Use the PRINT command to check the value placed into the host variable by the function.

5. Now place the function in a package. Also, add needed code to create and populate the needed packaged variables outlined earlier. Name the package **LOGIN_PKG**.

6. Use a host variable to test the packaged procedure using a user name of "gma1" and password of "goofy" to verify the procedure works properly.

7. Use DBMS_OUTPUT statements in an anonymous block to display the values stored in the packaged variables.

Assignment 6-5: Package Overloading

In this assignment, you will create packaged procedures to retrieve shopper information. Brewbean's is adding a screen to the application in which customer service agents can retrieve shopper information using either the shopper id or the last name. Create a package named SHOP_QUERY_PKG containing overloaded procedures to facilitate the lookup outlined. The procedures should return the shopper's name, city, state, phone number, and e-mail address. Test the package twice. First, invoke using shopper id 23. Second, invoke using last name of "Ratman". Both of the test values refer to the same shopper and, therefore, should return the same shopper information.

Assignment 6-6: Create a Package with a Specification Only

In this assignment, you will create a package consisting of a specification only. The Brewbean's lead programmer noticed there are only a few states that require Internet sales tax and the rates do not change often. Create a package named TAX_RATE_PKG to hold the following tax rates in packaged variables for reference: PV_TAX_NC = .035, PV_TAX_TX = .05, and PV_TAX_TN = .02. Code the variables to prohibit the rates from being modified. Use an anonymous block with DBMS_OUTPUT statements to display the value of each of the packaged variables.

Assignment 6-7: Use a Cursor in a Package

In this assignment, we need to address the sales tax computation because the Brewbean's lead programmer anticipates the rates and states applying the tax to undergo a series of changes. The tax rates are currently being held in packaged variables but now need to be more dynamic to handle the expected changes. The lead programmer has requested that a package be developed that will hold the tax rates by state in a packaged cursor. The BB_TAX table will be updated as needed to reflect which states are applying sales tax and at what rates. This package will need to contain a function that can receive a two-character state abbreviation (the shopper's state) as an argument, and it will need to find a match in the cursor and return the appropriate tax rate. Use an anonymous block to test the function with a state value of 'NC'.

✓ Assignment 6-8: Use a One Time Only Procedure in a Package

The Brewbean's application currently contains a package that is used in the shopper logon process. However, one of the developer's wants to be able to reference at what time the user logged on to be able to determine when the session should be timed out and entries rolled back. Modify the LOGIN_PKG package, which is in the Assignment06-08.txt file in the Chapter.06 folder. Use a one time only procedure to populate a packaged variable with the date and time upon user logon. Use an anonymous block to verify the one time procedure works and populates the packaged variable.

CASE PROJECTS

Case 6-1: Brewbean's Order Checkout Package

At the end of Chapter 5, you created a procedure and a number of functions to handle the updating of basket columns during a shopper checkout process. Create a package named SHOP_PROCESS_PKG that will contain all the program units created in Chapter 5. Modify the BASK_CALC procedure so that the subtotal, tax, shipping, and total amounts are placed into packaged variables rather than into the database. This is to allow the application to display a confirm purchase page for the shopper. Test this procedure using basket 3.

The lead programmer has requested that all packaged program units be in alphabetical order to make them easy to locate. Use forward declarations if needed to enable the alphabetization of program units.

Case 6-2: More Movies Program Unit Packaging

In the Chapter 4 More Movies case project, you created two procedures to support the check out and check in of rented movies. In Chapter 5, you created a function to allow inquiries as to movie availability. Create a package named MM_RENTALS_PKG to contain all these program units. Make all the program units public. Test the MOVIE_INFO function using movie ids of 6 and 7.

7

PROGRAM UNIT DEPENDENCIES

In this chapter, you will:

- Identify local program unit dependencies
- Determine direct and indirect dependencies
- View data dictionary information concerning dependencies
- Run the dependency tree utility
- Identify the unique nature of package dependencies
- Understand remote object dependency actions
- Use remote dependency invalidation methods
- Avoid recompilation errors
- Grant program unit privileges

In this chapter, you learn about **program unit dependencies**, which are the interrelationships of objects as they relate to procedures, functions, and packages. A developer must cultivate an appreciation for program unit dependencies to avoid complications that can be raised from object references.

Note that relationships or dependencies determine the validity of any program unit after modifications to database objects that the program unit references. It is this validity that determines the need for recompilation. That is, if we have a procedure named ORDER_TOTAL_SP that reads particular columns from a table, what happens when the table is modified? Is the procedure still valid? Does the procedure need to be recompiled? In this chapter, we discover the answers to these questions and more as we explore the nature of dependencies and the information and tools available to assist us in managing these dependencies.

The Current Challenge in the Brewbean's Application

The Brewbean's manager has been reviewing user feedback regarding application performance and wants to take any steps possible to make the execution more efficient. Also, recently, users have been hitting some unexpected errors while using the Brewbean's application. The programmers noticed that many of these errors were related to recent modifications to the database and program units. The lead programmer recognized that both of these issues involved object dependency concerns and that the development group needs to become aware of database dependencies and their impact. The lead programmer gathered the group together for a brief workshop to explore dependency issues.

Rebuilding Your Database

Complete the following steps to rebuild the Brewbean's database for this chapter.

To rebuild the Brewbean's database:

1. Locate the **c7Dbcreate.sql** file in the Chapter.07 folder to ensure that it exists. This file contains the script to create the database.

2. Open SQL*Plus.

3. Create a spool file so that SQL*Plus keeps a copy of all the messages received from running the file that creates the database. On the main menu in SQL*Plus, click **File**, point to **Spool**, and then click **Spool File**. A Select File dialog box appears.

4. Browse to a folder used to contain your working files, and in the File name text box type **DB_log7**, and then click **Save**. Now, whatever text is seen in our SQL*Plus session is saved to this file for future reference.

5. Now, let's create the database. In SQL*Plus, type the following command, which runs all the statements contained in the c7Dbcreate.sql file. Messages verifying the creation and data insertion steps scroll on the SQL*Plus screen. This may take a couple minutes to complete.

    `@<pathname to PL/SQL files>\Chapter.07\c7Dbcreate`

6. Now, we need to turn the spooling off so that we can review the results in the spool file created. On the main menu, click **File**, point to **Spool**, and then click **Spool Off**.

7. Open a text editor.

8. Open the file named **DB_log7.lst** in your working folder.

9. Review the messages for any errors.

LOCAL DEPENDENCY ACTIVITY

First, let's take a look at an individual procedure and the dependencies of a procedure. The ORDER_TOTAL_SP procedure is used in the Brewbean's ordering application to calculate order totals, including item subtotal, shipping cost, and overall total. In this scenario, as displayed in Figure 7-1, the procedure references two database objects: a table and another procedure.

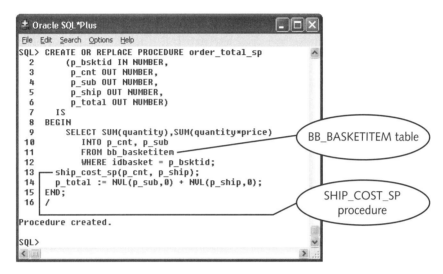

Figure 7-1 The ORDER_TOTAL_SP procedure

The table named BB_BASKETITEM and the procedure named SHIP_COST_SP are both considered as being the referenced objects. The procedure named ORDER_TOTAL_SP is considered as the dependent object. If we modify one of the referenced objects, table BB_BASKETITEM or procedure SHIP_COST_SP, the status of the dependent object, procedure ORDER_TOTAL_SP, is flagged as invalid. What does this mean? It means that the program unit needs to be recompiled. The status of each database object can be checked via the data dictionary view USER_OBJECTS.

To take a look at the current status of the procedures of immediate concern in the Brewbean's application:

1. Open or return to SQL*Plus.

2. One column of the data dictionary view we use is too long; therefore, type the following code to avoid the wrapping of the output:

   COLUMN object_name FORMAT A20

3. Type the query listed in Figure 7-2 to review the status of all the procedures in your schema. If you have additional procedures listed, it is not a problem. Depending on your experimentation, you may have created additional procedures. You should at least find all those listed in Figure 7-2.

Figure 7-2   Query status of all procedures

The results display a list of all the procedures in your schema and their current status. After you successfully save or compile a procedure, the status is marked as VALID. The status column value is automatically changed to INVALID if one of the procedures' referenced objects is modified. For example, the ORDER_TOTAL_SP procedure contains a call to the SHIP_COST_SP procedure. If we modify the SHIP_COST_SP procedure, the ORDER_TOTAL_SP procedure then has a status of INVALID until it is recompiled. The system provides various methods to recompile the procedure to return it to a status of VALID. These methods include manual and automatic database actions that can take place.

To confirm the behavior of the program unit status, let's make a modification and review the impact on the status.

To make a modification:

1. Open or return to SQL*Plus. Note that the earlier query in Figure 7-2 confirmed that the status of the ORDER_TOTAL_SP and SHIP_COST_SP procedures are currently VALID. Rebuild the SHIP_COST_SP procedure using the status07.sql file. Click **File** and then **Open** from the main menu.

2. Navigate to and select the **status07.sql** file.

3. Click **Open** and a copy of the file contents appears in the SQL*Plus screen, as shown in Figure 7-3. Notice that a declaration for a variable named LV_JUNK_NUM has been added to the procedure. It is added only for this demonstration and is not actually used in the procedure.

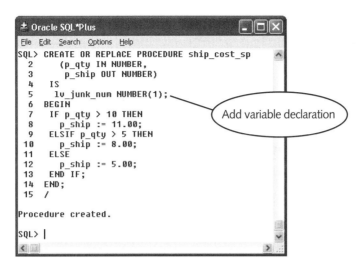

Figure 7-3   A variable declaration added to the SHIP_COST_SP procedure

4. To run the code and re-create the procedure, click **File** and then **Run** from the main menu. You receive a "Procedure created" message.

5. Query the USER_OBJECTS view to check the status with the query shown in Figure 7-4. Note that the status of the ORDER_TOTAL_SP procedure is now INVALID because one of the referenced objects, the SHIP_COST_SP procedure, has been changed.

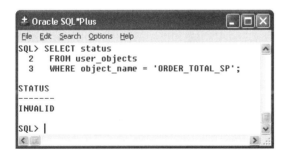

Figure 7-4   Check the status of procedure ORDER_TOTAL_SP

6. The INVALID status indicates that the program unit must be recompiled before it is run again. We can manually recompile a procedure in two ways. First, we can use the CREATE OR REPLACE statement to rebuild the procedure, which completes a compile action. Use the status_b07.sql file. Click **File** and then **Open** from the main menu.

7. Navigate to and select the **status_b07.sql** file from the Chapter.07 folder. Click **Open** and a copy of the file contents appears in the SQL*Plus screen. To run the code and re-create the procedure, click **File** and then **Run** from the main menu. You receive a "Procedure created" message.

8. Query the USER_OBJECTS view again, as previously done in Figure 7-4, to confirm the status of the ORDER_TOTAL_SP procedure is now VALID.

9. A second way to manually recompile entails using the ALTER COMPILE command. Let's alter the SHIP_COST_SP procedure again to make the ORDER_TOTAL_SP procedure INVALID once again. Rebuild the SHIP_COST_SP procedure using the status_c07.sql file. Click **File** and then **Open** from the main menu.

10. Navigate to and select the **status_c07.sql** file from the Chapter.07 folder.

11. Click **Open** and a copy of the file contents appears in the SQL*Plus screen. Notice that a declaration for the variable named LV_JUNK_NUM has been removed.

12. Run the statements in Figure 7-4. The ORDER_TOTAL_SP procedure is once again INVALID.

13. In SQL*Plus, enter the ALTER COMPILE command, as shown in Figure 7-5, to recompile the ORDER_TOTAL_SP procedure.

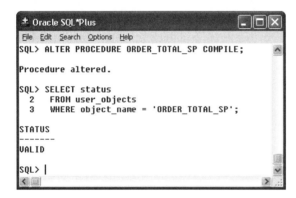

Figure 7-5 ALTER COMPILE command

14. Run the query shown in Figure 7-5 to confirm that the status of ORDER_TOTAL_SP is now VALID.

The preceding steps demonstrate the impact of modifications to referenced objects on the status of a program unit dependent on that object. We saw that we can recompile invalid program units manually with the ALTER COMPILE command or by re-creating the program unit. However, what happens if we do not perform either of these actions and a program unit with a status of INVALID is called during runtime? The Oracle system automatically compiles the next time the invalid program unit is invoked. This automatic recompilation only applies to local objects or dependent objects that are on the same database. If dependent objects are located on remote databases, the behavior is different. (**Remote databases** are connections to other Oracle database servers.)

There are a couple of drawbacks to letting the system handle this recompilation automatically:

- First, the compile operation is occurring during the runtime operation of the user application and, therefore, the compile time is added to the user wait time for the program processing.

- Second, what if the referenced object is changed in such a way that the recompile is not successful? The end user receives an error on this program unit during runtime of the application.

If we had changed the number of parameters in the SHIP_COST_SP procedure in the example in the preceding steps, the ORDER_TOTAL_SP procedure would not successfully recompile. We would need to make modifications to the ORDER_TOTAL_SP procedure for it to operate properly. If the manual recompile methods are used, we have essentially tested our modifications to be sure the dependent object is not adversely affected by the changes.

In the example in the preceding steps, the SHIP_COST_SP procedure was the referenced object modified to determine the effect on the ORDER_TOTAL_SP procedure. Note that the ORDER_TOTAL_SP procedure also queries or references the BB_BASKETITEM table; therefore, the ORDER_TOTAL_SP procedure is also dependent on the BB_BASKETITEM table. If the BASKETITEM table is altered in any way, such as increasing the length of a column, the ORDER_TOTAL_SP procedure status changes to INVALID. In this scenario, if the column named PRICE is dropped from the table, the ORDER_TOTAL_SP procedure could not recompile successfully without modification.

IDENTIFYING DIRECT AND INDIRECT DEPENDENCIES

A **direct dependency** occurs when one database object, such as the ORDER_TOTAL_SP procedure, directly references other objects, such as the BB_BASKETITEM table and the SHIP_COST_SP procedure. In this case, we can review the procedure code and identify the referenced items fairly simply. However, what if one of the object names referenced is actually a public synonym for an object? We may not easily recognize what schema houses this object. Recall that when we create objects on the database, these objects are by default stored in our own schema. However, if granted appropriate rights, we can call database objects from other schemas.

For example, let's say a table named BB_BASKETITEM is created in the schema named DBA1 and assigned a public synonym called ITEMS for developers to reference in their program units. Without the public synonym, a reference to this table would be included as "DBA1. BB_BASKETITEM " to instruct the system to go to the DBA1 schema and look for the BB_BASKETITEM table. However, the reference using the public synonym would be included only as ITEMS, which does not clearly indicate the schema or object name.

In addition, a program unit may contain a number of **indirect dependencies**, which involve references to database objects that in turn reference other database objects making the chain of dependencies less obvious to track. For example, a procedure could reference a database view, which in turn references a database table. In this scenario, the procedure has a direct dependency on the view and an indirect dependency on the table through this view. If the underlying table is modified, the procedure status changes to INVALID due to the indirect dependency. Therefore, indirect dependencies are just as important to consider as direct dependencies.

DATA DICTIONARY VIEWS FOR DEPENDENCIES

The USER_DEPENDENCIES data dictionary view is a most helpful view in providing information on direct object dependencies. If you run a DESCRIBE command on this view, you find the columns listed in Table 7-1.

Table 7-1   Columns in the USER_DEPENDENCIES data dictionary view

Column Name	Description
NAME	Object name
TYPE	Object type
REFERENCED_OWNER	Schema name of the referenced object
REFERENCED_NAME	Name of the referenced object
REFERENCED_TYPE	Type of the referenced object
REFERENCED_LINK_NAME	Name of database link if being used to access the referenced object
SCHEMAID	Internal id assigned to the schema
DEPENDENCY_TYPE	Ref or Hard

Let's list the dependency information for the ORDER_TOTAL_SP procedure. Figure 7-6 illustrates a query on the USER_DEPENDENCIES view for this procedure. A COLUMN FORMAT command was used first to make the output more readable.

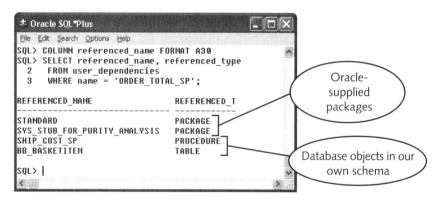

Figure 7-6   Query on the USER_DEPENDENCIES view

The output listing shows four items on which the ORDER_TOTAL_SP procedure is dependent. The first two objects are related to common, built-in, Oracle-supplied packages to allow the operation of PL/SQL code and program units. The last two items are objects from our schema that the procedure references. One is a table and the other is a procedure.

Another way to use this data dictionary view is to list all the objects that are dependent on a particular object or, in other words, search based on the referenced name column. This could answer the question, "The SHIP_COST_SP procedure has been changed; which objects are dependent on this procedure?" Figure 7-7 displays a query on the USER_DEPENDENCIES view to answer this question.

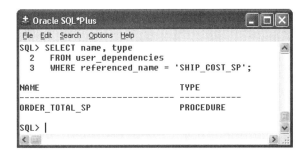

Figure 7-7   Another query on the USER_DEPENDENCIES view

If you have object dependencies across different schemas, in addition to the data listed in Figure 7-7, you can view the owner of each object. Another data dictionary view named DBA_DEPENDENCIES is available. It contains the same columns as USER_DEPENDENCIES with the addition of an owner column.

> The DBA_DEPENDENCIES view is not accessible by the default user of SCOTT that is automatically created during the Oracle install. You must log onto Oracle with a DBA account (such as the default SYSTEM user), and grant privileges to users to list data from this view.

THE DEPENDENCY TREE UTILITY

In this section, we identify the lack of information in the data dictionary views regarding indirect dependencies and then look at the dependency tree utility, which provides a mechanism to track indirect dependencies. Let's modify the ORDER_TOTAL_SP procedure to introduce an indirect dependency situation. We modify the procedure to SELECT from a view of the BB_BASKETITEM table rather than the table itself. This results in a direct dependency of the ORDER_TOTAL_SP procedure to the view and an indirect dependency to the table. After this modification, we review what dependency data is contained in the data dictionary views.

To modify the ORDER_TOTAL_SP procedure:

1. In a text editor, open the **ordview07.txt** file from the Chapter.07 folder.

2. Modify the SELECT statement FROM clause to state **FROM bb_basketitem _vu** instead of "FROM bb_basketitem." The BB_BASKETITEM_VU object is a view created on the BASKETITEM table.

3. Copy and paste the CREATE PROCEDURE code into SQL*Plus.

4. Run the query shown in Figure 7-8 to list all the objects that the ORDER_TOTAL_SP procedure references. Notice the BB_BASKETITEM table is not included in the list. Why? It is an indirect dependency.

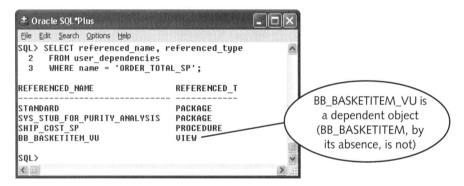

Figure 7-8 A query of the USER_DEPENDENCIES view, which does not show indirect dependencies

5. While in SQL*Plus, do a second query to list all the objects that are dependent on the BB_BASKETITEM table, as shown in Figure 7-9. Notice the ORDER_TOTAL_SP procedure is not included in this list because it is indirectly dependent on the BB_BASKETITEM table via the BB_BASKETITEM _VU view.

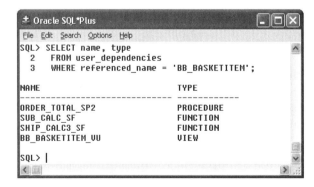

Figure 7-9 A second query of the USER_DEPENDENCIES view, which does not show indirect dependencies

The **dependency tree utility** is a mechanism available in the Oracle system to provide a map to visualize both direct and indirect dependencies within the database. For this utility to be operable, a script that builds the necessary views, tables, sequences, and procedures must be run first. Under the Oracle home directory where the database is installed, you find the subdirectory \rdbms\admin, which contains the file utldtree.sql. You run this script in SQL*Plus by using the following code:

```
@<path to Oracle Home directory>\rdbms\admin\utldtree
```

 To become familiar with the objects the utldtree.sql script is creating, open the file in a text editor and review the statements. A number of helpful comments are included to explain parts of the script.

Some of the statements (such as the DROP commands) fail if this is the first time you have run the utldtree.sql script. Why? Because the script contains DROP statements for the same objects created in case you have executed the script previously. This script needs to be run only once to create the needed objects. This script creates a procedure named DEPTREE_FILL that can fill the table created with data regarding the dependencies of a given object. Then, we use the views created to display a listing of the dependencies. The messages from the initial execution of the utldtree.sql script match the following listing:

```
drop sequence deptree_seq
              *
ERROR at line 1:
ORA-02289: sequence does not exist

Sequence created.

drop table deptree_temptab
              *
ERROR at line 1:
ORA-00942: table or view does not exist

Table created.
Procedure created.
drop view deptree
*
ERROR at line 1:
ORA-00942: table or view does not exist

SQL>
SQL> REM This view succeed if current user is sys. This view shows
SQL> REM which shared cursors depend on the given object. If the current
SQL> REM user is not sys, then this view get an error either about lack
SQL> REM of privileges or about the non-existence of table x$kglxs.
SQL>
SQL> set echo off
  from deptree_temptab d, dba_objects o
      *
ERROR at line 5:
ORA-00942: table or view does not exist
```

7

```
SQL>
SQL> REM This view succeed if current user is not sys.  This view
SQL> REM does *not* show which shared cursors depend on the given object.
SQL> REM If the current user is sys then this view get an error
SQL> REM indicating that the view already exists (since prior view create
SQL> REM have succeeded).
SQL>
SQL> set echo off
View created.
drop view ideptree
*
ERROR at line 1:
ORA-00942: table or view does not exist
View created.
```

Let's try using the utldtree utility to review all the dependencies of the BB_BASKETITEM table:

1. Open or return to SQL*Plus.

2. Execute the utldtree script to build the necessary objects to use the utility by typing the following code, and then pressing **Enter**.

   @<*path to Oracle Home directory*>\rdbms\admin\utldtree

3. Run the procedure DEPTREE_FILL, as shown in Figure 7-10, to populate the DEPTREE_TEMPTAB table with dependency data for a specific object, the BB_BASKETITEM table. The DEPTREE_FILL procedure has three parameters: object type, object schema, and object name.

4. Now, we have two choices in listing the dependency information. First, use the DEPTREE view to list the dependencies using a numeric level scheme. Run the query, as shown in Figure 7-11. Notice that the number in the nested level column represents the relation to the object being analyzed.

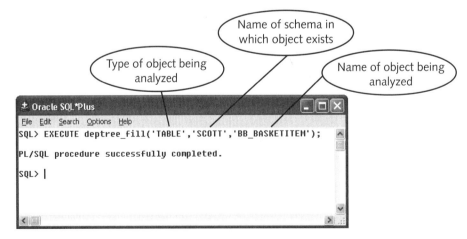

Figure 7-10   Run the DEPTREE_FILL procedure

5. Using the seq# sort assists in identifying the dependency line. Study Figure 7-11. (Note that the SEQ# column output may differ.) Note that the ORDER_ TOTAL_SP procedure is directly related to BB_ BASKETITEM_VU view, which is directly related to the BB_BASKETITEM table. In other words, the ORDER_TOTAL_SP procedure is indirectly dependent on the BB_BASKETITEM table via the BB_BASKETITEM_VU view.

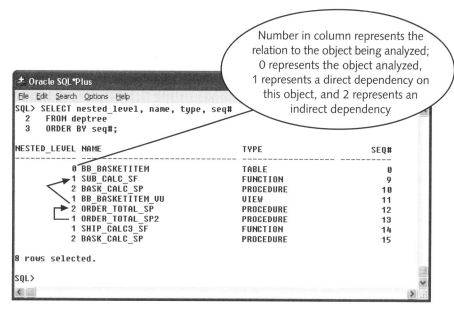

Figure 7-11   List of all dependencies using the DEPTREE view

6. Another view named IDEPTREE is also available, and this view displays the same dependency information in a different format. Run the SELECT on this view, as shown in Figure 7-12.

As we have now seen, the data dictionary view of USER_DEPENDENCIES is quite helpful in identifying direct dependencies; however, the dependency tree utility provides a much broader picture of all dependencies, both direct and indirect, and the path of dependencies through the database objects.

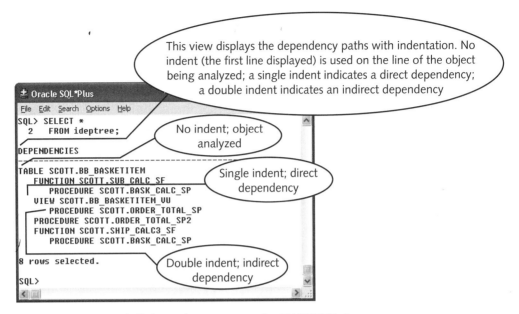

Figure 7-12　List of all dependencies using the IDEPTREE view

PACKAGE DEPENDENCIES

One of the advantages identified with packages is the improved handling of dependencies. This ease of handling is driven by the fact that the package structure separates the program unit code (which is in the body), from the program unit header (which is in the specification). If modifications are made to the code in the package body, the status of dependent objects is not changed to INVALID, as seen with stand-alone program units.

Only modifications to the package specification or the program unit header section raise the INVALID status of dependent objects. This can prove quite beneficial in terms of not having to accomplish the amount of recompilation that would be necessary if all the program units were stand-alone.

Let's take a look at how this works in the Brewbean's application. We can use the ORDER_INFO_PKG package to test the package dependencies. This package contains a public procedure and a private function. The packaged procedure uses the packaged function. A stand-alone procedure named PKG_DEPTEST_SP calls the procedure in the ORDER_INFO_PKG package and is, therefore, dependent on the package. We can check the status of the PKG_DEPTEST_SP procedure (the dependent object) after various modifications to the ORDER_INFO_PKG package. The code and dependency are shown in Figure 7-13.

Recall that you can use the USER_SOURCE data dictionary view to review the source code of a procedure, function, or package.

To work with the package dependencies:

1. Open or return to SQL*Plus. Enter the PKG_DEPTEST_SP procedure shown on the right side of Figure 7-13. You now need to check the status of the procedure PKG_DEPTEST_SP. Enter the query shown in Figure 7-14 to confirm the procedure status is VALID.

2. Let's first make a modification to the package body to determine the effect on the dependent procedure. Brewbean's wants to list the shopper name with an asterisk if the basket contains multiple items. Open the **pkgdep07.txt** file in a text editor to make the modification to the package body.

7

ORDER_INFO_PKG Package

```
CREATE OR REPLACE PACKAGE order_info_pkg IS
PROCEDURE basket_info_pp
 (p_basket IN NUMBER,
  p_shop OUT NUMBER,
  p_date OUT DATE,
  p_name OUT VARCHAR2);
END;
/
CREATE OR REPLACE PACKAGE BODY order_info_pkg IS
FUNCTION ship_name_pf
 (p_basket IN NUMBER)
 RETURN VARCHAR2
 IS
  lv_name_txt VARCHAR2(25);
BEGIN
  SELECT shipfirstname||''||shiplastname
  INTO lv_name_txt
  FROM bb_basket
  WHERE idBasket = p_basket;
  RETURN lv_name_txt;
  EXCEPTION
   WHEN NO_DATA_FOUND THEN
   DBMS_OUTPUT.PUT_LINE('Invalid basket id');
END ship_name_pf;
PROCEDURE basket_info_pp
 (p_basket IN NUMBER,
  p_shop OUT NUMBER,
  p_date OUT DATE,
  p_name OUT VARCHAR2)
  IS
 BEGIN
  SELECT idshopper, dtordered
  INTO p_shop, p_date
  FROM bb_basket
  WHERE idbasket = p_basket;
  p_name := ship_name_pf(p_basket);
 EXCEPTION
  WHEN NO_DATA_FOUND THEN
  DBMS_OUTPUT.PUT_LINE('Invalid basket id');
  END basket_info_pp;
END;
/
```

PKG_DEPTEST_SP Procedure

```
CREATE OR REPLACE PROCEDURE pkg_deptest_sp
 (p_bask NUMBER)
 IS
 lv_shop_num NUMBER(4);
 lv_bask_dat DATE;
 lv_name_txt VARCHAR2(25);
BEGIN
 order_info_pkg.basket_info_pp(p_bask,lv_shop_num,
 lv_bask_dat,lv_name_txt);
 DBMS_OUTPUT.PUT_LINE(lv_shop_num);
 DBMS_OUTPUT.PUT_LINE(lv_bask_dat);
 DBMS_OUTPUT.PUT_LINE(lv_name_txt);
END;
/
```

The PKG_DEPTEST_SP procedure calls the BASKET_INFO_PP packaged procedure, and the PKG_DEPTEST_ SP procedure is dependent on the ORDER_INFO_PKG package

Figure 7-13   Dependency of the PKG_DEPTEST_SP procedure

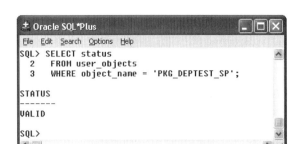

Figure 7-14   Check the status of the PKG_DEPTEST_SP procedure

3. Modify the package body by making the changes to the BASKET_INFO_PP procedure indicated in Figure 7-15 in the package.

```
CREATE OR REPLACE PACKAGE BODY order_info_pkg IS
FUNCTION ship_name_pf
 (p_basket IN NUMBER)
 RETURN VARCHAR2
IS
 lv_name_txt VARCHAR2(25);
BEGIN
 SELECT shipfirstname||' '||shiplastname
  INTO lv_name_txt
  FROM bb_basket
  WHERE idBasket = p_basket;
  RETURN lv_name_txt;
 EXCEPTION
  WHEN NO_DATA_FOUND THEN
  DBMS_OUTPUT.PUT_LINE('Invalid basket id');
 END ship_name_pf;
 PROCEDURE basket_info_pp
  (p_basket IN NUMBER,
   p_shop OUT NUMBER,
   p_date OUT DATE,              Add variable declaration
   p_name OUT VARCHAR2)            for LV_QTY_NUM
  IS                                                    Retrieve order quantity
   lv_qty_num NUMBER(2);                                     in SELECT
 BEGIN
   SELECT idshopper, dtordered, quantity
    INTO p_shop, p_date, lv_qty_num
    FROM bb_basket
    WHERE idbasket = p_basket;
   p_name := ship_name_pf(p_basket);         Add IF statement to
   IF lv_qty_num > 1 THEN                      check quantity
    p_name := '*'||p_name;
   END IF;
```

Figure 7-15   Package body modifications

4. Copy and paste all the modified code into SQL*Plus.

5. Now, you can check the status of the dependent object, the PKG_DEPTEST_SP procedure, to see if the modifications had an impact on the dependent object status. Enter the query shown in Figure 7-16 to confirm the procedure status is still VALID.

6. Now, let's test a modification of the package specification. Brewbean's decided that the asterisk indicating multiple items in an order should be output as a separate value from the procedure rather than attached to the shopper name. Open the pkgdepB07.txt file in a text editor to make the modification to the package specification.

Figure 7-16   Confirming the procedure status

7. Modify the package specification by making the changes to the BASKET_INFO_PP procedure header indicated in Figure 7-17.

```
CREATE OR REPLACE PACKAGE order_info_pkg
IS
 PROCEDURE basket_info_pp
 (p_basket IN NUMBER,
  p_shop OUT NUMBER,
  p_date OUT DATE,
  p_name OUT VARCHAR2,
  p_mult OUT CHAR);          Add parameter
END;
```

Figure 7-17   Package specification modifications

8. Copy and paste all the modified code into SQL*Plus.

Of course, adding a parameter means we also need to modify this procedure code in the package body as well. However, we skip this for now because it is not required for this example.

9. Now, you can check the status of the dependent object, the PKG_DEPTEST_SP procedure, to see if the modifications had an impact on the dependent object status. Enter the query shown in Figure 7-18 to confirm the procedure status is now INVALID.

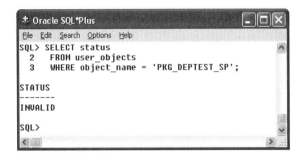

Figure 7-18   Check status of procedure again

The tasks in the preceding step sequence demonstrate the benefit of packages in handling object dependencies in that only program unit header modifications require a recompilation of dependent objects. This makes sense, as we know any changes to the parameters of a procedure require modification to associated procedure calls because the arguments in the call must match with the parameters in the program unit. This separation of the specification and body allows developers to create code with calls to packaged program units without having the package code completed. This is quite useful for larger development projects in that only package specifications need to exist to create dependent objects.

REMOTE OBJECT DEPENDENCIES

Remote database connections are used to link to another database and use or call objects that exist on that database. The object dependencies are handled in a slightly different manner when remote database dependencies are involved. This different handling is rooted in the fact that the data dictionary does not track remote dependencies and, therefore, if an object on one database is changed, any remote program units that use or depend on that object are not immediately changed to an INVALID status. Oracle has made this decision because this operation could be very expensive in regards to processing time and raise additional problems if the remote databases are not currently available at the time of modifications. Remote dependencies are checked at runtime rather than instantly as with local database objects. This causes a failure on the first run of a dependent object if the referenced object has been altered.

To work with remote dependencies within Brewbean's:

1. Open or return to SQL*Plus. You need to create a database link so that the system treats a call to an object as a remote connection. Type and execute the command shown in Figure 7-19.

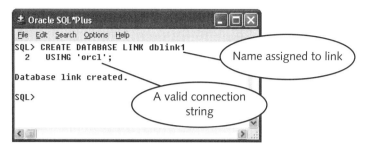

Figure 7-19   Create a database link

2. You use the SHIP_COST_SP procedure as the referenced object and the ORDER_TOTAL_SP procedure, which calls SHIP_COST_SP, as the dependent object. You need to modify the call to SHIP_COST_SP to use the database link and simulate a remote connection. Open the remote07.txt file in a text editor to make the modification to the procedure code.

3. Modify the ORDER_TOTAL_SP procedure by making the changes indicated in Figure 7-20.

```
CREATE OR REPLACE PROCEDURE order_total_sp
 (p_bsktid IN NUMBER,
  p_cnt OUT NUMBER,
  p_sub OUT NUMBER,
  p_ship OUT NUMBER,
  p_total OUT NUMBER)
IS
BEGIN
  SELECT SUM(quantity),SUM(quantity*price)
   INTO p_cnt, p_sub
   FROM bb_basketitem
   WHERE idbasket = p_bsktid;
 ship_cost_sp@dblink1(p_cnt, p_ship);
 p_total := NVL(p_sub,0) + NVL(p_ship,0);
END;
/
```

(annotation: Add the database link reference)

Figure 7-20   Add the database link reference to the procedure call

4. Copy and paste all the modified code into SQL*Plus.

5. Query the USER_OBJECTS data dictionary view to confirm the current status of both procedures is VALID, as shown in Figure 7-21. Notice the use of the column format command to make the output more readable.

6. Now, you need to make a change to this referenced object to test how the status of the dependent object reacts. Open the remote_b07.txt file in a text editor to make the modification to the procedure code.

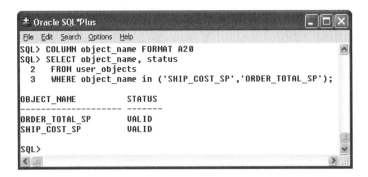

Figure 7-21 Check the status of the two procedures

7. Modify the SHIP_COST_SP procedure by making the shipping cost change indicated in Figure 7-22.

```
CREATE OR REPLACE PROCEDURE ship_cost_sp
 (p_qty IN NUMBER,
  p_ship OUT NUMBER)
IS
BEGIN
 IF p_qty > 10 THEN
  p_ship := 12.00;
 ELSIF p_qty > 5 THEN
  p_ship := 8.00;
 ELSE
  p_ship := 5.00;
 END IF;
END;
 /
```

Change the shipping cost from $11 to $12 for orders with more than 10 items

Figure 7-22 Modifying the SHIP_COST_SP procedure

8. Copy and paste all the modified code into SQL*Plus.

9. In SQL*Plus, check the status of the two procedures once again, as displayed in Figure 7-23. Notice that the dependent object, procedure ORDER_TOTAL_SP, is still VALID. If this procedure did not use the database link to call the SHIP_COST_SP procedure, it would now have an INVALID status; however, this is treated as a remote connection and, therefore, is not immediately updated.

10. Run the ORDER_TOTAL procedure, as shown in Figure 7-24. What happens? The dependency of ORDER_TOTAL_SP on SHIP_COST_SP is checked at runtime because a remote connection is used. The first run after the modification

of the referenced object fails because it recognizes that a change has been made and ORDER_TOTAL_SP needs to be recompiled. Notice that one of the error messages mentions that the timestamp of SHIP_COST has been changed.

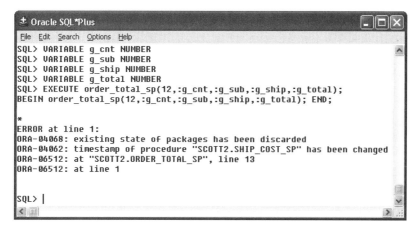

Figure 7-23   Verify that the ORDER_TOTAL_SP procedure status is still VALID

7

```
± Oracle SQL*Plus
File  Edit  Search  Options  Help
SQL> VARIABLE g_cnt NUMBER
SQL> VARIABLE g_sub NUMBER
SQL> VARIABLE g_ship NUMBER
SQL> VARIABLE g_total NUMBER
SQL> EXECUTE order_total_sp(12,:g_cnt,:g_sub,:g_ship,:g_total);
BEGIN order_total_sp(12,:g_cnt,:g_sub,:g_ship,:g_total); END;

*
ERROR at line 1:
ORA-04068: existing state of packages has been discarded
ORA-04062: timestamp of procedure "SCOTT2.SHIP_COST_SP" has been changed
ORA-06512: at "SCOTT2.ORDER_TOTAL_SP", line 13
ORA-06512: at line 1

SQL> |
```

Figure 7-24   First run of dependent procedure following referenced object modification

11. In SQL*Plus, check the current status of the procedures, as shown in Figure 7-25. Notice the status of the ORDER_TOTAL_SP procedure is now INVALID.

12. In SQL*Plus, run the ORDER_TOTAL_SP procedure again, as shown in Figure 7-26. Note that the second run accomplishes the automatic recompile of ORDER_TOTAL_SP and the procedure now runs successfully.

The preceding example highlights the fact that we must give extra consideration when working with remote database connections, because it is less forgiving. The first run does not automatically recompile dependent objects; instead, it raises an error. In these cases, the ALTER COMPILE command should be used to recompile any remote dependencies to avoid an error at runtime.

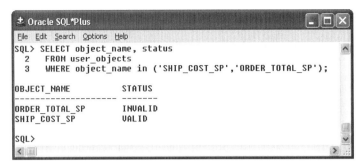

Figure 7-25   Verify the ORDER_TOTAL procedure status is now INVALID

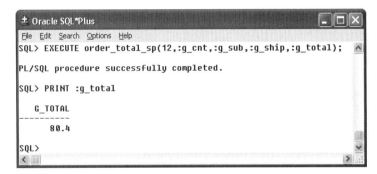

Figure 7-26   The second run of ORDER_TOTAL_SP processes successfully

REMOTE DEPENDENCY INVALIDATION METHODS

Oracle offers two models or methods to determine invalidation of remote dependent objects at runtime: timestamp and signature.

The **timestamp model** compares the last modification date and time of objects to determine if invalidation occurs. If the dependent object has an older timestamp than the referenced object when called, the dependent object is flagged as INVALID and needs to be recompiled. A column named TIMESTAMP is maintained in the data dictionary and is included in the USER_OBJECTS view. Do you recall the remote dependency example in the previous section? Calling the dependent procedure ORDER_TOTAL_SP raised an error after the referenced procedure was modified. Figure 7-27 shows a query from the USER_OBJECTS VIEW, which displays the timestamp information for both procedures involved.

The timestamp model is the default method, however, it does have a couple issues to consider. First, if the databases are in different time zones, the timestamp method does not recognize this and may cause unnecessary recompilation. Second, this method is also used if the dependent procedure is a client-side procedure, while the referenced object is on the server. If the client-side procedure is part of an Oracle forms application, the source code may not be available in the runtime environment and, therefore, cannot be recompiled.

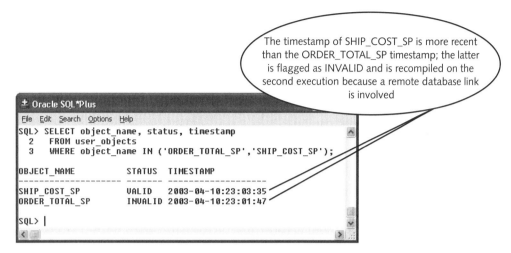

The timestamp of SHIP_COST_SP is more recent than the ORDER_TOTAL_SP timestamp; the latter is flagged as INVALID and is recompiled on the second execution because a remote database link is involved

Figure 7-27   Timestamp information

The **signature model** compares the mode, data type, and the order of parameters and the RETURN data type (for functions) when a remote object call is accomplished to determine invalidation. When a procedure is compiled, it stores the signature of any referenced objects. When the procedure runs, it compares the signature information stored about the referenced object to the actual parameters in the referenced object. If anything has changed in regards to the parameters, the dependent object is flagged as INVALID and needs recompilation. The signature model is not the default method and, if desired, needs to be set in the system using one of the methods listed in Table 7-2.

Table 7-2   Setting the signature mode

Setting or Command	Setting Affects
REMOTE_DEPENDENCIES_MODE= SIGNATURE (set this parameter in the database initialization file, typically init.ora)	All user sessions from database startup
ALTER SYSTEM SET REMOTE_ DEPENDENCIES_MODE = SIGNATURE;	All user sessions after command issued
ALTER SESSION SET REMOTE_ DEPENDENCIES_MODE = SIGNATURE;	The user session who issued the command

The signature model can resolve the time issues of using the timestamp model; however, it does have some considerations of its own. There are two situations in which parameter modifications are not viewed as a signature modification and, therefore, do not prompt a recompilation:

- First, if an IN parameter has a default value setting and this value is changed, the dependent object continues to refer to the old default value until the dependent object is manually recompiled.

- Second, if the dependent object calls a packaged procedure for which a new overloaded version is added, the dependent object uses the old version and does not see the new overloaded version until a manual recompilation of the dependent object is completed.

In both cases, the dependent object is not flagged as INVALID and, therefore, does not indicate that recompilation is needed.

TIPS TO AVOID RECOMPILATION ERRORS

The need for application maintenance or modifications always arises, and as developers, we need to attempt to write code as flexible as possible to reduce maintenance and possible recompilation errors. Oracle documentation suggests several techniques to use in coding program units to avoid errors associated with program changes:

- Use the '%TYPE' and '%ROWTYPE' attributes in variable declarations. These declarations instruct the system to assign the variable data type based on the underlying table column data type at runtime. If changes are made to tables, such as column length or data type, program units using these columns in variables or parameters do not have to be edited. Keep in mind this does not avoid other logic errors that can be raised by data type modifications. For example, if we change a variable data type from NUMBER to VARCHAR2 and the program unit code uses a numeric function, such as round on this variable, it now raises an error.

- Use 'SELECT *' when querying data from the database rather than a named column list. If the table changes, such as with a column drop, the query does not fail if it is using this notation. There is some debate among Oracle users as to whether the benefit is greater than the downside of this technique. If the notation is used, the program unit may be retrieving far more data than is needed and this adds to the processing time.

- Use a column list in INSERT statements so that changes to a table, such as dropping and adding columns, do not necessarily raise an error in the INSERT. If a column list is not used, then the INSERT expects data to match the physical order of the columns and requires a value for each column, which raises an error for any table column deletions or additions.

GRANTING PROGRAM UNIT PRIVILEGES

As we know from our SQL experience, to access any object on the database, a user must have appropriate privileges granted, typically by the DBA. For example, if user SCOTT needs query and modification rights on the BB_BASKET table, then the following GRANT command needs to be issued by the DBA:

```
GRANT SELECT, INSERT, UPDATE ON bb_basket TO scott;
```

The first items listed in a GRANT command are the privileges being granted which are SELECT, INSERT, and UPDATE in this case. The ON clause indicates a specific object if applicable. The TO clause indicates which user or role should receive these privileges.

The same form of the GRANT command is used to assign privileges regarding program units. Table 7-3 contains a list of privileges related to procedures, functions, and packages. The term "program units" in this table equates to all of these objects (procedures, functions, and packages). The keyword PROCEDURE is used in privilege grants to represent all of the program unit objects.

7

Table 7-3 Program unit privileges

System Privilege	Explanation
CREATE PROCEDURE	Allows a user to create, modify, and drop program units within their own schema
CREATE ANY PROCEDURE	Allows a user to create program units in any schema; does not allow the modification or dropping of the program units
ALTER ANY PROCEDURE	Allows a user to modify program units in any schema
DROP ANY PROCEDURE	Allows a user to drop program units in any schema
EXECUTE ON *program_unit_name*	Allows a user to execute a specific program unit
EXECUTE ANY PROCEDURE	Allows a user to execute program units in any schema

Again, notice the table does not show a CREATE FUNCTION or CREATE PACKAGE privilege because all the program unit types are encompassed in the CREATE PROCEDURE privilege. Also, note that the EXECUTE privilege cannot be granted for individual package constructs; EXECUTE on the entire package must be granted. Just as with all privileges, these can be cancelled with the REVOKE command.

Figure 7-28 displays examples of GRANT statements executed within a DBA account (SYS). Note that the creator of the program unit can grant EXECUTE privileges to other users for this program unit.

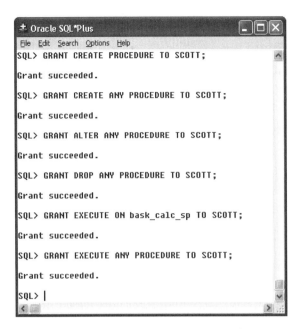

Figure 7-28 GRANT program unit privileges

Recall that privileges can also be granted to roles and PUBLIC, not just to specific users, as shown in the examples.

The data dictionary contains information regarding user privileges. As always, a variety of views are available to review this information. Table 7-4 lists the name, description, and column details of important data dictionary views regarding privileges.

Table 7-4 Data dictionary views on privileges

View Name	Description	Column Information
SESSION_PRIVS	Shows all privileges of the current schema, direct and indirect	PRIVILEGE, which is the name of the privilege granted
SESSION_ROLES	Shows all roles granted to the current schema	ROLE, which is the name of the role granted
USER_SYS_PRIVS	Shows only direct privileges of the current schema	• USERNAME, which is the recipient of the privilege • PRIVILEGE, which is the name of the privilege granted • ADMIN_OPTION, which is Yes or No, indicating if the privileges were granted WITH ADMIN OPTION

Table 7-4   Data dictionary views on privileges (continued)

View Name	Description	Column Information
USER_ROLE_PRIVS	Shows only direct roles granted to the current schema	• USERNAME, which is the recipient of the privilege • GRANTED_ROLE, which is the name of the role granted • ADMIN_OPTION, which is Yes or No, indicating if the role was granted WITH ADMIN OPTION • DEFAULT_ROLE, which is Yes if it is the user's default role; otherwise, it is No • OS_GRANTED, which is Yes if the operating system manages the roles; otherwise, it is No

Figure 7-29 shows a query on the SESSION_PRIVS view to list the privileges of the current user.

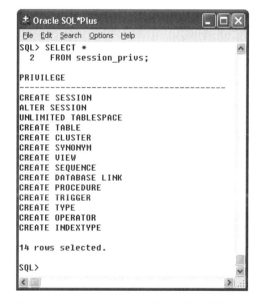

Figure 7-29   Query on the SESSION_PRIVS view

CHAPTER SUMMARY

❏ When a program unit refers to other database objects, the program unit is considered dependent on that object. If the referenced object is modified, the dependent object status changes to INVALID, indicating a need to be recompiled.

❐ With local database dependencies or for database objects residing on the same database, the dependent object can be compiled manually or it automatically compiles the next time it is run.

❐ A program unit can be recompiled manually by using the ALTER COMPILE command.

❐ Certain changes may call for a manual recompilation of dependent program units to ensure proper operation versus raising an error at runtime. Example situations include: 1) if we have dependent program units on a table and we modify the table structure, and 2) if we have dependent program units on a program unit in which we changed the parameters.

❐ A direct dependency is where object 1 uses or calls object 2. An indirect dependency is where object 1 calls object 2, which calls object 3; the dependency of object 1 on object 3 is indirect.

❐ The data dictionary USER_DEPENDENCIES view allows a user to review the object status and referenced objects. However, this view is not as helpful in identifying indirect dependencies.

❐ The dependency tree utility allows the user to review direct and indirect dependencies of an object in either a numeric order or indented format. The utldtree.sql file is provided by Oracle and is run to set up the necessary table, view, and procedure to use the dependency tree. The DEPTREE_FILL procedure is used to populate the dependency data for the object of interest.

❐ Package dependency action differs from stand-alone program units in that only a modification to the package specification changes dependent object status to INVALID. A modification to the body does not change the status of dependent objects. This is an advantage of packages in that less recompilation is required.

❐ Remote database object dependencies are not automatically tracked when modifications are accomplished. These dependencies are not checked until runtime and result in an error on the first run unless a manual recompilation of dependent objects is accomplished.

❐ Remote dependencies are determined by using one of two available models. The timestamp model is the default and it checks the last modification date to the referenced object, and if it is more recent than the last modification date on the dependent object, the dependent object is flagged with a status of invalid. The signature model compares the type and order of parameters to determine invalidation.

❐ To minimize recompilation errors, Oracle suggests using %ROWTYPE and %TYPE variable declarations, select with the ALL notation for data queries, and use a column list with INSERT statements.

❐ To enable a user to be able to run a program unit, the user must be granted execution privileges.

REVIEW QUESTIONS

1. Which data dictionary view can be used to check the status of a program unit?

   a. USER_DEPENDENCIES

   b. USER_OBJECTS

   c. USER_ERRORS

   d. USER_STATUS

2. Procedure A includes a call to function A. If the code in function A is modified, then the status of procedure A is _____ .

   a. ON

   b. OFF

   c. INVALID

   d. VALID

3. A program unit that has an INVALID status can be recompiled by _____ .

   a. using the ALTER COMPILE command

   b. using the RECOMPILE command

   c. executing it

   d. all of the above

4. A query on the USER_DEPENDENCIES view for a given object displays all the associated _____ dependencies.

   a. formal

   b. direct

   c. indirect

   d. informal

5. The dependency tree utility allows users to view _____ dependencies.

   a. indirect

   b. direct

   c. valid

   d. invalid

6. To review the dependencies of a particular object with the dependency tree utility, the _____ procedure must first be run.

   a. LOAD_DEPTREE

   b. FILL_DEPTREE

   c. DEPTREE_LOAD

   d. DEPTREE_FILL

7. The IDEPTREE view created by the dependency tree utility displays dependencies in a(n) _____ format.

   a. numbered level

   b. indented

   c. matrix

   d. tabular

8. Review the dependency listing in Figure 7-30. The procedure BASK_CALC_SP has a(n) _____ dependency on table BB_BASKETITEM.

   a. direct

   b. indirect

   c. no dependency

   d. cannot be determined

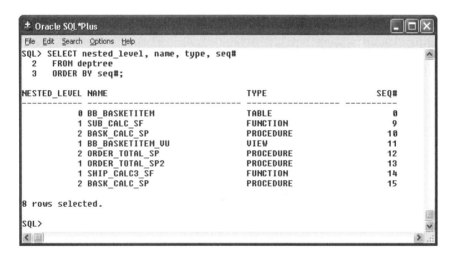

Figure 7-30   DEPTREE view listing

9. Review the dependency listing in Figure 7-31. The BB_BASKETITEM_VU view has a(n) _____ dependency on BB_BASKETITEM.

   a. direct

   b. indirect

   c. no dependency

   d. cannot be determined

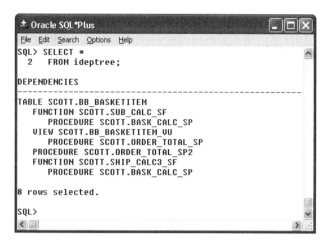

Figure 7-31   Determining dependency

10. A database link is used to accomplish _____ database connections.

    a. direct

    b. indirect

    c. remote

    d. a database link does not exist

11. If you have object dependencies on remote databases with one server in Virginia and the other in California, which invalidation model should *not* be used?

    a. tracking

    b. timestamp

    c. signature

    d. check

12. Currently, procedure A has a VALID status. Procedure A calls function B, which is in a package. A programmer just added an IF statement to the code in function B. Now, the status value of procedure A is _____.

    a. invalid

    b. valid

    c. check

    d. cannot be determined

7

13. Procedure A calls function C. If a parameter is added to function C, procedure A
    _____.

    a. could recompile successfully

    b. does not recompile successfully

    c. does not need to be recompiled

    d. will not be affected

14. The signature model of checking remote dependencies compares the
    _____.

    a. time of last compile

    b. object status

    c. parameter names

    d. type and order of parameters

15. Oracle recommends which of the following to reduce program unit recompila-
    tion errors following database modifications? (Choose all that apply.)

    a. use of %TYPE variable declarations

    b. use of numeric variables if possible

    c. use of a column list with inserts

    d. use of a column list with queries

16. Explain how dependencies on packaged program units are handled differently
    than those of stand–alone program units.

17. Review the dependency listing in Figure 7-31. List all the objects that are indi-
    rectly dependent on the BB_BASKETITEM table.

18. Why should developers be concerned with dependencies?

19. Explain the different handling of local versus remote dependencies.

20. If we are about to modify an object, what is the best way to view both direct and
    indirect dependencies of this object? What steps must be taken to make this feature
    available?

ADVANCED REVIEW QUESTIONS

1. You have a stored procedure named CALC_COST, which was created in the
   SCOTT schema. Which of the following statements executed in the SCOTT
   schema allow user TESTER to execute the procedure?

   a. `GRANT PROCEDURE calc_cost TO TESTER;`

   b. `GRANT EXECUTE ON calc_cost TO TESTER;`

   c. `GRANT EXECUTE calc_cost TO TESTER;`

   d. none of the above

2. Which of the following data dictionary views contains information regarding the status of program units?

a. USER_TABLES

b. USER_SOURCE

c. USER_STATUS

d. USER_OBJECTS

3. Review the following CREATE PROCEDURE and ALTER TABLE statements. Given these, what occurs if the next statement you issue invokes the LOAD_HOURS procedure?

```
CREATE PROCEDURE load_hours
  (p_id IN NUMBER,
  p_hours IN NUMBER)
IS
BEGIN
INSERT INTO work_track (e_id, hours)
  VALUES (p_id, p_hours);
END;
/
ALTER TABLE work_track ADD job_id NUMBER(5);
```

a. The execution raises an error because the ALTER TABLE statement made the procedure status INVALID.

b. The system automatically attempts to recompile the procedure.

c. The procedure runs with no compilation action.

d. The procedure fails due to the table modification.

4. Which data dictionary view contains information on all your privileges, direct and indirect?

a. SESSION_PRIVS

b. USER_SYS_PRIVS

c. SESSION_ROLES

d. USER_ROLE_PRIVS

5. You create a package specification that currently has a status of VALID. You execute a CREATE OR REPLACE statement to modify the associated package body. This statement produces errors and the package body has a status of INVALID. Now, what is the status of the package specification?

a. INVALID

b. VALID

c. NULL

d. cannot determine

HANDS-ON ASSIGNMENTS

Assignment 7-1: Review Dependency Information in the Data Dictionary

There are two data dictionary views that store information associated with dependencies: USER_OBJECTS and USER_DEPENDENCIES. Let's take a look at both of these to determine the relevant information.

1. Open SQL*Plus and issue a DESCRIBE command on the view USER_OBJECTS, as shown in Figure 7-32. Which columns are particularly relevant to dependencies? The STATUS column indicates if the object is VALID or INVALID. The TIMESTAMP column is used in remote connection to determine invalidation.

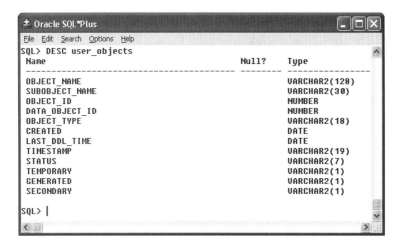

Figure 7-32   Describe the USER_OBJECTS view

2. Query USER_OBJECTS selecting the OBJECT_NAME, STATUS, and TIMESTAMP columns for all procedures. Recall that you can use a WHERE clause to look for object types of 'PROCEDURE' to list only procedure information.

3. Now, let's perform a DESCRIBE command on the USER_DEPENDENCIES view to review the available columns, as shown in Figure 7-33. Notice if you query this table for the name of a specific object, you receive a list of all the objects it references. However, if you query for a specific referenced name, you receive a list of objects that are dependent on this particular object.

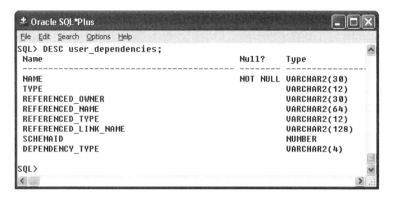

Figure 7-33   Column listing of the USER_DEPENDENCIES view

4. Let's say we intend to make a modification to the BB_BASKET table and need to identify all the dependent program units to complete recompilation. Execute the following query to determine all the objects that are dependent on the BB_BASKET table.

```
COLUMN name FORMAT A15
SELECT name, type
 FROM user_dependencies
 WHERE referenced_name = 'BB_BASKET';
```

Assignment 7-2: Test Dependencies on Stand-alone Program Units

In this assignment, you verify the effect of object modifications on the status of dependent objects. You work with a procedure and a function.

1. In a text editor, open the **assignment07–02.txt** file that is located in the Chapter.07 folder. This file contains statements to create the procedure STATUS_CHECK_SP and function STATUS_DESC_SF. Review the code and note that the procedure includes a call to the function.

2. Copy and paste the code to create the function and the procedure.

3. Enter the following query to verify the status of both objects.

```
SELECT object_name, status
 FROM user_objects
 WHERE object_name IN
('STATUS_CHECK_SP','STATUS_DESC_SF');
```

4. The function identifies a description to the numeric value for the idstage. The company needs to add another order status stage for situations in which the credit card approval failed. Return to the text file and add the following ELSIF clause:

```
ELSIF p_stage = 6 THEN
   lv_stage_txt := 'Credit Card Not Approved';
```

5. Does the modification in Step 5 affect the status of the STATUS_CHECK_SP procedure? Verify by returning to SQL*Plus and repeat the query in Step 4. The procedure is dependent on the function and, therefore, is now INVALID and requires recompilation.

6. Invoke the procedure for basket 13, as shown in the following code:

```
VARIABLE g_stage NUMBER
VARIABLE g_desc VARCHAR2(30)
EXECUTE status_check_sp(13,:g_stage,:g_desc);
```

7. Repeat the query in Step 4 to verify the status of the STATUS_CHECK_SP procedure. The procedure now shows a status of VALID as a result of automatic recompilation when the procedure was invoked.

Assignment 7-3: Test Dependencies on Packaged Program Units

In this problem, you verify the effect of object modifications on the status of dependent objects. You work with a procedure and a packaged function.

1. In a text editor, open the **assignment07-03.txt** file that is located in the Chapter.07 folder. This file contains statements to create the procedure STATUS_CHECK_SP and package LOOKUP_PKG. Review the code and note that the procedure includes a call to the packaged function STATUS_DESC_PF.

2. Copy and paste the code to create the procedure and the package.

3. Run the code.

4. Use the following query to verify the status of the procedure.

```
SELECT status
  FROM user_objects
 WHERE object_name = 'STATUS_CHECK_SP';
```

5. The function identifies a description to the numeric value for the idstage. The company needs to add another order status stage for situations in which the credit card approval failed. Return to the text file and add the following ELSIF clause to the packaged function:

```
ELSIF p_stage = 6 THEN
   lv_stage_txt := 'Credit Card Not Approved';
```

6. Copy and paste the package code to SQL*Plus to rebuild with the modifications.

7. Does the modification in Step 5 affect the status of the STATUS_CHECK_SP procedure? Verify it is still VALID by repeating the query in Step 4. The procedure is dependent on the function; however, if the referenced program unit is in a package, only changes to the program unit header (package specification) cause an invalidation of the dependent object.

Assignment 7-4: Test Remote Object Dependencies

At times, you may encounter program unit calls that use a database link to another database. These are called remote dependencies and act differently in regards to program unit invalidation, as you see in the following steps.

1. Create a database link named dblink2. If you have a second Oracle database running, then use a valid connection string for that database. Otherwise, use a connection string for the database you are connected to.

2. In a text editor, open the **assignment07-04.txt** file that is located in the Chapter.07 folder. This file contains statements to create the procedure STATUS_CHECK_SP and function STATUS_DESC_SF. Notice the procedure uses a database link when calling the function, which is treated as a remote database connection. Execute the code to create the objects.

If your database link connects to another database, be sure to create the function on that database.

3. Check the status of the procedure with a query to the data dictionary.

4. The function identifies a description to the numeric value for the idstage. The company needs to add another order status stage for situations in which the credit card approval failed. Return to the text file and add the following ELSIF clause to the packaged function:

```
ELSIF p_stage = 6 THEN
   lv_stage_txt := 'Credit Card Not Approved';
```

5. Copy and paste the package code to SQL*Plus to rebuild with the modifications. Does the modification in Step 4 affect the status of the STATUS_CHECK_SP procedure? Verify that it is still VALID. The procedure is dependent on the function; however, because it is a remote dependency, the status is not checked at the time of modification on the referenced object.

6. Test invoking the procedure.

7. Verify the status of the procedure again.

8. Invoke the procedure a second time. What happens?

Assignment 7-5: Identify All Dependencies Using the Dependency Tree Utility

At this point, you have created a variety of database objects in your schema. Use an appropriate tool to identify all the direct and indirect dependencies on the BB_BASKET table. Produce dependency lists in the two different formats available. Identify each object as a direct or indirect dependency. Also, identify the path of dependency of each of the indirectly dependent objects.

Assignment 7-6: Review the utldtree.sql Script

In Windows, search for the file named utldtree.sql. This file should be in the database directory under the rdbms\admin subdirectory. Open the file in a text editor and review the script. Identify all the objects created (name and type) and a brief description on how each object is used for tracking dependencies.

Assignment 7-7: Avoid Recompilation Errors

Any application inevitably is modified, so we must attempt to produce code that assists in minimizing maintenance. Identify two coding techniques that help avoid recompilation errors following referenced object modification and briefly describe how this technique achieves this goal.

Assignment 7-8: Identify the Types of Dependencies

In this chapter, we discussed direct, indirect, and remote dependencies. Define each of these and describe how they differ in regards to program unit invalidation and recompilation.

CASE PROJECTS

Case 7-1: The Brewbean's Application Maintenance

To avoid unexpected application errors experienced by end users, the Brewbean's head programmer wants you to automate the process of recompiling any program units that have an INVALID status. This procedure will be run nightly to recompile any program units that did not get recompiled following modifications. This procedure could reference the USER_OBJECTS view to determine which program units are INVALID.

Case 7-2: The More Movies Rental Application

In previous chapters, we have created procedures, functions, and packages to support the rental process. As application modifications are made in the future, we need to be able to identify all object dependencies to test our changes. Use the data dictionary and/or the dependency tree utility to complete a list of dependencies for all the More Movies database objects. Do so in the format shown in Table 7-5. The dependency type should be listed as direct or indirect.

Table 7-5 More Movies database dependency list

Object Name	Dependent Object	Dependency Type

8

DATABASE TRIGGERS

In this chapter, you will:

♦ Learn about database triggers and syntax

♦ Know how to create and test a DML trigger in SQL*Plus

♦ Know how to create and test an Instead-Of database trigger

♦ Use system triggers

♦ Identify when triggers should be used

♦ Identify trigger restrictions

♦ Use the ALTER TRIGGER statement

♦ Delete a trigger

♦ Use data dictionary information relevant to triggers

A database trigger is a block of PL/SQL code that runs automatically when a particular database event occurs. It is tied to a DML action on a specific table and does not care where the action was generated, such as from an application or an SQL command entered directly into SQL*Plus.

This chapter explores the creation and potential uses of database triggers. We first identify the structure of database triggers and concentrate on the frequently used DML event triggers. Then, after we have a fundamental understanding of a trigger, we explore further, looking in particular at system event triggers, the application of triggers, trigger limitations, and relevant data dictionary information.

THE CURRENT CHALLENGE IN THE BREWBEAN'S APPLICATION

The Brewbean's company coffee ordering application is coming along; we have already addressed a number of the order processing issues such as shipping cost calculations. However, the owner described the product inventory process to the application development group today and you have been given the task of developing PL/SQL blocks to handle updating the product inventory when a sale is completed.

The in-stock level of products needs to be updated in real time so that a customer can be apprised at the time of ordering if an item is currently out of stock. After a shopper has selected all the desired items and views the basket listing, as shown in Figure 8-1, the Check Out link is clicked and the order confirmation page appears. This page asks the customer to enter credit card, name, address, and contact information, as partially displayed in Figure 8-2. When the Submit button on the bottom of this page is clicked, the order is confirmed and the inventory data in the BB_PRODUCT table needs to be updated.

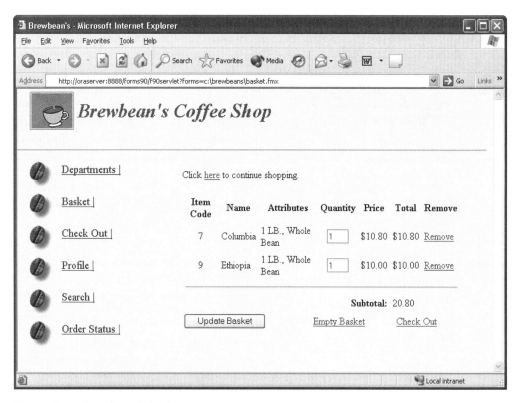

Figure 8-1   Brewbean's basket screen

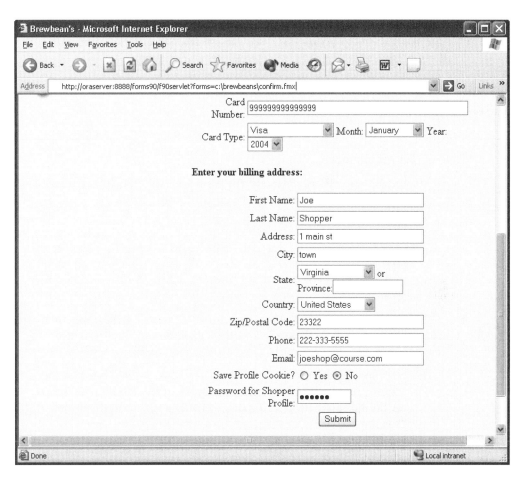

Figure 8-2   Brewbean's order confirmation screen

The BB_PRODUCT table includes three columns involving inventory: STOCK holds the current number of items available, ORDERED holds the current number on request or to be purchased to replenish stock, and REORDER holds the number at which the stock should be replenished. The BB_PRODUCT_REQUEST table holds information regarding each stock replenishment request.

Note that Brewbean's uses the term request instead of order because they roast and package all their own coffee products. Only the equipment items must actually be purchased from wholesalers. In addition, at this time, the coffee stock and replenishment amounts are in terms of whole pounds. Therefore, if a customer purchases a half-pound indicated by the OPTION1 column in the BB_BASKETITEM table, the stock level must be reduced only by .5.

The company also has a similar application process included in their in-store ordering system for walk-in customers. Thus, regardless of whether the order is being processed by the in-store application or the Web site application, the product inventory needs to be updated at the time of sale. In this scenario, a database trigger would be an appropriate mechanism to accomplish this task because the sales confirmation may occur in different applications.

REBUILDING YOUR DATABASE

Complete the following steps to rebuild the Brewbean's database for this chapter.

To rebuild the database:

1. Locate the **c8Dbcreate.sql** file in the Chapter.08 folder to ensure it exists. This file contains the script to create the database.

2. Open SQL*Plus.

3. Create a spool file so that SQL*Plus keeps a copy of all the messages received from running the file that creates the database. On the main menu in SQL*Plus, click **File**, point to **Spool**, and then click **Spool File**. A Select File dialog box appears.

4. Browse to a folder used to contain your working files, and in the File name text box, type **DB_log8**, and then click **Save**. Now, whatever text is seen in our SQL*Plus session is saved to this file for future reference.

5. Now let's create the database. In SQL*Plus, enter the following command, which runs all the statements contained in the c8Dbcreate.sql file. Messages verifying the creation and data insertion steps scroll on the SQL*Plus screen. This may take a couple minutes to complete.

   `@<pathname to PL/SQL files>\Chapter.08\c8Dbcreate`

6. Now, we can turn the spooling off so that we can review the results in the spool file created. On the main menu, click **File**, point to **Spool**, and then click **Spool Off**.

7. Open a text editor.

8. Open the file named **DB_log8.lst** in your working folder.

9. Review the messages for any errors.

INTRODUCTION TO DATABASE TRIGGERS

A database trigger is a block of PL/SQL code that runs automatically when a particular database event occurs. The event may be a DML action of INSERT, UPDATE, and DELETE, or an Oracle system action such as a user logging on. The trigger contains a PL/SQL block with a header, which instructs Oracle as to which event this trigger is associated with on the database. For example, if the trigger is associated with an UPDATE statement on the BB_BASKET table, then the trigger code runs or fires automatically whenever an UPDATE statement on the BB_BASKET table is processed—regardless of its source.

Note that it does not matter whether the Brewbean's Web site application issues the UPDATE or the in-store application issues the UPDATE; in either case, the trigger fires. Even if we logged onto the database with SQL*Plus and entered an UPDATE for the BB_BASKET table, the trigger still fires. A database trigger is tied to a database table or view and is implicitly fired by the database system when that table or view is affected with the associated DML action. This is quite a change from procedure and functions that are explicitly called by our program code.

DATABASE TRIGGER SYNTAX AND OPTIONS

Let's take a quick peek at the basic syntax layout of a CREATE TRIGGER statement in Figure 8-3. Notice that the clauses related to the trigger timing and events are labeled.

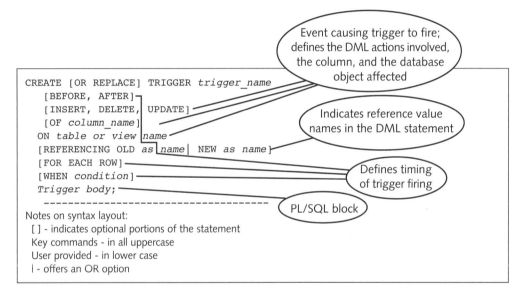

```
CREATE [OR REPLACE] TRIGGER trigger_name
    [BEFORE, AFTER]
    [INSERT, DELETE, UPDATE]
    [OF column_name]
    ON table or view name
    [REFERENCING OLD as name| NEW as name]
    [FOR EACH ROW]
    [WHEN condition]
    Trigger body;
    ------------------------------------
Notes on syntax layout:
    [ ] - indicates optional portions of the statement
    Key commands - in all uppercase
    User provided - in lower case
    | - offers an OR option
```

Event causing trigger to fire; defines the DML actions involved, the column, and the database object affected

Indicates reference value names in the DML statement

Defines timing of trigger firing

PL/SQL block

Figure 8-3   CREATE TRIGGER statement

The trigger timing and events must indicate what events cause this trigger to fire and whether the firing occurs before or after the event occurs. Additional options can indicate whether to fire the trigger only one time for the DML firing event or for each row affected by the DML event. In addition, a condition can be included that is checked when the event occurs and causes the trigger to fire only when the condition is true.

The trigger body is a PL/SQL block that contains the actions that take place when the trigger fires. Additional important features available with triggers include conditional predicates and correlation identifiers. Many options exist for each area of the CREATE TRIGGER statement; the next sections explore these options, but first, we need to explore a related code example.

Database Trigger Code Example

Understanding a database trigger is easier when you have an actual code example. The following code sample is a completed trigger statement creating a trigger named PRODUCT_ INVENTORY_TRG. We use this code to solve our updating inventory problem presented in the Brewbean's application. Line numbers are listed to the left for easier reference. This example is used to familiarize you with the CREATE TRIGGER statement and to serve as the basis for discussion in the subsequent sections.

```
1 CREATE OR REPLACE TRIGGER product_inventory_trg
2  AFTER UPDATE OF orderplaced ON bb_basket
3  FOR EACH ROW
4  WHEN (OLD.orderplaced <> 1 AND NEW.orderplaced = 1)
5 DECLARE
6  CURSOR basketitem_cur IS
7   SELECT idproduct, quantity, option1
8   FROM bb_basketitem
9   WHERE idbasket = :NEW.idbasket;
10  lv_chg_num NUMBER(3,1);
11 BEGIN
12  FOR basketitem_rec IN basketitem_cur LOOP
13   IF basketitem_rec.option1 = 1 THEN
14    lv_chg_num := (.5 * basketitem_rec.quantity);
15   ELSE
16    lv_chg_num := basketitem_rec.quantity;
17   END IF;
18   UPDATE bb_product
19   SET stock = stock - lv_chg_num
20   WHERE idproduct = basketitem_rec.idproduct;
21  END LOOP;
22 END;
```

Each of the following sections explains the various parts of the CREATE TRIGGER statement. Do not worry about understanding or running the statement yet. A text file named c8invent.sql containing this code can be found in the Chapter.08 folder. You may want to print this file and make notes throughout the subsequent sections.

Trigger Timing and Correlation Identifiers

The correlation and timing features are so intertwined that they are discussed together in this text. The timing of the trigger encompasses a number of issues, including firing before or after the associated DML event, firing only once for the entire DML event or for each row the DML event affects, and firing only if a particular condition is true.

First, the timing must be indicated as either BEFORE or AFTER. Notice Line 2 of our trigger includes the key word "AFTER," indicating the code in the body of the trigger should be executed after the DML statement, which fired the trigger, is completed. In this case, we want the trigger code to execute only after the BB_BASKET table has been updated, setting the ORDERPLACED column to 1 and confirming the order has been completed.

After we know the order has been completed, we need to make the inventory changes. If we used BEFORE instead of AFTER, the inventory would be updated prior to the order being confirmed. What if a problem occurs, such as a failed credit card verification, when the order information is confirmed? We would end up with an inaccurate stock level in the BB_PRODUCT table because the order was not completed but we already updated the inventory as if it were completed.

Another consideration in trigger timing is whether to execute the trigger code for every row the DML event affects or just one time. This is called row level and statement level triggers, respectively. **Row level** indicates firing the trigger code for each row affected in the DML statements, whereas **statement level** indicates firing the trigger only once for the event regardless of the number of rows affected by the DML statement. The BEFORE and AFTER options combine with either row or statement level options to produce a trigger firing sequence, as shown in Figure 8-4, for a DELETE statement that affects two rows. Keep in mind that numerous triggers can be constructed on one table and, therefore, a DML action on that table can fire off more than one trigger.

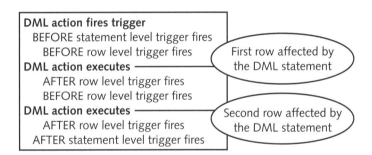

Figure 8-4   Trigger firing sequence

 If multiple triggers exist on a table, there is no guarantee on the order in which the triggers will be fired. A BEFORE statement level trigger fires before a BEFORE row level trigger; however, if we have two statement level triggers, the order in which they are fired is random.

Note that the row level option is applicable only for UPDATE and DELETE events. Why? Because each of these DML statements can affect more than one row. An INSERT can affect only one row at a time and is, therefore, not relevant to the row level feature. In our case, the application completes only one order at a time, so a statement level trigger should be all that is needed. The default timing of a trigger is statement level and no code is included to achieve this timing. However, notice line 3 of the PRODUCT_INVENTORY_TRG trigger contains the clause "FOR EACH ROW," which makes this a row level trigger. Why did we do this? We must be working with a row level trigger to use the features of the WHEN clause. In addition, we need correlation identifiers in our inventory trigger.

Consider what tasks each of these features can accomplish. The WHEN clause allows us to include a condition so that the trigger executes only if this condition is true. For example, in our inventory trigger, we want the inventory update activity to occur only when the order is being confirmed. Our code needs to check if the UPDATE on the BB_BASKET table is changing the ORDERPLACED column from 0 to 1, indicating the initial placement of the order.

Of course, other update actions can occur on the BB_BASKET table, but we want to perform this particular inventory update only with the initial order confirmation. Notice Line 4 of the PRODUCT_INVENTORY_TRG trigger contains a WHEN clause. So, what do the OLD and NEW references refer to? These are **correlation identifiers**, or qualifiers, which are special PL/SQL bind variables that refer to values associated with the DML action that fires a trigger. The correlation identifiers are powerful in that they actually allow us to refer to and use the row data values of the DML action.

As an example of the usefulness of correlation identifiers, consider our inventory trigger. We can check the original or old value of the ORDERPLACED column for the row being updated for a value of 0 and the new value of ORDERPLACED for a value of 1. An UPDATE involves an existing row with original values and a resulting or new row with the updated values. If an UPDATE fires a trigger, the Oracle server actually holds the original and new row column values in these special bind variables that can be referenced. The correlation identifier name (OLD or NEW), a period, and then the column name are the elements in the proper notation, as seen in Line 4 of the example trigger.

 A common point of confusion is when to precede a correlation identifier with a colon. Notice that in the WHEN clause of the example trigger, no preceding colon is used to check OLD.orderplaced and NEW.orderplaced. However, as you will see in later examples, a preceding colon must be used with correlation identifiers in code in the trigger body.

Note that both the OLD and NEW identifiers are not available for all DML actions, as listed in Table 8-1. Only an UPDATE statement has both an original row and a new row, so both identifier values are available.

Table 8-1 Correlation identifier availability

DML Event	OLD Identifier	NEW Identifier
INSERT	Not available	Contains insert values
UPDATE	Contains values of the original row	Contains new values for any columns updated and original values for any columns not updated
DELETE	Contains values of the original row	Not Available (Note that "Not Available" indicates any references retrieve a NULL value)

The REFERENCING clause shown in the trigger syntax layout in Figure 8-3 can be used to change the OLD and NEW qualifier names to something of your choice. However, the OLD and NEW names are so logical they are typically used as is by most PL/SQL developers.

Remember these three critical pieces of information regarding the WHEN clause of a trigger: First, a WHEN clause can be used only in row level triggers. Second, the values checked in the WHEN clause must reference correlation identifiers. Third, the condition must be enclosed in parentheses.

Trigger Events

The trigger event information defines which DML statements on which table cause the trigger to fire. The first item to be indicated is the DML event or events. This includes the options of INSERT, UPDATE, and DELETE of which we can include only one, two, or all of them. Notice the PRODUCT_INVENTORY_TRG in line 2 is as follows:

```
AFTER UPDATE OF orderplaced ON bb_basket
```

The only DML event that can cause this trigger to fire is an UPDATE. Also, notice the OF clause that follows the UPDATE keyword. The OF clause is optional, can only be used in conjunction with the UPDATE event, and allows a column name to be specified. In our example, the trigger fires only if the UPDATE affects the ORDERPLACED column. If the SET clause of the UPDATE statement does not affect the ORDERPLACED column, this trigger code does not fire. If we want the trigger to fire for both an UPDATE or DELETE action, then the events can be listed as:

```
AFTER DELETE OR UPDATE OF orderplaced ON bb_basket
```

The ON clause indicates which table is being affected by the DML events. The PRODUCT_INVENTORY_TRG trigger fires when the UPDATE is occurring on the BB_BASKET table, which contains the ORDERPLACED column. Unlike procedures and functions, a database trigger is tied to the table that is named in the ON clause. When a DML action occurs on this table, the database system first checks through all the triggers associated with this table and determines if any should fire.

Trigger Body

The trigger body is a PL/SQL block. The header section of the trigger is not followed by an IS keyword as in procedures and functions; therefore, the DECLARE SECTION keyword must be used if any variables need to be declared for the trigger body code block. In our example, the DECLARE keyword is used because we need to declare a cursor to accomplish our task of updating the inventory.

8

The block retrieves the items ordered data from the BB_BASKETITEM table into a cursor so that the stock amount of each product can be updated. To retrieve the correct rows associated with the basket being ordered, we need to reference the IDBASKET value from the UPDATE statement that fires this trigger by using the correlation identifiers. Notice the cursor declaration in the trigger is as follows:

```
CURSOR basketitem_cur IS
  SELECT idproduct, quantity, option1
  FROM bb_basketitem
  WHERE idbasket = :NEW.idbasket;
```

The WHERE clause references the value of the IDBASKET from the UPDATE statement using the ":NEW.idbasket" notation. Notice that a colon now precedes the correlation identifier. Recall that the correlation identifiers are considered to be host or bind variables and, therefore, the colon is used in the PL/SQL block just as we do with other host variables. In this example, it does not matter if we reference the NEW or OLD IDBASKET value as both hold the same value because the UPDATE statement did not modify this column. The OPTION1 column value is retrieved, because this indicates if the order size of coffee is a half-pound or a whole pound. If it is a half-pound, the quantity is multiplied by .5 to determine the change in the inventory amount because the coffee inventory is in terms of whole pounds.

The executable or BEGIN section of the trigger body uses a CURSOR FOR loop to enable looping through all the products ordered and update the appropriate product stock value in the BB_PRODUCT table. An IF clause is included in the CURSOR FOR loop to check if the order size of coffee is for a half-pound. If it is, the adjustment is made to the quantity to properly update the stock amount. Note that the trigger issues DML or UPDATE statements on the BB_PRODUCT table to modify the stock amount. Consider the flow of database activity for this trigger: An UPDATE statement on the BB_BASKET table causes the trigger to fire, the trigger retrieves the products ordered information by accomplishing a query on the BB_BASKETITEM table, and UPDATEs are issued on the BB_PRODUCT table to modify the stock amount of the applicable products. This flow is diagramed in Figure 8-5.

In our trigger example, the PL/SQL block code is entered right into the trigger. However, triggers can also use a CALL statement to invoke a stored subprogram (procedure or function) or a wrapper for a C or Java routine. What if we already have a procedure named UPDATE_STOCK_SP available that contains the code we want to use in our trigger body? We would use a CALL statement in our trigger, as shown in the following code. If the procedure required parameters, then these would be included as well, just as we would normally invoke the procedure. The OLD and NEW values can also be used to supply parameter values.

```
CREATE OR REPLACE TRIGGER product_inventory_trg
  AFTER UPDATE OF orderplaced ON bb_basket
  FOR EACH ROW
  WHEN (OLD.orderplaced <> 1 AND NEW.orderplaced = 1)
  CALL update_stock_sp(:NEW.idbasket)
```

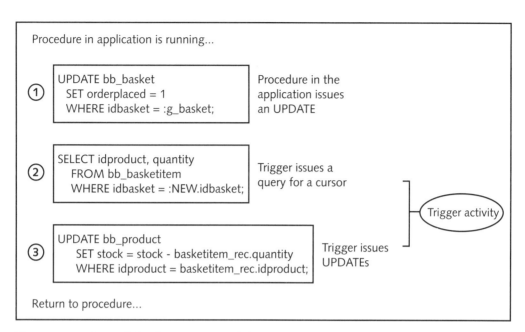

Figure 8-5   Flow of database activity of the PRODUCT_INVENTORY_TRG trigger

Conditional Predicates

What if we want to accomplish different activities in the trigger based on the DML statement issued? **Conditional predicates** are clauses available within a trigger to allow separate handling of different database triggering events within a single trigger. The predicates consist of an IF (establishing the conditional part) word and a predicate, which is a verb indicating the DML action. Recall that we can list multiple DML events in a trigger, as shown in the following code:

```
AFTER DELETE OR UPDATE ON bb_basket
```

In this trigger clause, either a DELETE or UPDATE action on the BB_BASKET table causes the trigger to fire. What if we want to accomplish something different in the trigger for each of these events? This can be achieved by using conditional predicates in the trigger body.

Let's simplify our inventory trigger example for a moment to demonstrate how these work. The following code reduces the product stock level if an UPDATE occurs but increases the stock level if the order is deleted or cancelled. It also assumes all coffee orders are for whole pounds. Notice the IF clauses in Lines 10 and 17.

```
1 CREATE OR REPLACE TRIGGER product_inventory_trg
2  AFTER DELETE OR UPDATE ON bb_basket
3  FOR EACH ROW
4 DECLARE
5  CURSOR basketitem_cur IS
6    SELECT idproduct, quantity
```

```
7    FROM bb_basketitem
8    WHERE idbasket = :NEW.idbasket;
9 BEGIN
10 IF UPDATING THEN
11   FOR basketitem_rec IN basketitem_cur LOOP
12   UPDATE bb_product
13    SET stock = stock - basketitem_rec.quantity
14    WHERE idproduct = basketitem_rec.idproduct;
15   END LOOP;
16 END IF;
17 IF DELETING THEN
18   FOR basketitem_rec IN basketitem_cur LOOP
19   UPDATE bb_product
20    SET stock = stock + basketitem_rec.quantity
21    WHERE idproduct = basketitem_rec.idproduct;
22   END LOOP;
23 END IF;
22 END;
```

The predicates of INSERTING, DELETING, and UPDATING are available to indicate particular actions based on an event. Also, the UPDATING conditional predicate can be further refined to indicate a particular column. For example, if you want to accomplish a particular action if the UPDATE modifies the ORDERPLACED column, you could use the following code:

```
IF UPDATING ('orderplaced') THEN
```

 We would want to use the column refinement in an UPDATING predicate if the trigger fires regardless of which columns are being updated. In this case, we can use the column refinement to indicate different actions depending on the column updated. However, if we have already indicated that the trigger should fire only after an update of a particular column with the UPDATE OF syntax, it is redundant to use the column refinement in the conditional predicate in this manner.

CREATING AND TESTING A DML TRIGGER IN SQL*PLUS

Now that we have a basic understanding of database trigger coding, let's create the PRODUCT_INVENTORY_TRG trigger that we have discussed in this chapter thus far.

To create the trigger:

1. Open or return to SQL*Plus.

2. In SQL*Plus, navigate to and select the **c8invent.sql** file from the Chapter.08 folder.

3. Click **Open** and a copy of the file contents appears in the SQL*Plus screen. Notice the code matches the PRODUCT_INVENTORY_TRG trigger code we reviewed earlier in the chapter.

4. To run the code and create the trigger, click **File** and then **Run** from the main menu. (Press **Enter** if necessary to make the code execute.) You receive a "Trigger created" message. If you receive the message "Trigger created" as in Figure 8-6, then no errors occurred. If you receive an error message, recopy the code from the text file and be sure to capture all of the code.

```
Oracle SQL*Plus
File  Edit  Search  Options  Help
SQL> CREATE OR REPLACE TRIGGER product_inventory_trg
  2      AFTER UPDATE OF orderplaced ON bb_basket
  3      FOR EACH ROW
  4      WHEN (OLD.orderplaced <> 1 AND NEW.orderplaced = 1)
  5  DECLARE
  6      CURSOR basketitem_cur IS
  7        SELECT idproduct, quantity, option1
  8         FROM bb_basketitem
  9         WHERE idbasket = :NEW.idbasket;
 10      lv_chg_num NUMBER(3,1);
 11  BEGIN
 12      FOR basketitem_rec IN basketitem_cur LOOP
 13        IF basketitem_rec.option1 = 1 THEN
 14          lv_chg_num := (.5 * basketitem_rec.quantity);
 15        ELSE
 16          lv_chg_num := basketitem_rec.quantity;
 17        END IF;
 18        UPDATE bb_product
 19          SET stock = stock - lv_chg_num
 20          WHERE idproduct = basketitem_rec.idproduct;
 21      END LOOP;
 22  END;
 23  /

Trigger created.————(  Message indicating
                      successful trigger creation )

SQL> |
```

Figure 8-6   Creating the PRODUCT_INVENTORY_TRG trigger in SQL*Plus

Recall that you can use the SHOW ERRORS command in SQL*Plus to view error messages.

Now that we have a trigger, we can confirm it is operating properly by doing some testing. In the following steps, we use basket 15 to test the trigger because the order has not yet been completed and the ORDERPLACED column currently holds a value of 0.

To test the trigger with basket 15:

1. First, let's do a query on the BB_BASKETITEM table to confirm which products are involved with basket 15. Open or return to SQL*Plus and enter the query shown in Figure 8-7.

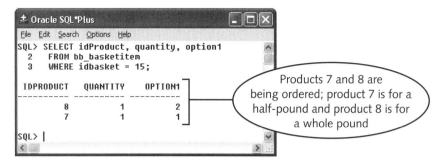

Figure 8-7 Query the database to view the basket product list

2. Verify the current stock amounts for these two products to determine if the correct modifications are being accomplished. Enter the query as shown in Figure 8-8.

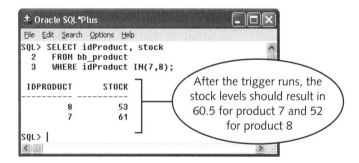

Figure 8-8 Query the database to confirm the stock amounts

3. Now that we know what should happen when the trigger fires, let's submit an UPDATE statement that should cause the trigger to fire. Enter the UPDATE as seen in Figure 8-9 to modify the ORDERPLACED column value from a 0 to a 1 to indicate the order has been placed. Also, issue a COMMIT statement to save the transaction because we are issuing DML actions.

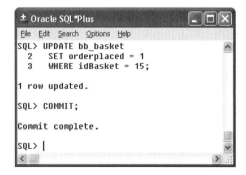

Figure 8-9 Issue an UPDATE to fire the trigger

4. Now, we need to verify the trigger actually fired and acted as we expected. Query the BB_PRODUCT table, as shown in Figure 8-10. This verifies that the trigger fired and performed the expected modification to the stock value of each product.

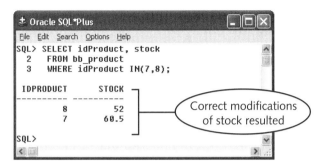

Figure 8-10   Querying the BB_PRODUCT table to check the new stock values

CREATING AND TESTING AN INSTEAD-OF TRIGGER

An **Instead-Of trigger** is a PL/SQL block that executes in place of a DML action on a database view. This type of trigger is typically used to allow the modification of data through a view on a view that is nonmodifiable. Recall from SQL that a view that contains any of the following is typically nonmodifiable:

- Joins
- Set operators
- Aggregate functions
- GROUP BY, CONNECT BY, START WITH clauses
- DISTINCT operator

We may want to allow developers to use a nonmodifiable view in DML statements to simplify data modifications that affect more than one table. Or, we may want to process DML actions via a view but need to affect each of the base tables of the view differently. For example, let's suppose Brewbean's develops an application screen that allows store employees to enter shipping information for orders, as shown in Figure 8-11.

In this case, the SHIPFLAG column needs to be updated in the BB_BASKET table, and the DTSTAGE, SHIPPER, SHIPPINGNUM, and NOTES columns need to be updated in the BB_BASKETSTATUS table. Let's assume when an order is approved and sent to shipping, a new row is inserted into the BB_BASKETSTATUS table at that time with an IDSTAGE of 3 to hold shipping information for this order. A view including

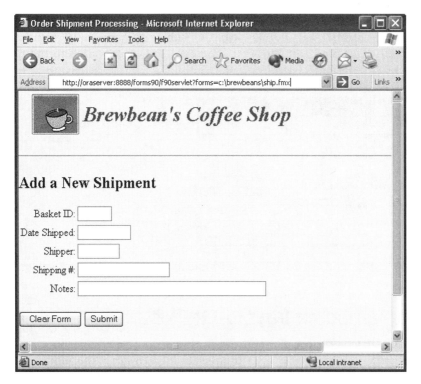

Figure 8-11   Brewbean's shipping information screen

columns from the BB_BASKET and BB_BASKETSTATUS tables was created in the database creation script for this text and looks like the following statement:

```
CREATE OR REPLACE VIEW bb_ship_vu
  AS SELECT b.idbasket, b.shipflag, bs.idstage, bs.dtstage,
       bs.notes, bs.shipper, bs.shippingnum
    FROM bb_basket b, bb_basketstatus bs
    WHERE b.idBasket = bs.idBasket;
```

Notice that this view does not include the primary key column of IDSTATUS in the BB_BASKETSTATUS table. What happens if we attempt an UPDATE using this view? Review the attempt in Figure 8-12 and notice the error refers to a "non key-preserved table."

A **key-preserved** table is one that is involved in a join and the keys of the original table are included in the keys of the resultant join. In this case, the primary key column of IDSTATUS is not included in the view, yet we are attempting to insert a new row into the BB_BASKETSTATUS table, which would require a value for this column. Before we create the trigger, let's review the Instead-Of trigger code needed to accomplish the shipping data update task, as shown in Figure 8-13. Notice the key words INSTEAD OF are used and the UPDATE in this trigger execute instead of the UPDATE on the BB_SHIP_VU view that will fire this trigger.

```
± Oracle SQL*Plus
File  Edit  Search  Options  Help
SQL> UPDATE bb_ship_vu
  2      SET shipflag = 'Y', idstage = 3, dtstage = '20-FEB-03', shipper = 'UPS',
  3            shippingnum = 'ZH49KXK70MFC3M0'
  4      WHERE idBasket = 12;
  SET shipflag = 'Y', idstage = 3, dtstage = '20-FEB-03', shipper = 'UPS',
      *
ERROR at line 2:
ORA-01779: cannot modify a column which maps to a non key-preserved table

SQL>
```

Figure 8-12   UPDATE action via a view containing a join

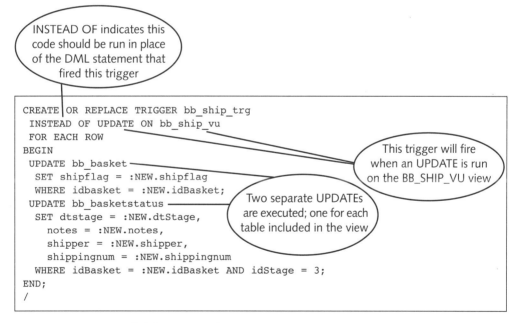

Figure 8-13   Instead-Of trigger code

Now, we create this Instead-Of trigger to support the Brewbean's order shipping application screen.

To create the Instead-Of trigger:

    1. Open or return to SQL*Plus.

    2. Navigate to and select the **c8ship.sql** file from the Chapter.08 folder.

    3. Click **Open** and a copy of the file contents appears in the SQL*Plus screen.

    4. To run the code and create the trigger, click **File** and then **Run** from the main menu.

Now, we have an Instead-Of trigger on the BB_SHIP_VU view. To test it, we issue the UPDATE on the view that we attempted earlier.

To test the Instead-Of trigger:

    1. Go to SQL*Plus.

    2. Enter the UPDATE statement as shown at the top of Figure 8-14.

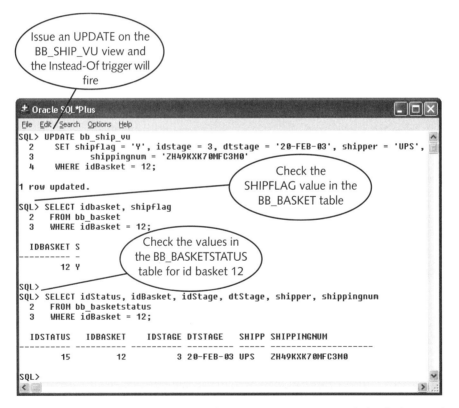

Figure 8-14   Issue an UPDATE on the BB_SHIP_VU view and check the results

3. Query the BB_BASKET table, as shown in Figure 8-14, to verify the SHIPFLAG has been set to 'Y' for idBasket 12.

4. Query the BB_BASKETSTATUS table, as shown in Figure 8-14, to verify that the DTSTAGE, SHIPPER, and SHIPPINGNUM columns have been updated for basket 12.

Note that the goal of an Instead-Of trigger is to replace a DML statement that is run on a view. This is to not only simplify code by using views, but also to enable a variety of different actions to occur if needed.

SYSTEM TRIGGERS

System triggers refer to database triggers that are fired by Data Definition Language (DDL) statements or database system events rather than DML actions. DDL events include CREATE, ALTER, DROP, and GRANT. A more comprehensive list includes the following:

- CREATE
- ALTER
- DROP
- GRANT
- REVOKE
- RENAME
- TRUNCATE

- ANALYZE
- AUDIT
- NOAUDIT
- COMMENT
- ASSOCIATE STATISTICS
- DISASSOCIATE STATISTICS

Database system events include LOGON, LOGOFF, STARTUP, SHUTDOWN, and SERVERERROR. The STARTUP and SHUTDOWN are concerned with the starting and stopping of the database. The syntax layout of the system trigger, as listed in the following syntax example, is basically the same as other triggers in regards to timing and the list of events. The ON clause indicates the trigger is either a database or schema level trigger rather than a specific table or view as used in DML triggers.

```
CREATE [OR REPLACE] TRIGGER trigger_name [BEFORE, AFTER]
[List of DDL or Database System Events] [ON DATABASE |
SCHEMA] Trigger body;
```

A database level trigger fires whenever the event occurs, regardless of the schema in which it occurs. For example, the following code creates a trigger that fires when anyone logs onto the schema in which the trigger was created. If we log on to Oracle as user SCOTT and create a trigger with the header code indicated in the following code, this trigger fires each time we log on to the Scott schema:

```
CREATE OR REPLACE TRIGGER logon_trg AFTER logon ON SCHEMA
Trigger body;
```

If we use ON DATABASE rather than ON SCHEMA, then this trigger fires when any-one logs onto the database, regardless of the schema. This would be used to maintain a typical audit file containing data of who and when the database has been accessed. Notice the timing of the logon event is AFTER. There are some common sense limi-tations on which timings work properly with which events. For example, the system could not entertain a BEFORE logon trigger because it cannot anticipate this event—unless of course the next version of Oracle software builds in psychic powers! In addition, the STARTUP and SHUTDOWN events must be used in a database level trigger. Let's review the necessary code to create a system trigger that captures the user name and current date upon each logon, as shown in Figure 8-15.

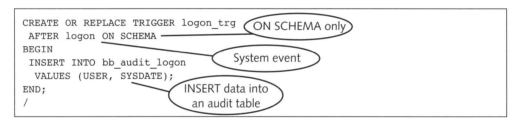

Figure 8-15   System trigger creation

Now, let's create this trigger to track logon information for your own schema.

To build a system trigger:

1. Open or return to SQL*Plus.

2. Create the BB_AUDIT_LOGON table with the following command:

```
CREATE TABLE bb_audit_logon (
    userid VARCHAR2(12),
    logdate DATE );
```

3. Navigate to and select the **c8systrig.sql** file from the Chapter.08 folder.

4. Click **Open** and a copy of the file contents appears in the SQL*Plus screen.

5. To run the code and create the trigger, click **File** and then **Run** from the main menu.

We now have a trigger and need to test it. We need to log off and log back on to ver-ify that the trigger fires.

To test the system trigger:

1. In SQL*Plus, type **Exit** and then press **Enter**.

2. Using your Windows Start menu, start up SQL*Plus.

3. Log back on to the database.

4. Verify that the trigger fired by querying the BB_AUDIT_LOGON table, as shown in Figure 8-16. A row has been inserted into the table indicating that you just logged on.

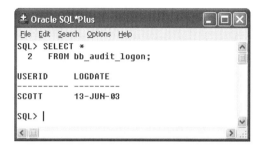

Figure 8-16   Query the BB_AUDIT_LOGON table to verify the system trigger fired

Note that system triggers typically fall under the responsibility of a Database Administrator (DBA) versus a developer. The DBA usually has the role regarding database security, and the system triggers assist in tracking database access. In addition, the DBA may want to perform a number of system maintenance tasks before a database shutdown that could be handled via a BEFORE SHUTDOWN system trigger. However, it is important for a developer to be aware of the potential capabilities offered by system triggers.

APPLYING TRIGGERS TO ADDRESS PROCESSING NEEDS

Now that you have an understanding of the types of triggers available and their basic operation, you may wonder about the common uses of triggers. A variety of tasks are well-suited for triggers. These tasks are discussed in Table 8-2.

Table 8-2   Possible trigger uses

Task Type	How a Trigger May be Applied
Auditing	Log files of database activity are widely used, for example, the tracking of sensitive data modifications such as employee payroll data. A trigger could be used to write the original and new values of the employee salary update to an audit table. If any questions arise concerning the change, a record of the original values and new values assigned is now available.

Table 8-2 Possible trigger uses (continued)

Task Type	How a Trigger May be Applied
Data integrity	Simple data validity checks can be accomplished with CHECK constraints. However, more complex CHECKS that require comparison to a live data value from the database can be accomplished using triggers. A trigger could be used to ensure that any changes to the regular price of a product do not allow a decrease from the current price. The NEW and OLD price values can be compared in a trigger.
Referential integrity	Foreign key constraints are used to enforce relationships between tables. If a parent key value, such as a department number, is modified, a foreign key error occurs if we still have products assigned to that department. Triggers provide a way to avoid this error and accomplish a cascade update action. An example of this is demonstrated after this table.
Derived data	We may have columns that hold values that are derived from using other columns in a calculation. For example, Brewbean's may have a product sales summary table that holds the total quantity and dollar sales by product. If this table needs to be updated in real time, then a trigger could be used. Every time a new sale is recorded, the trigger would fire and add the new sales amounts to the totals in the sales summary table.
Security	Additional checks on database access, such as a simple check on the time of user logon, can be accomplished. Some companies use a trigger to determine if it is a weekend day; if so, access is denied. In this case, the company identifies any weekend access as suspicious. (Don't we wish all companies were like this?!!)

Let's look at an example of the referential integrity enforcement using triggers. At times, Brewbean's changes the department numbers and wants to be sure that products assigned to the department are changed accordingly. Currently, because there is an existing foreign key constraint on the department values between the product and department tables, any department number changes require the addition of the new department number, modification of the department number on each of the associated product rows, and then a deletion of the original department row. This process is a pain and a trigger is requested to simplify this department number modification process.

To create the referential integrity trigger:

1. Open or return to SQL*Plus. First, attempt the UPDATE as shown in Figure 8-17 to verify the foreign key error issue.

2. Navigate to and select the **c8refint.sql** file from the Chapter.08 folder.

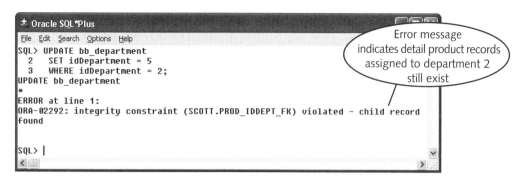

Figure 8-17   Verify foreign key constraint issue

3. Click **Open** and a copy of the file contents appears in the SQL*Plus screen. Review this trigger code, as shown in Figure 8-18.

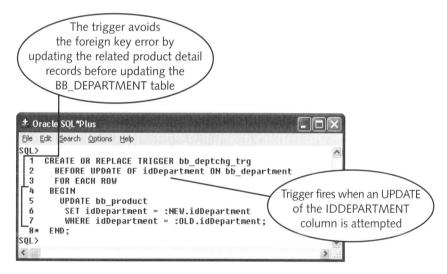

Figure 8-18   Trigger to avoid foreign key error

4. To run the code and create the trigger, click **File** and then **Run** from the main menu.

5. Now try the UPDATE statement we attempted earlier to see if the error is resolved. Type the UPDATE statement and queries to verify the data changes, as shown in Figure 8-19.

6. Be sure to do a rollback to undo these changes. Type **ROLLBACK;** and then press **Enter**. Verify the data is returned to the original state of IDDEPARTMENT 2 with the same queries we used to check the data changes.

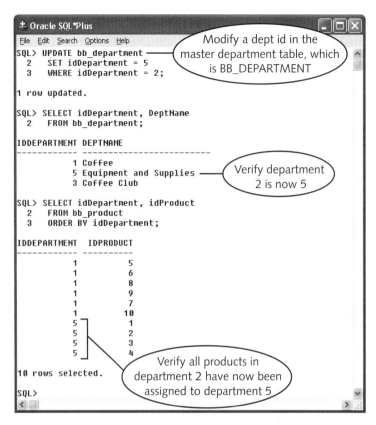

Figure 8-19   Testing the referential integrity trigger

RESTRICTIONS OF TRIGGER USE INCLUDING MUTATING TABLES

A developer needs to be aware of a few limitations that exist in the use of triggers. One prominent limitation is that triggers cannot issue transaction control statements of COMMIT, ROLLBACK, and SAVEPOINT. If you do include a transaction control statement, the trigger compiles fine but then produces a runtime error when the trigger is fired. This limitation also includes any subprogram code called by a trigger with the exception of a subprogram that is declared as an autonomous transaction.

Two other limitations that exist deal with handling specific data types. No LONG or LONG RAW variables can be declared in triggers and the NEW and OLD qualifiers cannot refer to these types of columns. LOB and OBJECT columns can be referenced in a trigger but cannot be modified.

Probably one of the most common problems we hit when we begin creating triggers is the mutating table issue. A **mutating table** is a table that is being modified by a DML action when a trigger is fired. For example, let's say the Brewbean's application has some screens available that allow the manager to enter in sales periods for products. The BB_PRODUCT table contains the SALESTART, SALEEND, and SALEPRICE columns that indicate when the current or last sale period of the item occurred. When new sales period information is entered with an UPDATE to the BB_PRODUCT table, we need a trigger to check to make sure that the sales period being entered does not overlap the time period of the current sale period information for that product. Thus, if the current information for an item indicates a sale running from 6/1/04 to 6/15/04 and we attempt to enter a new sales period of 6/10/04 to 6/20/04, we would want to stop the transaction because we already have a special running for that item on 6/10/04.

What does this trigger need to do to accomplish the task of checking the current sales period? We need to query the BB_PRODUCT table to retrieve the current SALESTART and SALEEND dates to use for comparison. This is where the trouble begins. The trigger is fired by an UPDATE on the BB_PRODUCT table and, therefore, the BB_PRODUCT table is considered to be a mutating table by the trigger. That is, the table modification process has already started. Yet, we want to do a query on this same table. Let's create this trigger and see what happens.

8

To create the trigger:

1. Open or return to SQL*Plus.

2. Navigate to and select the **c8salestrig.sql** file from the Chapter.08 folder.

3. Click **Open** and a copy of the file contents appears in the SQL*Plus screen. Review this trigger code, as shown in Figure 8-20.

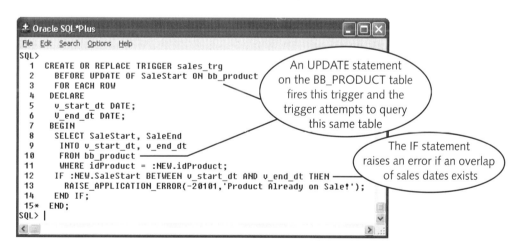

Figure 8-20   Trigger to test the mutating table issue

4. To run the code and create the trigger, click **File** and then **Run** from the main menu.

Now that we have the trigger in place, let's test it.

To test the trigger:

1. Open or return to SQL*Plus and query the BB_PRODUCT table, as shown in Figure 8-21 to list the current sales information for product 6.

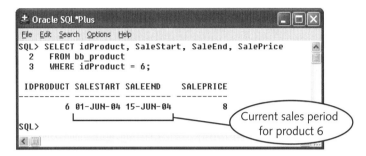

Figure 8-21 Current sales information for product 6

2. Now attempt an UPDATE to enter a new sales period for product 6, as shown in Figure 8-22. Note the error regarding table mutation.

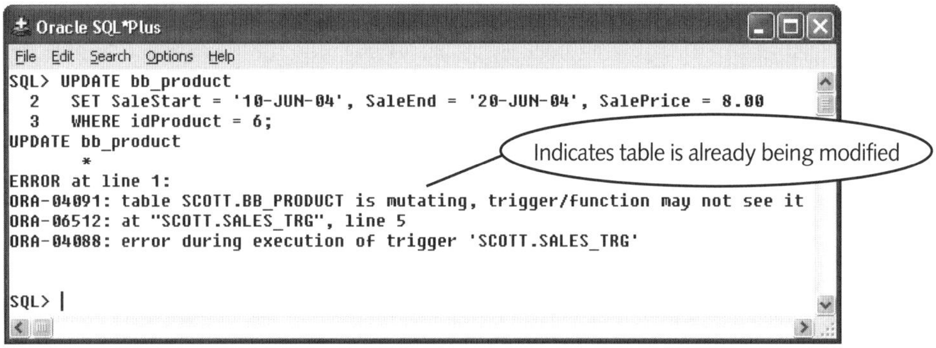

Figure 8-22 DML action resulting in a mutating table error

Because the system was currently in the midst of an UPDATE on the BB_PRODUCT table when the trigger fired and the trigger attempted to query the same table, the table mutation error was raised. This issue is specific to row level triggers only because a statement level trigger occurs in its entirety either before or after the triggering DML statement.

So, is there a way to work around this issue? We can use two triggers to avoid the table mutation problem in our example. Because we could do a query on the BB_PRODUCT table in a statement level trigger, we could add a statement level trigger with this query and put the values retrieved into package variables. Then, the row level trigger, which checks for sales period date overlapping, could reference these package variables rather than doing the query itself. Let's try it.

To work around the table mutation error:

1. Open or return to SQL*Plus. Enter the following command to execute the **c8mutate.sql** file that will create the objects needed. Figure 8-23 outlines the object creation accomplished with this file.

   `@<pathname to PL/SQL files>\Chapter.08\c8mutate`

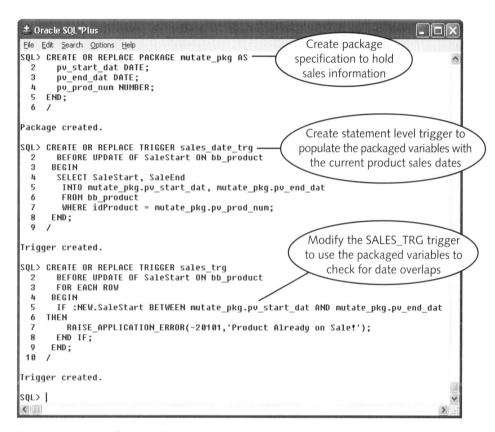

Figure 8-23   Code execution

2. Before the application code submits the UPDATE on sales dates, the packaged variable named MUTATE_PKG.PV_PROD_NUM needs to be populated with the product id so that the statement level trigger can retrieve the proper data. Enter the following code to initialize the packaged variable:

```
BEGIN
 mutate_pkg.pv_prod_num := 6;
END;
/
```

3. Now, we are ready to test our work. Enter the DML statement, as shown in Figure 8-24.

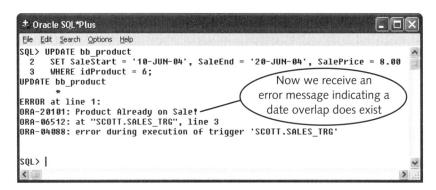

Figure 8-24 Testing the sales date triggers

In Figure 8-24, notice that the table mutation issue is no longer a problem. Both the statement level and row level triggers fired and were able to check for a sales date overlap. We received the error message "ORA-20101: Product Already on Sale!" as coded in our trigger because we have a date overlap situation. The RAISE_APPLICATION_ ERROR statement is used in the trigger to stop the processing of the UPDATE statement. In this case, the UPDATE to product 6 was stopped as we intended.

You may wonder why the :NEW.IDPRODUCT correlation identifier from the UPDATE was not used in the statement level trigger to query the correct sales dates versus initializing the packaged variable identifying the product id as we did in the example. Recall that correlation identifiers are only available with a row level trigger, not a statement level trigger.

Here are two final notes regarding the mutating table issue:

- First, a table in an INSERT statement that fires a row level trigger affecting the same table is not considered a mutating table unless the INSERT includes a subquery.

- Second, there are restrictions regarding constraining tables. A **constraining table** is a table that is referenced via a foreign key constraint on the table that a trigger is modifying. These tables may need to be queried by the trigger to enforce the referential integrity constraint. Therefore, the trigger body cannot read or modify the primary, unique, or foreign key columns of a constraining table.

THE **ALTER TRIGGER** STATEMENT

The ALTER TRIGGER statement is used to recompile, enable, or disable a trigger. If a trigger body references a database object that is modified, the trigger's status is set to INVALID and must be recompiled. The following statement recompiles a trigger named SALES_TRG:

```
ALTER TRIGGER sales_trg COMPILE;
```

Another use of the ALTER TRIGGER statement is disabling and enabling a trigger. Disabling a trigger puts it on hold. It remains stored on the database but it does not fire. Why would we disable a trigger? Sometimes, a situation arises that you may need to import a large amount of data into your database. If you know the data is already clean, or in other words, complies with all your business rules, you can save a lot of processing time by disabling triggers—and constraints for that matter! Keep in mind that each row inserted is checked by every applicable constraint and fires off associated triggers. If you are inserting 500,000 rows, this takes a lot of processing.

To disable or enable a specific trigger, use the following statement:

```
ALTER TRIGGER trigger_name DISABLE|ENABLE;
```

If you want to enable or disable all triggers associated to a table, you can use the ALTER TABLE statement:

```
ALTER TABLE table_name DISABLE|ENABLE ALL TRIGGERS;
```

8

DELETE A TRIGGER

As with all database objects, the DROP statement can be used to delete an object from the database. The DROP statement in the following code is used to delete a trigger.

```
DROP TRIGGER trigger_name;
```

There is a striking difference in regards to how dropping associated database objects affects triggers versus subprograms. Recall that a subprogram may use database objects but is not tied or associated to a specific object. This is unlike how DML triggers are tied to either a table or view. If a table that a procedure used is dropped, the procedure remains on the system but is marked with a status of INVALID. However, if a table or view that a DML trigger is built on is dropped, the trigger is dropped as well.

DATA DICTIONARY INFORMATION FOR TRIGGERS

We can refer to many of the same data dictionary tables we use with subprograms to retrieve information regarding triggers. These tables include USER_OBJECTS, USER_DEPENDENCIES, and USER_ERRORS. However, the source code of your triggers is not found in the USER_SOURCE table that is used for subprograms. Instead, for triggers, a data dictionary table named USER_TRIGGERS exists and maintains details regarding all your triggers. Table 8-3 lists some of the columns and descriptions of the USER_TRIGGERS table.

Table 8-3   Columns in the USER_TRIGGERS table

Column Name	Description
TRIGGER_NAME	Name assigned
TRIGGER_TYPE	Timing (such as BEFORE) and row or statement level
TRIGGERING_EVENT	Applicable DML, DDL, or system events
TABLE_NAME	Table or view name with which the trigger is associated
BASE_OBJECT_TYPE	Type of object associated with the trigger: table, view, schema, or database
WHEN_CLAUSE	Code if a WHEN clause is used
STATUS	Enabled or disabled
DESCRIPTION	Summary of type and event information
TRIGGER_BODY	PL/SQL block in the trigger

A simple query retrieving the DESCRIPTION and TRIGGER_BODY columns, as shown in Figure 8-25, produces a nice listing of details for a trigger. Be sure to set the LONG value in SQL*Plus as shown because the TRIGGER_BODY column is of type LONG and is cut off on the screen if not lengthened.

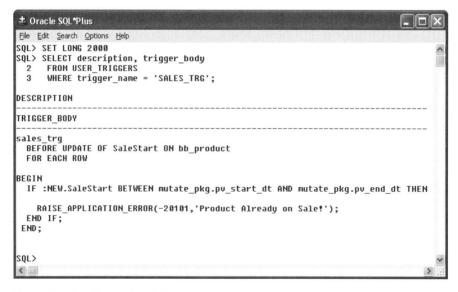

Figure 8-25   Query USER_TRIGGERS

CHAPTER SUMMARY

❒ A database trigger is a PL/SQL block that is executed based on a defined event, such as an INSERT or LOGON action.

❑ Database triggers fire automatically based on an event.

❑ The CREATE TRIGGER statement includes timing options, such as BEFORE or AFTER, and identifies row level versus statement level.

❑ Row level triggers fire for each row affected by a DML event. A statement level trigger fires only once for a DML event, regardless of the number of rows affected by the DML event.

❑ The CREATE TRIGGER statement includes an optional WHEN clause that can check a condition to determine if the trigger should be executed.

❑ Correlation identifiers of OLD and NEW allow original and new row values of a DML action to be referenced.

❑ Conditional predicates of INSERTING, UPDATING, and DELETING allow different PL/SQL code to be processed within a trigger for different DML events.

❑ An Instead-Of trigger is based on a view and enables DML activity through the view.

❑ System triggers are fired on DDL statements and system events.

❑ A mutating table error is raised if a DML statement fires a trigger that attempts to reference the same table that is involved in the DML event that fired the trigger.

❑ The ALTER TRIGGER statement allows a trigger to be compiled or ENABLED/DISABLED.

❑ To remove a trigger from the system, use the DROP TRIGGER statement.

❑ The USER_TRIGGERS data dictionary view contains all the trigger header information and source code.

8

REVIEW QUESTIONS

1. Which of the following is a trigger? (Choose all that apply.)

   a. SYSTEM

   b. TRANSACTION

   c. DML

   d. INSTEAD-OF

2. If a trigger needs to fire when an UPDATE is executed on the CK_DATE column of the table named BB_JUNK and needs to perform an INSERT operation on another table only if the UPDATE that fires the trigger succeeds, which event would be needed in this trigger?

   a. `BEFORE UPDATE ON bb_junk`

   b. `AFTER UPDATE ON bb_junk`

   c. `BEFORE UPDATE OF ck_date ON bb_junk`

   d. `AFTER UPDATE OF ck_date ON bb_junk`

3. An Instead-Of trigger can be attached to a _____ database object.

    a. view

    b. table

    c. sequence

    d. trigger

4. A table mutation error can occur only with which type of trigger?

    a. statement level

    b. row level

    c. system level

    d. none of the above

5. Which statement would we use to disable a trigger?

    a. `ALTER TABLE`

    b. `MODIFY TRIGGER`

    c. `ALTER TRIGGER`

    d. None of the above; the trigger must be dropped to be disabled.

6. What data dictionary view contains the source code for a trigger?

    a. USER_OBJECTS

    b. USER_TRIGGERS

    c. USER_SOURCE

    d. USER_TRIGGER_CODE

7. Which events are available in a DML trigger? (Choose all that apply.)

    a. INSERT

    b. UPDATE

    c. DELETE

    d. SELECT

8. Which timing option is available in an Instead-Of trigger? (Choose all that apply.)

    a. AFTER

    b. BEFORE

    c. Instead-Of

    d. none of the above

9. Conditional predicates allow triggers to _____.

    a. identify different actions for different events

    b. refer to original and new values in an UPDATE

    c. add a WHEN clause to the trigger

    d. add a condition that must be met before the trigger fires

10. Correlation identifiers allow triggers to _____ .

  a. identify different actions for different events

  b. refer to original and new values in an UPDATE

  c. add a WHEN clause to the trigger

  d. add a condition that must be met before the trigger fires

11. What clause can be used to change the names of correlation identifiers?

  a. INSTEAD-OF

  b. REFERENCING

  c. WHEN

  d. None of the above; these names cannot be changed.

12. System triggers fire on _____ . (Choose all that apply.)

  a. database events such as LOGON

  b. DDL statements such as ALTER TABLE

  c. DML statements such as UPDATE

  d. transaction control statements such as COMMIT

13. A trigger is different from a procedure in that it _____ .

  a. does not need to be explicitly executed

  b. needs to be explicitly executed

  c. contains a PL/SQL block

  d. can refer to packaged variables

14. An AFTER UPDATE trigger can be attached to a _____ .

  a. table

  b. view

  c. procedure

  d. DDL event

15. Which correlation identifier is available and contains values in an INSERT trigger?

  a. OLD

  b. NEW

  c. REFERENCE

  d. none of the above

16. Explain the mutating table issue as it relates to database triggers.

17. Describe two typical uses or applications of triggers.

18. What are the fundamental differences between a program unit (procedure and function) and a database trigger?

19. The NEW correlation identifiers contain no values in association with a DELETE event. Explain why this is true.

20. Explain what happens to the program units and triggers associated with a table when the table is dropped.

ADVANCED REVIEW QUESTIONS

1. If the following statement is executed, what happens to the triggers that exist for this table?

   ```
   DROP TABLE bb_junk;
   ```

   a. The triggers are disabled.

   b. The triggers are dropped.

   c. The triggers are not affected.

   d. The triggers' status is marked as INVALID.

2. The following two triggers have been created. When an UPDATE statement is issued on the BB_JUNK table, which trigger fires first?

   ```
   CREATE TRIGGER trig_1
     BEFORE UPDATE OF ck_date ON item_sale
   FOR EACH ROW
   BEGIN
    CALL proc_1
   END;
   CREATE TRIGGER trig_2
     BEFORE UPDATE OF ck_date ON item_sale
   BEGIN
    CALL proc_2
   END;
   ```

   a. TRIG_1

   b. TRIG_2

   c. PROC_2

   d. cannot be determined

3. Review the following trigger header. Select all of the following statements that delete the trigger from the database.

   ```
   CREATE TRIGGER trig_1
     BEFORE DELETE ON item_sale
   ```

   a. ALTER TABLE item_sale DELETE ALL TRIGGERS;

   b. ALTER TABLE item_sale DROP ALL TRIGGERS;

   c. DROP TRIGGER trig_1;

   d. DELETE TRIGGER trig_1;

4. The row level trigger is the only type of trigger that can include _____.
   (Choose all that apply.)

   a. a WHEN clause

   b. conditional predicates

   c. correlation identifiers

   d. an OF column clause

5. If a trigger were fired from an INSERT statement, which situations would raise a
   mutating table error if the trigger includes a DML action on the same table used
   in the INSERT statement?

   a. The table is in the midst of being modified.

   b. The trigger is a statement level trigger.

   c. The trigger is a row level trigger.

   d. The trigger is a row level trigger and the INSERT includes a subquery.

HANDS-ON ASSIGNMENTS

Assignment 8-1: Create a Trigger to Address Product Restocking

Brewbean's has a couple columns in the product table to assist in inventory tracking.
REORDER is one of the columns and contains the stock level at which the product
should be reordered. If the stock falls to this level, Brewbean's wants the application to
automatically insert a row into the BB_PRODUCT_REQUEST table to alert the
ordering clerk that additional inventory is needed. Brewbean's currently uses the reorder
level amount as the quantity that should be ordered. This task can be accomplished using
a trigger.

1. Take out a piece of scrap paper and pencil. Think about the tasks the triggers
   needs to accomplish, including checking if the new stock level falls below the
   reorder point. If so, then check if the product is already on order by checking the
   product request table; if not, then enter a new product request. Try to write out
   the trigger code on the scratch paper. Even though we learn a lot by reviewing
   code, we really become skilled when we can create the code on our own.

2. Open or return to SQL*Plus.

3. Navigate to and select the **c8reorder.sql** file from the Chapter.08 folder.

4. Click **Open** and a copy of the file contents appears in the SQL*Plus screen.
   Review this trigger code; how does it compare to your code?

5. To run the code and create the trigger, click **File** and then **Run** from the
   main menu.

6. Let's test the trigger on product 4. First, run the query shown in Figure 8-26 to
   verify the current stock data for this product. Notice a sale of one more should
   initiate a reorder.

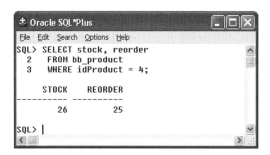

Figure 8-26   Checking the stock data

7. Run an UPDATE, as shown in Figure 8-27. This should cause the trigger to fire. Notice the query to check if the trigger fired and if a product stock request was inserted into the BB_PRODUCT_REQUEST table.

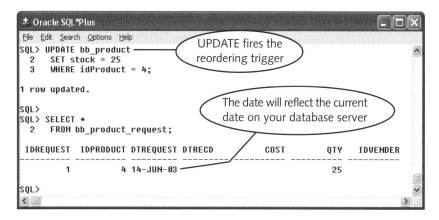

Figure 8-27   UPDATE the stock level for product 4

8. Be sure to issue a rollback to undo these DML actions to restore data to the original state for use in later projects. Type **ROLLBACK;** in SQL*Plus and then press **Enter**.

9. Also, be sure to run the following statement to disable this trigger so that it does not affect other projects:

```
ALTER TRIGGER bb_reorder_trg DISABLE;
```

Assignment 8-2: Update Stock Information When a Product Request Is Filled

Brewbean's has a BB_PRODUCT_REQUEST table in which requests to refill stock levels are automatically inserted via a trigger. After the stock level falls below the reorder level, this trigger fires and enters a request into the table. This works great; however, when a store clerk records that the product request has been filled by updating the DTRECD and COST columns of the request, he or she wants the stock level in the product table to be updated. Create a trigger named BB_REQFILL_TRG to accomplish this task using the following steps as a guideline:

1. Open or return to SQL*Plus and run the following INSERT to create a product request that we can use in this project:

```
INSERT INTO bb_product_request (idRequest, idProduct,
                 dtRequest, qty)
   VALUES (3, 5, SYSDATE, 45);
   COMMIT;
```

2. Create the trigger (BB_REQFILL_TRG) so that it fires when a date that is received is entered into the product request. This trigger needs to modify the STOCK column in the BB_PRODUCT table to reflect the increased inventory.

3. Now test the trigger. First, query the stock and reorder data for product 5, as shown in Figure 8-28.

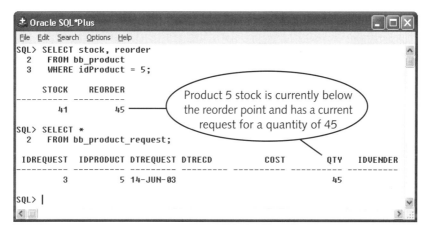

Figure 8-28   Query the data for product 5 stock and reorder amount

4. Now update the product request to record it as fulfilled using the UPDATE, as shown in Figure 8-29.

5. Perform data queries to verify that the trigger did fire and the stock level of product 5 has been appropriately modified. Then, issue a ROLLBACK statement to undo the data modifications.

6. If you are not doing Assignment 8-3, then disable the trigger so that it does not affect other assignments.

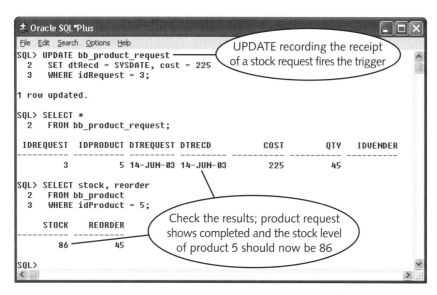

Figure 8-29   UPDATE the product request

Assignment 8-3: Update the Stock Level If a Product Fulfillment Is Cancelled

Brewbean's has really made some progress on their inventory handling processes; however, they hit a snag when a store clerk incorrectly recorded a product request as fulfilled. When the product request was updated to record a DTRECD value, the stock level of the product was automatically updated via an existing trigger named BB_REQFILL_TRG. If the clerk empties out the DTRECD column to indicate the product request has not been actually filled, then the product stock level needs to be corrected or reduced as well. Modify the BB_REQFILL_TRG trigger to resolve this problem.

1. Open or return to SQL*Plus. Modify the trigger code as needed. Add code to check if the DTRECD column already has a date in it and it is now being set to NULL.

2. Submit the following DML actions to create and update rows that we can use to test the trigger:

```
INSERT INTO bb_product_request (idRequest, idProduct,
            dtRequest, qty, dtRecd, cost)
  VALUES (4, 5, SYSDATE, 45, '15-JUN-03',225);
UPDATE bb_product
  SET stock = 86
  WHERE idProduct = 5;
COMMIT;
```

3. Execute the following UPDATE statement to test the trigger. Also, perform the appropriate queries to verify the data has been correctly modified.

```
UPDATE bb_product_request
SET dtRecd = NULL
WHERE idRequest = 4;
```

4. Be sure to run the following statement to disable this trigger so that it does not affect other projects:

```
ALTER TRIGGER bb_reqfill_trg DISABLE;
```

Assignment 8-4: Update Stock Levels When an Order Is Cancelled

At times, customers make mistakes in submitting their orders and call to cancel the order. Brewbean's wants to create a trigger that automatically updates the stock level of all products associated with a cancelled order and that updates the ORDERPLACED column of the BB_BASKET table to zero, reflecting that the order was not completed. Create a trigger named BB_ORDCANCEL_TRG to accomplish this task by considering the following items:

❑ The trigger needs to fire when a new status record is added to the BB_BASKETSTATUS table and when the IDSTAGE column is set to 4, which indicates order cancellation.

❑ Each basket can contain multiple items in the BB_BASKETITEM table, so a CURSOR FOR loop may be an appropriate mechanism to update the stock level of each item.

❑ Keep in mind that coffee can be ordered in half or whole pounds.

❑ Use basket 6 for testing. Note that this basket contains two items.

1. Open or return to SQL*Plus. Execute the following INSERT to test the trigger:

```
INSERT INTO bb_basketstatus (idStatus, idBasket, idStage,
                             dtStage)
VALUES (bb_status_seq.NEXTVAL, 6, 4, SYSDATE);
```

2. Execute queries to confirm that the basket order status and product stock levels have been appropriately modified by the trigger.

3. Be sure to run the following statement to disable this trigger so that it does not affect other projects:

```
ALTER TRIGGER bb_ordcancel_trg DISABLE;
```

Assignment 8-5: Process Discounts

Brewbean's is implementing a new discount for return shoppers—every fifth completed order should receive a 10% discount. The count of orders for a shopper is placed in a packaged variable named PV_DISC_NUM during the ordering process. This count needs to be tested upon check out to determine if a discount should be applied. Create a trigger named

BB_DISCOUNT_TRG so that when an order is confirmed (the ORDERPLACED value is set from 0 to 1), the PV_DISC_NUM PACKAGED variable is checked. If it is equal to 5, then set a second host variable named PV_DISC_TXT to 'Y'. This variable is used in calculating the order summary so that a discount is applied if appropriate.

Create a package specification named DISC_PKG that contains the needed packaged variables. Use an anonymous block to initialize the packaged variables to use for testing the trigger. Test the trigger by using the following UPDATE:

```
UPDATE bb_basket
  SET orderplaced = 1
  WHERE idBasket = 13;
```

Reset the ORDERPLACED to zero for the basket used (this is done as needed to complete trigger testing). Also, disable this trigger when completed so that it does not affect other assignments.

Assignment 8-6: Use Triggers to Maintain Referential Integrity

At times, Brewbean's has changed the id number for existing products. In the past, they have had to add a new product row with the new id to the BB_PRODUCT table, modify all the corresponding BB_BASKETITEM and BB_PRODUCTOPTION table rows, and then delete the original product row. Can a trigger be developed to avoid all these steps and handle the UPDATE of the BB_BASKETITEM and BB_PRODUCTOPTION table rows automatically for a given change in product id? If so, create the trigger and test by issuing an UPDATE statement, which changes the IDPRODUCT of 7 to 22. Do a rollback to return the data back to its original state. Also, disable the new trigger after you have completed the assignment.

Assignment 8-7: Update Summary Data Tables

The Brewbean's owner uses several summary sales data tables on a daily basis to monitor business activity. The table BB_SALES_SUM holds the product id, total $ sales, and total quantity sold for each product. A trigger is needed so that every time an order is confirmed, or the ORDERPLACED column is updated to 1, the BB_SALES_SUM table is updated accordingly. Create a trigger named BB_SALESUM_TRG that accomplishes this task. Before testing, go to SQL*Plus and reset the ORDERPLACED column to 0 for basket 3, as shown in the following code, and use this basket to test the trigger.

```
UPDATE bb_basket
  SET orderplaced = 0
  WHERE idBasket = 3;
```

Note that the BB_SALES_SUM table already contains some data. Test the trigger by executing the following UPDATE in SQL*Plus. Also, confirm that the trigger is working correctly.

```
UPDATE bb_basket
  SET orderplaced = 1
WHERE idBasket = 3;
```

Complete a rollback and disable the trigger when completed so that it does not affect other assignments.

Assignment 8-8: Maintain an Audit Trail of Product Table Changes

The accuracy of product table data is critical and the Brewbean's owner wants to have an audit file that contains information regarding all DML activity on the BB_PRODUCT table. This information should include the user id of the user issuing the DML action, the date, the original values of the changed row, and the new values. This audit table needs to track specific columns of concern, including PRODUCTNAME, PRICE, SALESTART, SALEEND, and SALEPRICE. Create a table named BB_PRODCHG_AUDIT that can hold the relevant data. Then create a trigger named BB_AUDIT_TRG that fires an update to this table whenever one of the specified columns in the BB_PRODUCT table is changed.

Be sure to issue the following command. If you created the SALES_DATE_TRG trigger in the chapter, it will conflict with this assignment.

```
ALTER TRIGGER sales_date_trg DISABLE;
```

Use the following UPDATE statement to test your trigger. Then, complete a rollback and disable the trigger when completed so that it does not affect other assignments.

```
UPDATE bb_product
  SET salestart = '05-MAY-03',
    Saleend = '12-MAY-03'
    saleprice = 9
WHERE idProduct = 10;
```

 Multiple columns can be listed in an OF clause of a trigger by simply listing them separated by commas.

CASE PROJECTS

Case 8-1: Map the Flow of Database Triggers

One thing we learn soon after beginning to create triggers is that we must be careful to recognize that one statement can end up firing off a number of triggers. Not only can we have multiple triggers attached to a single table, but also a trigger itself can perform DML operations that may affect other tables and fire off additional triggers. At times, we may be able to combine triggers to make our code more clear and maintainable. For example, if we have one trigger for an INSERT on the BB_PRODUCT table and another for an UPDATE on the BB_PRODUCT table, we may want to combine these into one trigger and use conditional predicates. In addition, triggers that execute DML statements that could fire off additional triggers could lead to a table mutation error. However, this error may not be apparent unless we identify the flow of trigger processing and objects involved.

Review the CREATE TRIGGER statements in the c8case1.txt file in the Chapter.08 folder. Prepare a flowchart of which triggers will fire and the firing sequence if a user logs on and issues the following code:

```
UPDATE bb_product
  SET stock = 50
  WHERE idProduct = 3;
```

Case 8-2: More Movies Inventory Processing

The More Movies company needs to be sure the number on hand of each movie is updated for each new rental recorded and returned. The MM_MOVIE table contains a column named MOVIE_QTY that reflects the current number of copies in stock for that movie. Create triggers that update the MOVIE_QTY column for rentals and returns. Name the triggers MM_RENT_TRG and MM_RETURN_TRG, respectively. Test your trigger with the following two DML actions, which add a new rental and update that rental to reflect it has been returned. Perform needed queries to confirm the triggers fired and operated correctly.

```
INSERT INTO mm_rental (rental_id, member_id, movie_id,
payment_methods_id)
 VALUES (13,10, 6, 2);
UPDATE mm_rental
 SET checkin_date = SYSDATE
 WHERE rental_id = 13;
```

ORACLE-SUPPLIED PACKAGES

> **In this chapter, you will:**
> ♦ Use communications packages
> ♦ Generate output via packages
> ♦ Include large objects in the Oracle database
> ♦ Explore dynamic SQL and PL/SQL
> ♦ Identify other important built-in packages

Oracle includes many complete packages called Oracle-supplied or built-in packages that extend the functionality of PL/SQL. To become efficient, every PL/SQL developer needs to become familiar with the built-in packages and take advantage of the capabilities that are already provided.

Each package comes with parameters, just like the ones we construct ourselves. It is helpful to review the individual scripts that create each package, which can be found in the rdbms\admin directory of the Oracle9*i* database software. Each script contains comments documenting the code and can be opened with a text editor. This chapter introduces a handful of commonly used packages. A few are covered in some detail to give you a good foundation on how to analyze and use built-ins. The packages are presented in the following categories: communications, output, large objects, dynamic SQL, and miscellaneous.

THE CURRENT CHALLENGE IN THE BREWBEAN'S APPLICATION

The Brewbean's lead programmer has compiled a list of the most important functionality needs that must be resolved for their coffee Web site application. The list of needs includes: a communication mechanism to assist submitting and verifying credit card information, real-time messages to notify appropriate personnel when product inventory falls below the restocking level, automatic creation of e-mails to provide customer purchase confirmations, ability to read external text files to import product information submitted by vendors, ability to integrate product images into the database to be displayed on the Web site, and automatic submission of programs that can be scheduled to execute during low usage hours.

In talking with colleagues at other companies, the programmers determined much of this functionality already exists within the Oracle-supplied packages. The lead programmer narrowed down a list of packages to be explored. Table 9-1 lists information on each of the packages presented in this chapter. Notice all the built-in package names begin with a prefix of either DBMS_ or UTL_.

Table 9-1   Built-in packages

Chapter Section	Built-in Package Name	Description	Script Filename
"Communications"	DBMS_PIPE	Allows different database sessions to communicate	dbmspipe.sql
"Communications"	DBMS_ALERT	Enables notification of database events	dbmsalrt.sql
"Communications"	UTL_SMTP	Enables e-mail features	utlsmtp.sql
"Communications"	UTL_HTTP	Enables HTML retrieval	utlhttp.sql
"Communications"	UTL_TCP	Enables TCP/IP communications	utltcp.sql
"Generating Output"	DBMS_OUTPUT	Displays data to the screen	dbmsotpt.sql
"Generating Output"	UTL_FILE	Reads and writes data to external files	utlfile.sql
"Large Objects"	DBMS_LOB	Creates and manipulates LOBs within the database	dbmslob.sql
"Dynamic SQL and PL/SQL"	DBMS_SQL	Constructs and parses statements at runtime	dbmssql.sql
"Miscellaneous Packages"	DBMS_JOB	Schedules jobs	dbmsjob.sql
"Miscellaneous Packages"	DBMS_DDL	Enables access to DDL statements not allowed directly from within PL/SQL	dbmsdesc.sql

REBUILDING YOUR DATABASE

Complete the following steps to rebuild the Brewbean's database for this chapter.

To rebuild the Brewbean's database:

1. Locate the **c9Dbcreate.sql** file in the Chapter.09 directory to ensure it exists. This file contains the script to create the database.

2. Open SQL*Plus and log on.

3. Create a spool file so that SQL*Plus keeps a copy of all the messages received from running the file that creates the database. On the main menu in SQL*Plus, click **File**, point to **Spool**, and then click **Spool File**. A Select File dialog box will appear.

4. Browse to a folder used to contain your working files and in the File name text box, type **DB_log9**, and then click **Save**. Now, whatever text is seen in our SQL*Plus session is saved to this file for future reference.

5. Now let's create the database. In SQL*Plus, enter the following command, which runs all the statements contained in the c9Dbcreate.sql file. Messages verifying the creation and data insertion steps will scroll on the SQL*Plus screen. This may take a couple minutes to complete.

   ```
   @<pathname to PL/SQL files>\Chapter.09\c9Dbcreate
   ```

6. Now, we will turn the spooling off so that we can review the results in the spool file created. On the main menu, click **File**, point to **Spool**, and then click **Spool Off**.

7. Open a text editor.

8. Open the file named **DB_log9.lst** in your working folder.

9. Review the messages for any errors.

COMMUNICATIONS

Communications technology plays an increasingly critical role in systems as we depend more and more on real-time data. For example, online auctions have grown increasingly popular and depend on messaging capabilities to provide updated bid information to those participating. In addition, online merchants depend on generating order confirmation e-mails automatically to send to customers. The following sections cover built-in packages dealing with communications.

DBMS_PIPE Package

The DBMS_PIPE package lets two or more sessions in the same instance communicate. When the Oracle database is started, a system global area (SGA) or memory buffers are allocated and background processes are started; they both make up an Oracle instance. Oracle pipes are mechanisms that can send information through buffers in the SGA and, therefore, are lost when the instance is shut down.

An important aspect of the Oracle pipes is that they are asynchronous or, in other words, they operate independent of transactions. One session packs a message into a buffer and sends it, whereas another session receives and unpacks the message. DBMS_PIPE can transmit messages of the following data types: VARCHAR2, NUMBER, DATE, RAW, and ROWID.

The DBMS_PIPE package is used quite often to interface with service routines in the host operating environment or external to Oracle. Online credit card approval is one example of this type of use that Brewbean's might implement on their Web site. Other example uses of DBMS_PIPE, which are listed in the Oracle documentation, include the following:

- To complete independent transactions

- To communicate alerts such as database changes in a situation in which the data needs to be automatically refreshed when a modification occurs

- To enable stored program units to send debugging information through a pipe to be displayed or written to a file

- To concentrate groups of user transactions through a single session to improve performance

Sending a message through a pipe is a two-step process: (1) packing the message and then (2) sending the message. To accomplish these tasks, the DBMS_PIPE package contains the PACK_MESSAGE procedure and SEND_MESSAGE function.

Receiving a message is also a two-step process that includes the receiving and unpacking of the message. The package contains the RECEIVE_MESSAGE function and the UNPACK_MESSAGE procedure to receive the message from a pipe. For a simple demonstration of the Brewbean's credit card number submission process, let's send a message containing a number from one SQL*Plus session to another.

To use DBMS_PIPE:

1. Open or return to SQL*Plus. Log on as **SYSTEM**, **MANAGER**. We use this logon because the SYSTEM user automatically has privileges to all the built-in packages in the Oracle database installation.

Your instructor will indicate which logon to use depending on your configuration.

2. Type **SET SERVEROUTPUT ON**, and then press **Enter**.

3. Start a second SQL*Plus session by logging on as **SYSTEM**, **MANAGER** again.

4. Type **SET SERVEROUTPUT ON**, and then press **Enter**.

5. Return to the first SQL*Plus session. Click **File**, and then click **Open**.

6. Navigate to and select the **pipesend09.sql** file from the Chapter.09 folder.

7. Click **Open** and a copy of the file contents appears on the SQL*Plus screen. Notice the anonymous block uses PACK_MESSAGE and SEND_MESSAGE from the DBMS_PIPE package.

8. To run the code and create the procedure, click **File** and then **Run** from the main menu.

9. Return to the second SQL*Plus session. Click **File**, and then click **Open**.

10. Navigate to and select the **piperec09.sql** file from the Chapter.09 folder.

11. Click **Open** and a copy of the file contents appears in the SQL*Plus screen. Notice the anonymous block uses UNPACK_MESSAGE and RECEIVE_MESSAGE from the DBMS_PIPE package.

12. To run the code and create the procedure, click **File** and then **Run** from the main menu. Your screen should resemble Figure 9-1.

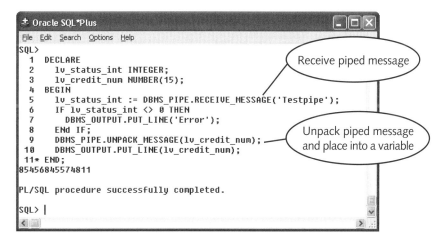

Figure 9-1   Receiving a message via an Oracle pipe

13. Close the second session of SQL*Plus.

In this example, the pipe is created implicitly with the first statement using the DBMS_PIPE package, which is the PACK_MESSAGE step. This creates a public pipe with few restrictions on access. If privacy is a concern with the message, then a private pipe can be explicitly created using the CREATE_PIPE function in the package. A private pipe restricts access to sessions running under the same user id as the creator of the pipe, stored subprograms executing in the same user id privilege domain as the pipe creator, and users connected as SYSDBA or INTERNAL. Table 9-2 lists some of the program units contained in the DBMS_PIPE package along with brief descriptions.

Table 9-2   DBMS_PIPE program units

Program Unit Name	Description
CREATE_PIPE	Creates a pipe; must be used to start a private pipe
PACK_MESSAGE	Puts message in buffer
SEND_MESSAGE	Sends message on a named pipe and creates a public pipe automatically if the named pipe does not exist
RECEIVE_MESSAGE	Copies message in a pipe to the buffer
REMOVE_PIPE	Deletes a named pipe
PURGE	Clears the contents of a named pipe
UNPACK_MESSAGE	Reads message in buffer

DBMS_ALERT Package

The DBMS_ALERT package allows notification of database events to be sent to interested users when the event occurs. For example, if you monitor an online auction, you might desire to see a new high bid when it is placed. Rather than having the system continuously poll the data to check for changes, an alert can be accomplished in a database trigger. The application allowing users to participate in the bid would register to receive alerts. Alerts are transaction based and are sent only if the transaction is committed.

The Brewbean's developers will use the DBMS_ALERT package to send alerts regarding product stock levels falling below the reorder point. This will be accomplished in Hands-on Assignment 9-2. Note that granting EXECUTE privileges to general users and a package synonym are not automatically accomplished in the package created during installation.

The basic mechanics of the alert process include registering an alert name, setting up when an alert should signal, and enabling the appropriate users to receive the alert. These tasks can be accomplished with the REGISTER, SIGNAL, and WAITONE procedures of the DBMS_ALERT package.

Let's look at some example statements to set up an alert. First, the following code uses the REGISTER procedure to set up an alert name. This procedure has only one IN parameter, which is a VARCHAR2 data type.

```
DBMS_ALERT.REGISTER('new_bid');
```

Now, we need to set up the firing of a signal for this alert. Normally, this would be in an AFTER INSERT or UPDATE database trigger to capture the data change. The SIGNAL procedure has two IN parameters of VARCHAR2 data type. The first parameter is the name of the alert that must match one that has been identified with the REGISTER procedure. The second parameter is the message to send with the alert, such as the new bid in our online auction example. The following statement is placed within a trigger and references the alert name we created earlier:

```
DBMS_ALERT.SIGNAL('new_bid', TO_CHAR(:new.bid));
```

Finally, our last step is to register the user to receive this particular alert. The WAITONE procedure can accomplish this task and has the following header containing four parameters:

```
DBMS_ALERT.WAITONE (
    name      IN    VARCHAR2,
    message   OUT   VARCHAR2,
    status    OUT   INTEGER,
    timeout   IN    NUMBER DEFAULT MAXWAIT);
```

The NAME parameter references a specific alert name or NEW_BID in our example. The MESSAGE parameter returns the message put into the alert signal. The STATUS parameter returns a 0 when an alert is received and a 1 for timing out. The TIMEOUT parameter is in terms of seconds and determines how long the session will wait for an alert signal before returning a timeout status. To indicate a desire to receive new bid messages, the user application needs to issue a statement, as shown in the following code:

```
DBMS_ALERT.WAITONE('new_bid', v_msg, v_status, 600);
```

Note that the V_MSG and V_STATUS variables need to be declared in this block because they will receive the information returned by the OUT parameters of the WAITONE procedure. The DBMS_ALERT package contains several more procedures, such as REMOVE to delete alerts from your system. If you have a need to periodically update viewed information, you should explore this package, because it is more efficient than continually polling data to determine when a change has occurred.

UTL_SMTP Package

The ability to send e-mails from within PL/SQL code has been simplified with the introduction of the UTL_SMTP package. It is easy to think of many reasons for which automatic e-mail notices from the database system would be quite useful. Brewbean's, of course, will use this package to automatically send e-mail order confirmations to customers. Other examples include:

- Sending customers notifications when airline ticket prices on a specified route have changed

- Sending customers notifications when a stock price reaches a particular price
- Notifying appropriate employees when inventory levels of items fall below the reorder point

Simple Mail Transfer Protocol (SMTP) is the protocol used to send e-mail across networks and the Internet. Within this protocol, e-mails basically have three parts to their structure. First, the **envelope** consists of the sender and recipient information or, in other words, the TO and FROM information. Second, the **header** consists of information about the message, such as the date and subject. Third, the **body** consists of the text in the message.

The process of the client communicating to the mail server is called a handshaking process. As the client sends parts of the e-mail, the server replies with a code to notify if the information was received successfully. This process becomes apparent in setting up a PL/SQL program to handle e-mail actions. Table 9-3 lists some of the functions available in the UTL_SMTP package.

Table 9-3   UTL_SMTP functions

Function Name	Description
HELO	Performs initial handshaking to identify the sender to the mail server
MAIL	Initiates a mail transaction that sends messages to mailbox destinations
RCPT	Identifies each of the recipients of an e-mail
DATA	Specifies the lines in the body of an e-mail
RSET	Aborts the current mail transaction
NOOP	Requests a reply from the mail server to verify connection is still alive
QUIT	Terminates the SMTP session and disconnects from the mail server
OPEN_CONNECTION	Opens a connection to the mail server
OPEN_DATA	Sends the DATA command
WRITE_DATA	Adds data to message body
CLOSE_DATA	Ends the message

The only other item for which we need to complete an example is a carriage-return line-feed to terminate each line as it is sent. The UTL_TCP package has a program named CRLF that issues a character sequence indicated to start a new line.

Now, let's review an example of a block that sends an e-mail to two recipients. Note that the following anonymous block contains values directly entered for many items, such as the sender and recipient e-mail addresses. This was done to make the code example easier to follow; however, most of these values will actually be IN parameters in a mail procedure constructed with this same code. Comments within the block of code identify the various steps in completing the e-mail.

```
DECLARE
  conn UTL_SMTP.connection;
  header VARCHAR2(1000);
  sender VARCHAR2(25) := 'jcasteel@tcc.edu';
  rec1 VARCHAR2(25) := 'jcasteel@tcc.edu';
  rec2 VARCHAR2(25) := 'sdavis@odu.edu';
  server VARCHAR2(25) := 'yoursmtpserver.com';
  subject VARCHAR2(35) := 'Testing email from PL/SQL';
  crlf VARCHAR2(2) := UTL_TCP.crlf;
  message VARCHAR2(500) := 'Here is the test email body!';
BEGIN
--Setup the header variable
  header:= 'Date: '||TO_CHAR(SYSDATE,'dd Mon yy
                           hh24:mi:ss')||crlf||
           'From: '||sender||''||crlf||
           'Subject: '||subject||crlf||
           'To: '||rec1||crlf||
           'CC: '||rec2;
--Start the connection
  conn := utl_smtp.open_connection(server);
--Complete handshake with SMTP server
  utl_smtp.helo(conn, server);
  utl_smtp.mail(conn, sender);
  utl_smtp.rcpt(conn, rec1);
  utl_smtp.rcpt(conn, rec2);
  utl_smtp.open_data(conn);
--Write the header info
  utl_smtp.write_data(conn,header);
--Write the body, crlf used to separate from header
  utl_smtp.write_data(conn, crlf || message);
--Close the connection and end transaction
  utl_smtp.close_data(conn);
  utl_smtp.quit(conn);
EXCEPTION
  WHEN UTL_SMTP.INVALID_OPERATION THEN
    dbms_output.put_line(' Invalid Op in transaction.');
  WHEN UTL_SMTP.TRANSIENT_ERROR THEN
    dbms_output.put_line(' Temporary prob - try later.');
  WHEN UTL_SMTP.PERMANENT_ERROR THEN
    dbms_output.put_line(' Errors in code.');
END;
```

Executing the previous block produces the e-mail shown in Figure 9-2. Keep in mind that the server variable must contain a valid and accessible SMTP server. In other words, the value of '*yoursmtpserver.com*' shown in the example needs to be replaced with a valid SMTP server address.

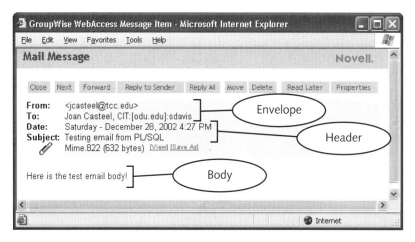

Figure 9-2   E-mail message produced with UTL_SMTP

Before using the UTL_SMTP package features, the Java Virtual Machine must be installed on the database server.

UTL_HTTP Package

The UTL_HTTP package provides the capability to make Hypertext Transfer Protocol (HTTP) calls from within PL/SQL programs. This package can be used to analyze the HTML source of Web sites to accomplish tasks such as tracking competitor pricing on products, gathering pricing for raw materials, and collecting relevant business news for a particular industry. For example, Brewbean's may wish to monitor some of the leading online coffee companies to compare product pricing.

Two functions named REQUEST and REQUEST_PIECES retrieve the HTML source from a specified URL. The REQUEST function retrieves only the first 2,000 bytes of data from the Web page cited. However, the REQUEST_PIECES function returns the entire page, regardless of size, in 2,000 byte segments. This is accomplished by storing each of the segments in a PL/SQL table variable of the HTML_PIECES data type that is also provided in the UTL_HTTP package. Therefore, the REQUEST_PIECES function is typically used to avoid possibly truncating information contained on the Web page.

Let's run an example that Brewbean's will use to track coffee press prices of a leading online competitor, *www.peets.com*. This example uses the UTL_FILE package to write the HTML returned by UTL_HTTP functions to a file to be viewed.

The *www.peets.com* site is an actual online coffee retailer at the time of publication. If this site no longer exists, substitute another available site.

To use UTL_HTTP to check competitor coffee press pricing:

1. Open a Web browser of your choice.

2. Enter the following address in the URL Address box of the browser to view the page listing the coffee presses available: **http://www.peets.com/stor/12/ essentials_items.asp?cat=3003**.

3. Open or return to SQL*Plus. Enter the following command:

```
CREATE DIRECTORY ora_files AS 'c:\oraclass';
```

Users must be granted the correct privileges to create directories.

4. Click **File**, and then click **Open**. Navigate to and select the **http09.sql** file from the Chapter.09 folder.

5. Click **Open** and a copy of the file contents appears in the SQL*Plus screen. Review the code. The code assumes you have a C:\oraclass directory.

6. To run the code and create the procedure, click **File** and then **Run** from the main menu.

7. Execute the procedure using the following statement (which needs to be entered in a single line):

```
EXECUTE
read_http_sp('http://www.peets.com/stor/12/essentials_items
.asp?cat=3003');
```

8. Open the **c:\oraclass\test_http.htm** file in a text editor. You will see the HTML source code for the page we originally viewed in the Web browser.

Additional options may need to be used if you are connected to the Internet through a firewall or desire to access pages via HTTPS. Refer to the *Oracle9i Supplied PL/SQL Packages and Types Reference* for more information.

UTL_TCP Package

Internet communications rely on the TCP/IP protocol, and the UTL_TCP package enables low-level calls using TCP/IP. As a matter of fact, the two packages that enable e-mail and HTTP communications rely on the UTL_TCP package to achieve their capabilities. An understanding of the TCP/IP protocol is necessary to take advantage of this package; however, this topic is beyond the scope of this book. Therefore, this section provides only a brief description of this package.

The UTL_TCP package contains functions to open and close connections with TCP/IP based servers. The package functions of OPEN_CONNECTION and

CLOSE_CONNECTION are used to establish and close connections. Included functions also enable the ability to read or write binary or text data from an open connection. The functions of READ_BINARY and WRITE_BINARY are used to read and write information over the connection. The package uses a socket (an Internet address plus a port number) to establish connections.

GENERATING OUTPUT

Built-in packages offer a variety of methods to create output from within a PL/SQL block. This section covers the two basic and widely used packages: DBMS_OUTPUT and UTL_FILE. DBMS_OUTPUT allows the display of data to the screen, whereas UTL_FILE allows the reading and writing of data to external files.

DBMS_OUTPUT Package

The DBMS_OUTPUT package is used to display messages to the screen from a stored program unit or an anonymous block run in SQL*Plus. We have used the PUT_LINE procedure of this package quite a bit already to display and check values within our PL/SQL block execution in constructing code for the Brewbean's application. As a matter of fact, debugging is probably the most popular use of this package.

When executing a stored program, any DBMS_OUTPUT lines are placed in an output buffer, which dumps its contents to the screen when the program has completed execution. The lines are displayed only upon completed execution and cannot be displayed in real time. However, numerous DBMS_OUTPUT statements can be used in a block, which is quite helpful for checking values at various points in the execution.

 Keep in mind that we must issue the command SET SERVEROUTPUT ON in SQL*Plus to see the display from DBMS_OUTPUT. If we do not set this to ON, no errors will occur and no lines are displayed to the screen.

Table 9-4 lists the procedures contained in the DBMS_OUTPUT package.

Table 9-4   DBMS_OUTPUT procedures

Procedure Name	Description
ENABLE	Allows message display (not necessary if you have SERVEROUTPUT set to ON in SQL*Plus)
DISABLE	Does not allow message display
PUT	Places information in the buffer
PUT_LINE	Places information in the buffer followed by an end-of-line marker
NEW_LINE	Places an end-of-line marker in the buffer
GET_LINE	Retrieves a line from the buffer
GET_LINES	Retrieves an array of lines from the buffer

Let's review each of these procedures briefly. The package is enabled by default, so the ENABLE procedure only needs to be run if a DISABLE procedure has been run previously. This procedure has only one parameter named buffer size and sets the size of the buffer in terms of bytes. A default value of 20000 is set, so no values are required to run the ENABLE procedure. The maximum buffer size allowed is 1 MB. The DISABLE procedure, of course, does the opposite and it turns off the ability to use any of the procedures in the package with the exception of ENABLE. DISABLE has no parameters.

The PUT and PUT_LINE procedures place information into the buffer to be displayed. These procedures accept one value that can be a data type of VARCHAR2, NUMBER, or DATE. If needed, you can use concatenation or data type conversion functions to create a single value from several different values. The significant difference between the two procedures is that PUT_LINE adds a new line marker in the buffer, whereas PUT continues placing information on the same line in the buffer. Let's experiment with using these two procedures.

To use PUT and PUT_LINE:

1. Open or return to SQL*Plus.

2. Type **SET SERVEROUTPUT ON**, if necessary, and then press **Enter** so that the output buffer can be displayed.

3. Enter the following anonymous block:

```
BEGIN
 DBMS_OUTPUT.PUT('This');
 DBMS_OUTPUT.PUT(' is a ');
 DBMS_OUTPUT.PUT('test');
 DBMS_OUTPUT.PUT_LINE('This is test 2');
 DBMS_OUTPUT.PUT('Working?');
END;
/
```

4. Compare the display to Figure 9-3.

5. Add the following line of code in two places: after the third DBMS_OUTPUT.PUT statement, and before the DBMS_OUTPUT.PUT_LINE statement, and as the last line in the block.

```
DBMS_OUTPUT.NEW_LINE;
```

6. Execute the block and compare the display to Figure 9-4. Note how the NEW_LINE procedure forces the display of lines in the buffer.

9

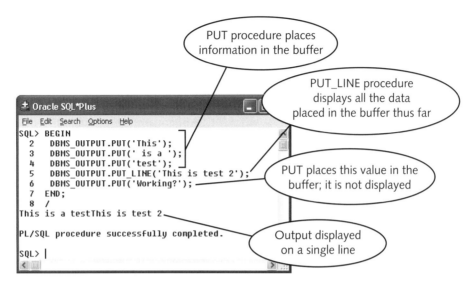

Figure 9-3 Using the PUT and PUT_LINE procedures

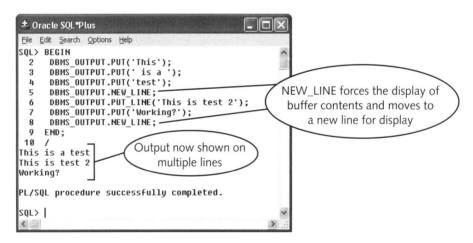

Figure 9-4 Using the NEW_LINE procedure

The GET_LINE and GET_LINES procedures retrieve values from the buffer; however, this is not necessary in SQL*Plus because the buffer display is handled automatically. Other environments such as the Enterprise Manager SQL Worksheet do require the use of these procedures. Also, you can use these procedures to extract information that was placed in the buffer by another program unit. The GET_LINE procedure has two OUT parameters, as the following header shows:

```
DBMS_OUTPUT.GET_LINE (
    line    OUT VARCHAR2,
    status  OUT INTEGER);
```

The LINE parameter holds the line retrieved from the buffer, which can be up to 255 bytes in length. The STATUS parameter contains a 0 if the procedure runs successfully and a 1 if not. If there are no lines in the buffer to retrieve, the status of 1 is returned.

The GET_LINES procedure retrieves multiple lines from the output buffer. This procedure has one OUT and one IN OUT parameter, as seen in the following header:

```
DBMS_OUTPUT.GET_LINES (
    lines    OUT  CHARARR,
    numlines   IN OUT  INTEGER);
```

The CHARARR data type is a table of VARCHAR2(255).

The LINES parameter holds the returned lines in which the maximum length of each line is 255 bytes. The NUMLINES parameter accepts a numeric value to indicate the number of lines to be retrieved. The number of lines actually retrieved is returned by this parameter. For example, you may provide a value of 40 for the NUMLINES parameter; however, there are only 35 lines in the buffer so these lines are retrieved and 35 is returned in the NUMLINES parameter.

Brewbean's is considering the use of the DBMS_OUTPUT package and the PUT and GET_LINES procedures to assist in logging debugging information. Exception handlers will include DBMS_OUTPUT.PUT_LINE statements to place information into a buffer and then invoke a procedure that will read the buffer and insert the messages into a log table. Let's try this in the following steps.

Your instructor will indicate which logon to use depending on your system's configuration.

To use DBMS_OUTPUT in exception handling:

1. Open or return to SQL*Plus as user **SCOTT**.

2. Type **SET SERVEROUTPUT ON**, and then press **Enter** so that the output buffer is available.

3. Enter the following command, which will run all the statements contained in the buffer09.sql file. See Figure 9-5.

   `@<pathname to PL/SQL files>\Chapter.09\buffer09`

4. Test the procedures by entering the following statement. Basket 88 does not exist and will raise the NO_DATA_FOUND error. You will see a "PL/SQL procedure successfully completed" message, as shown in Figure 9-6.

   `EXECUTE bsk_query_sp(88);`

9

5. Enter the following COLUMN FORMAT command to improve the read-ability of the query to check the BB_LOG_EXCEPTS table:

COLUMN descrip FORMAT A15

6. Query the BB_LOG_EXCEPTS table to determine if the values were inserted, as shown in Figure 9-6.

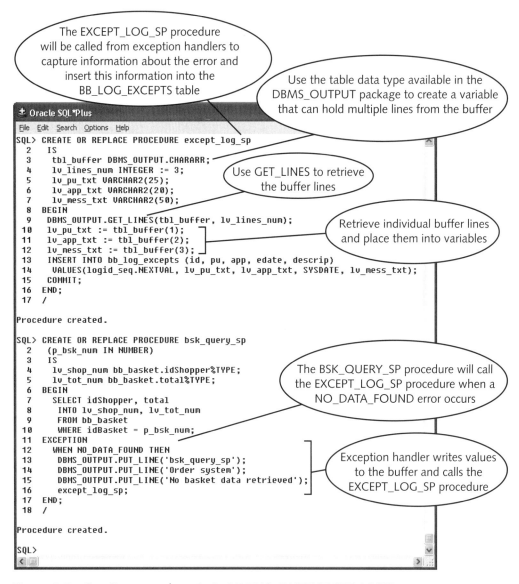

Figure 9-5 Creating procedures to test DBMS_OUTPUT.GET_LINES

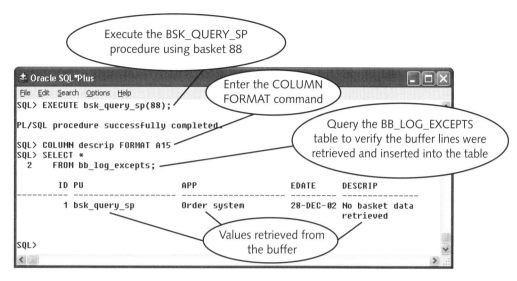

Figure 9-6 Testing the buffer procedures

In this example, the DBMS_OUTPUT.GET_LINES procedure was used to capture values from multiple lines of the buffer. Note that you do not see any values displayed from the DBMS_OUTPUT.PUT_LINE commands because the GET_LINES procedure retrieves and then clears the buffer.

UTL_FILE Package

The UTL_FILE package allows access to files that reside on an operating system and are accessible from the database server. This opens the possibility of reading from and writing to external files. This could be useful if another party provides data in the form of a text file. For example, let's say the Brewbean's company decides to add a large selection of teas from a particular vendor and the vendor can provide a text file containing product information, such as name and description. Using the UTL_FILE features can automate the process of adding these items to the product inventory table. In addition, the Brewbean's manager may request data exported from the database to be used in desktop analysis software, such as a spreadsheet program. Table 9-5 lists a handful of the programs available in the UTL_FILE package that we will use to test reading and writing to a text file.

Table 9-5 UTL_FILE partial list of programs

Program Unit Name	Description
FOPEN	Opens a file to write to or be read
PUT_LINE	Writes a line to a file
GET_LINE	Reads a line from a file
FCLOSE	Closes a file

In addition to the programs, the UTL_FILE package contains a FILE_TYPE data type that is used to declare a file variable. The FOPEN function must be provided with three IN parameter values: the file location, filename, and the open mode. The open modes available include r for read, w for write, and a for append. This function returns an identifier for this file that should be stored in a variable declared with the FILE_TYPE data type. This variable is then referenced in the remaining statements to refer to this particular file.

The PUT_LINE procedure needs IN parameter values to identify the file and the text that is written to the file. The GET_LINE procedure needs an IN parameter value to identify the file and an OUT parameter to hold the line data that is read. The FCLOSE procedure needs only one IN parameter, which identifies the file.

Before using the UTL_FILE features, access to the appropriate directories must be specified in the init.ora file using the UTL_FILE_DIR parameter or by using the CREATE DIRECTORY command, as we did earlier. The init.ora file contains all the settings for the database and is read when the database is started up. Even though the database administrator typically handles the init.ora parameter file, it is advantageous for a developer to be familiar with these settings because it contains environment settings, such as the UTL_FILE_DIR parameter, that determine if and how certain Oracle features are accessible from within PL/SQL code.

You can review a sample init.ora file from an Oracle9*i* installation to give you an idea of what the file contains. A sample init.ora file is included in the Chapter.09 folder and can be opened using a text editor. Notice that the UTL_FILE_DIR parameter is not included in the default init.ora file because this information is specific to a particular environment.

Now, we are ready to use the UTL_FILE package. Let's walk through an example in which we write a line to a file and then read that file from within PL/SQL code. Note that these steps assume that the directory object is already created.

To use UTL_FILE to write and read an external file:

1. Open or return to SQL*Plus, type **SET SERVEROUTPUT ON**, and then press **Enter**.

2. Click **File**, and then click **Open**.

3. Navigate to and select the **filew09.sql** file from the Chapter.09 folder.

4. Click **Open** and a copy of the file contents appears in the SQL*Plus screen. Review the code using the notes in Figure 9-7.

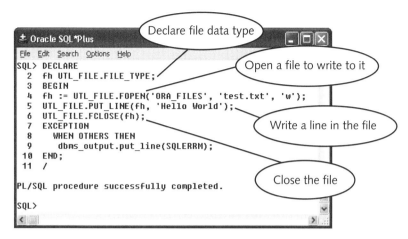

Figure 9-7   Procedure using UTL_FILE to write to a text file

5. To run the code and create the procedure, click **File** and then **Run** from the main menu.

6. In a text editor, open the **text.txt** file from the C:\oraclass directory and verify that the line 'Hello World' has been written into this file.

7. Now, let's read this same file. Click **File**, and then click **Open**.

8. Navigate to and select the **filer09.sql** file from the Chapter.09 folder.

9. Click **Open** and a copy of the file contents appears in the SQL*Plus screen. Review the code using the notes in Figure 9-8.

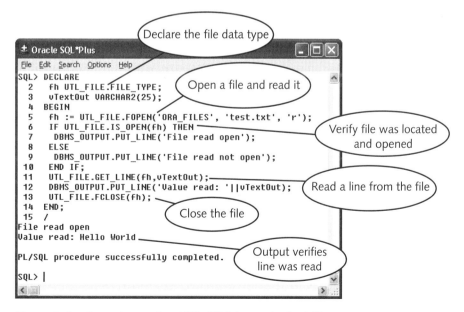

Figure 9-8   Procedure using UTL_FILE to read a text file

10. To run the code and create the procedure, click **File** and then **Run** from the main menu.

Beyond reading and writing to files, you should review a number of other useful file management procedures that have been added to the UTL_FILE package in Oracle9i, as listed in Table 9-6.

Table 9-6   New procedures in the UTL_FILE package

Program Unit Name	Description
FREMOVE	Deletes a file
FRENAME	Renames a file
FCOPY	Copies all or part of a file to another file
FGETATTR	Retrieves attributes of a file such as the size

LARGE OBJECTS

As media technology progressed, the types and size of data used in a database began to expand. Oracle8 introduced features to store and manipulate such data as large text data, graphic, and streaming video files. These types of data are referred to as large objects (LOBs) and take on various forms, such as those listed in Table 9-7.

Table 9-7   LOB types

LOB Type	Description
BLOB	Binary large object such as a photo file
CLOB	Character large object such as text documentation
BFILE	Binary large object such as a streaming video or movie file
NCLOB	Fixed-width character data for storing character data in other languages

A LOB can hold up to 4 GB of data, which well surpasses the 2 GB limit on the LONG data type. The LOB is actually a column in the database. Upon discovering the LOB capabilities, the technical staff at Brewbean's want to store an image of each product in the database so that it can be easily displayed when customers are shopping on the Web site.

One advantage of LOBs is that a table can have multiple LOB columns but only one LONG column. The table LOB column actually contains a locator that points to the actual object. The BLOB, CLOB, and NCLOB objects can all be stored internally (in the database) and are typically stored in a separate tablespace from database tables to enable optimization of the attributes of each tablespace for the type of data being stored.

A tablespace is similar to a partition or area of the physical disk. However, a BFILE object must be stored external to the database on the operating system. The DBMS_LOB package provides features to enable you to programmatically handle LOBs.

DBMS_LOB Package

The programs within the DBMS_LOB package can be grouped into two sets: mutators and observers. The mutators represent all the programs that add, change, or remove the LOBs. The observers represent all the programs that read the LOB or information about the LOB such as length. LOBs can be inserted and retrieved by SQL DML and query statements just as other types of data are manipulated. However, keep in mind that images will only be viewable in an environment such as Oracle Forms that can manage the display of graphical images. Even though we can manipulate LOBs in SQL*Plus, we cannot display them because the tool cannot handle graphical displays. Table 9-8 lists a number of the programs contained in the DBMS_LOB package to give you a feel for the type of activity accomplished with this package.

Table 9-8   List of the more useful DBMS_LOB programs

Program Unit Name	Description
LOADFROMFILE	Loads binary file into an internal LOB
WRITE	Writes data to a LOB
READ	Reads data from a LOB
ERASE	Deletes LOB data

Using DBMS_LOB to Manipulate Images

Let's return to Brewbean's desire to add images in the database to provide a product visual for Web site shoppers. A BLOB column needs to be added to the BB_PRODUCT table to house the images. The DBMS_LOB package will then be used to load the image into the database column.

To add product images:

1. Open or return to SQL*Plus.

2. Add a BLOB column to the BB_PRODUCT table using the following statement:

```
ALTER TABLE bb_product
  ADD pimage BLOB;
```

3. Initialize the BLOB column using the following UPDATE statement. The EMPTY_BLOB function prepares the column to hold image locator information.

```
UPDATE bb_product
  SET pimage = empty_blob();
```

9

4. Create a directory object that points to the folder containing the image file, using the following code:

```
CREATE directory bb_images as '<pathname to PL/SQL
    files>\Chapter.09';
```

5. If an error is raised due to not having the correct permission to create directories, then you will need to log on as SYSDBA and enter the following command to grant appropriate permissions (this code assumes that you are user SCOTT):

```
GRANT CREATE ANY DIRECTORY TO SCOTT;
```

6. Next you need to create a procedure to load the image file into the BLOB column. Click **File**, and then click **Open**.

7. Navigate to and select the **blob.sql** file from the Chapter.09 folder.

8. Click **Open** and a copy of the file contents appears in the SQL*Plus screen. Review the code using the notes in Figure 9-9.

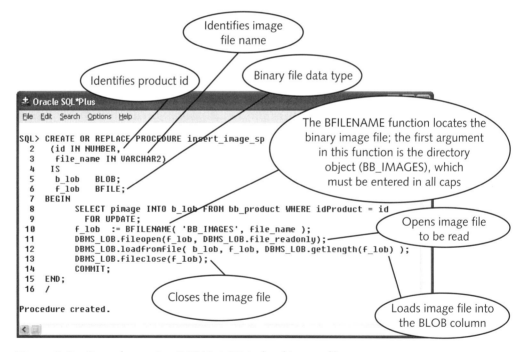

Figure 9-9   Procedure using DBMS_LOB to load image files

9. To run the code and create the procedure, click **File** and then **Run** from the main menu.

10. Use the following code to load the fpress.gif image file for product 3:

```
EXECUTE insert_image_sp(3,'fpress.gif');
```

11. To verify that the file has been loaded, we can use the query shown in Figure 9-10.

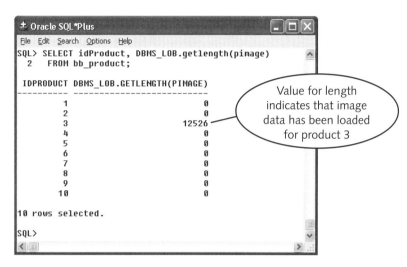

Figure 9-10   Listing the length of the BLOB column

Now that Brewbean's has product images stored in the database, when developing application screens, the image column can be queried for display just as any other column. Figure 9-11 displays the Brewbean's product screen containing the French press image we just loaded.

If a table contains any LOB columns, you will receive an error in SQL*Plus stating that "Column or attribute type cannot be displayed by SQL*Plus" if you attempt to select the LOB column or perform a SELECT * query.

DYNAMIC SQL AND PL/SQL

Thus far, the extent of our dynamic nature in SQL statements has been limited to supplying values into WHERE and HAVING clauses to potentially affect different rows each time the code is executed. However, note that we have had to supply all other parts of a statement, such as a table name, in the code at compile time.

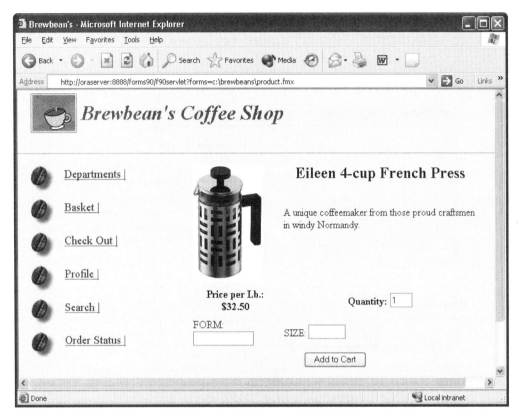

Figure 9-11   Brewbean's product screen

Fortunately, **dynamic SQL** and PL/SQL allow the construction and parsing of statements at runtime rather than compile time. (You'll recall that "parsing" refers to checking statement syntax during compilation.) Dynamic SQL features allow us to do tasks such as the following:

- Create utility programs to allow users to create or drop objects, such as tables and indexes.

- Create an ad-hoc query interface in which users can indicate such items as desired table columns and sort ordering.

- Allow users to select needed data manipulations that are actually procedures that are called. The front end could allow a user to select the desired functionality and input the necessary data values.

Prior to Oracle8i, the only mechanism available to accomplish dynamic code within our PL/SQL blocks was the DBMS_SQL package. With the introduction of Oracle8i, a feature named native dynamic SQL has simplified the process of building dynamic code. There is a great deal of already developed and deployed code containing DBMS_SQL so it is worth covering both types of dynamic SQL generation. These methods are covered in detail in Chapter 10.

MISCELLANEOUS PACKAGES

Developers discover quickly that a plethora of built-in packages exist in the Oracle server. This section covers two more popular packages: DBMS_JOB and DBMS_DDL. DBMS_JOB provides capabilities of a job scheduler to set up programs to run at specified times. DBMS_DDL provides access to DDL statements that cannot be included directly within PL/SQL code. We discuss both in the following sections.

DBMS_JOB Package

The **DBMS_JOB** package allows the scheduling of PL/SQL programs to process at some specified time or repeated interval. This job scheduling capability is more commonly referred to as the job queue and is available in most database systems.

Both application developers and database administrators schedule tasks using the features contained in DBMS_JOB. Developers, such as the Brewbean's application group, may need to schedule processing intensive jobs during off-hours such as updates to summary tables used for reporting purposes, which may only occur on a daily basis. Database administrators can use the job scheduler to process regular maintenance tasks automatically during low use hours.

Programs Within the DBMS_JOB Package

The DBMS_JOB package contains programs to submit jobs to the queue and execute jobs in the queue. This package allows adding job execution parameters, suspending jobs, and removing jobs from the queue. Table 9-9 lists the program components in the DBMS_JOB package.

Table 9-9   DBMS_JOB package programs

Program Unit Name	Description
BROKEN	Flags the job as broken so it is not executed
CHANGE	Alters job parameters set by a user
INTERVAL	Modifies execution interval for a job
ISUBMIT	Submits a job with a specified job number
NEXT_DATE	Modifies the next date of execution
REMOVE	Deletes a job from the queue
RUN	Executes the specified job immediately
SUBMIT	Adds a job to the queue
USER_EXPORT	Creates text of call to re-create the job
WHAT	Modifies the PL/SQL code to be executed

The dbmsjob.sql script creates the DBMS_JOB package. This script creates a public synonym for the package and grants the EXECUTE privilege to PUBLIC so that all Oracle users can access this package.

Parameters Within the DBMS_JOB Package

Each program in the DBMS_JOB package uses a set of parameters used to define a job, identify the code to process, indicate times and frequency of execution, and whether the job is valid. The **JOB parameter** is a unique integer that is assigned to each job in the queue. How this number is determined depends on the method used to add the job to the queue. If the SUBMIT procedure is used to add a job, the job number is assigned automatically by retrieving a value from the sequence SYS.JOBSEQ. If the ISUBMIT procedure allows a user to provide a job number, it is up to the user to make sure it is a unique value. The job numbers cannot be changed once entered into the job queue.

The **WHAT parameter** provides the PL/SQL code to be executed. Most commonly, this parameter contains a call to a PL/SQL stored program unit. However, supplying a literal string enclosed in single quotes for this parameter is valid. Also, a VARCHAR2 variable that contains valid code can be used for this parameter. Be aware of the 4,000-byte limit on the WHAT parameter if you intend to use large anonymous blocks as literal strings for this parameter. When providing stored program unit calls or invocations, it is recommended that the call be placed in an anonymous block in the WHAT parameter.

The **NEXT_DATE parameter** instructs the job queue as to when the job should be executed next. The default value of this parameter is SYSDATE. If a NULL value is provided for this parameter, then it is set to January 1, 4000, in an attempt to avoid the job being processed. This may be done to set up new jobs in the queue before the execution times have been determined. On the other hand, if it is desired to move a job to the top of the queue, a past date can be used because jobs will execute in the order of the oldest to newest NEXT_DATE value.

The **INTERVAL parameter** is an Oracle date expression in a character string that indicates how often the job should be executed. For example, if a job needs to be executed every day, the interval provided should be 'SYSDATE + 1.' If the job needs to run every hour, then the interval should be 'SYSDATE + 1/24.' The date expression in the interval is evaluated each time the job begins execution to determine when it should run next. The expression must evaluate to a future date or a NULL value. If the interval is NULL, then the job executes once and is automatically removed from the job queue. Therefore, for special one-time jobs, a NULL value for the interval achieves the needed execution.

The **BROKEN parameter** is a BOOLEAN value in which a TRUE indicates the job is broken and should not be executed by the job queue. In other words, if you have a job in the queue but need to discontinue its execution for a period, setting the BROKEN parameter to TRUE causes the job queue to not execute the job.

Checking Settings Within the DBMS_JOB Package

Before attempting to use the DBMS_JOB package, you need to check two parameter settings in the init.ora file.

For the first parameter setting, note that the Oracle job queue uses a dedicated background process and catalog tables to manage the execution of scheduled tasks automatically. The parameter JOB_QUEUE_PROCESSES must be set to a value between 1 and 36 to indicate the number of background processes to make available in the database instance. If this parameter is not set or is set to zero in the init.ora file, then no job in the queue is executed.

The second parameter, JOB_QUEUE_INTERVAL, must indicate how often the queue should be checked in terms of seconds for jobs that need to be run. Setting the interval too high could result in jobs running later than intended, whereas setting the interval too low could waste processing time as unnecessary checks are performed. The default setting is 60 seconds, but this may need to be modified depending on the job schedule you have planned.

By default, the JOB_QUEUE_PROCESSES and JOB_QUEUE_INTERVAL parameters are not included in the init.ora file and, therefore, must be added to use the job queue. Use the following steps to check the init.ora settings and add the job queue parameters, if necessary.

To set up the job queue parameters:

1. Open Windows Explorer.

2. In a text editor, open **init.ora** from the <*your Oracle database directory*>\ admin\orcl\pfile directory, which is on the drive where you installed Oracle. Depending on your installation, the file name may be appended with numbers.

3. Review the code to determine if the JOB_QUEUE_INTERVAL and JOB_QUEUE_PROCESSES are already in your init.ora file and have been set. The JOB_QUEUE_PROCESSES parameter must be set to at least 1. If you do not find these parameters in the init.ora file, continue with the following steps to add the parameters.

4. Make sure that the end of the init.ora file reads as follows:

```
###################################
# JOB QUEUE
###################################
JOB_QUEUE_PROCESSES=1
JOB_QUEUE_INTERVAL=60
###################################
```

5. Close SQL*Plus because you will be shutting down the database. Now you need to shut down the database and restart it so that the new settings in the init.ora file will take effect. Click the **Start** button, point to **All Programs**, point to the **Oracle OraHome92** entry, point to **Configuration and Migration Tools**, and then click **Administration Assistant for Windows NT**.

6. On the left side of the assistant pane initially shown, click the plus signs to expand the folders until the ORCL entry under the Databases item is exposed.

7. Right-click the **ORCL** item and click **Stop Service**. Click **OK** when the Service stopped successfully dialog box appears.

8. Wait a couple of minutes to allow Oracle to complete a shutdown. Right-click the **ORCL** item and then click **Start Service**. Click **OK** when the Service started successfully dialog box appears.

9. On the main menu of the assistant, click **File** and then click **Exit** to close the assistant. Save the console settings.

Before running an example, let's review the header information for the SUBMIT procedure of the DBMS_JOB package to review all the parameters. Review the following header code:

```
PROCEDURE submit
   ( job        OUT BINARY_INTEGER,
     what       IN  VARCHAR2,
     next_date  IN  DATE DEFAULT sysdate,
     interval   IN  VARCHAR2 DEFAULT 'null',
     no_parse   IN  BOOLEAN DEFAULT FALSE,
     instance   IN  BINARY_INTEGER DEFAULT 0,
     force      IN  BOOLEAN DEFAULT FALSE );
```

We have discussed all these parameters with the exception of the last three. The NO_PARSE indicates whether to parse the PL/SQL code when submitted, INSTANCE indicates the database instance number, and FORCE states whether the instance specified must be running. Notice that five of the seven parameters have DEFAULT values that are used if no value is provided for these parameters when invoking the SUBMIT procedure.

Now, let's run a simple example just to see the job queue in action. We will let all the DEFAULT values of the parameters be used because this will cause the job to run immediately and then be removed from the job queue.

To submit a job to the queue:

1. Open or return to SQL*Plus.

2. Click **File**, and then click **Open** from the main menu.

3. Navigate to and select the **jobq09.sql** file from the Chapter.09 folder.

4. Click **Open** and a copy of the file contents appears in the SQL*Plus screen. Review the code.

5. To run the code and create the procedure, click **File** and then **Run** from the main menu.

6. To provide the host variable that is needed to hold the job number returned by the queue when using the SUBMIT procedure, enter the following statement:

VARIABLE jobno NUMBER

7. Enter the block of code shown in Figure 9-12 to submit the BB_JOBTEST procedure into the job queue. The BB_JOBTEST procedure will insert one row into the BB_JOBQ table. Notice that the WHAT parameter is a text string; therefore, any literal strings within this parameter must be indicated with double quotes.

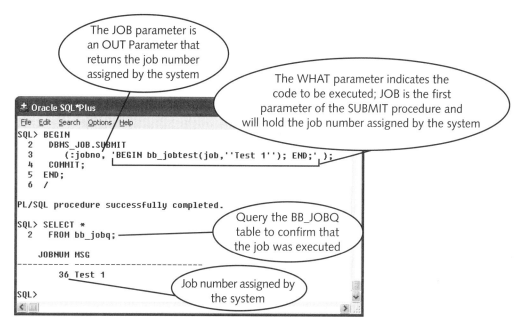

Figure 9-12   Submit a job into the queue

 Notice a COMMIT is inserted into the block immediately following the DBMS_JOB call. Some platforms do not place a job in the queue without the COMMIT action. You can test your platform by first excluding the COMMIT and then verifying if the job executes.

8. Now verify the job executed by entering the query displayed in Figure 9-12 to confirm if a row was inserted into the BB_JOBQ table. Note the job number will probably be different depending if the job queue has been used previously on your system.

The preceding step sequence verifies the job scheduler is actually operable; however, we really did not get to see jobs in the job queue because the job executed immediately. A data dictionary view named USER_JOBS is available to allow the viewing of information regarding jobs in the job queue. Figure 9-13 displays the results of performing a DESCRIBE on this view to list the available columns of data.

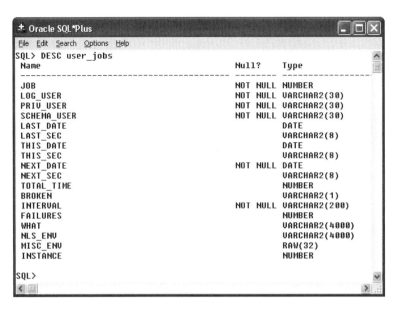

Figure 9-13   Column listing for USER_JOBS

    A DBA can also refer to two additional data dictionary views in regards to job
queue information: DBA_JOBS and DBA_JOBS_RUNNING. These views provide
more detailed execution information to assist the DBA in monitoring the system.

Let's submit another job into the queue that does not execute immediately so that we can
review the job queue information. Brewbean's wants to schedule a job to update the sales
summary table every evening. The existing procedure named PROD_SUM_SP accomplishes the updating task. We will use the Named Parameter Passing method, which makes
it easier to read and we do not have to be concerned with the order of the parameters.

To add a job to the queue:

1. Open or return to SQL*Plus.

2. Type the following code to schedule the job to run once a day at midnight.
   Note that the PROD_SUM_SP procedure contains no parameters.

```
BEGIN
 DBMS_JOB.SUBMIT
    (job => :jobno,
     what => 'BEGIN prod_sum_sp; END;',
     next_date => TRUNC(SYSDATE+1),
     interval => 'TRUNC(SYSDATE+1)' );
 COMMIT;
END;
/
```

3. To review the information in the job queue, enter the query shown in Figure 9-14. This information allows us to determine job details, such as when the job last ran, when it will next run, and what code the job executes.

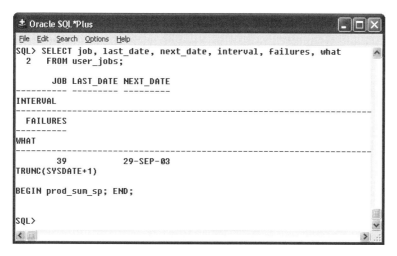

Figure 9-14   Viewing job queue information

4. Now that we have a job in the queue, let's try the REMOVE procedure to delete the job from the queue so that it will no longer be scheduled to run. Type the following code:

EXECUTE DBMS_JOB.REMOVE(*job number you placed in the queue*);

5. Type and run the same query used in Step 3 to confirm that the job is no longer in the queue.

Normally, if we have a job that fails, we want to know right away because it typically will continue to fail unless we make some modifications. Therefore, exception handling is important to immediately mark the job as broken and provide an alert to an appropriate person. Toward that end, one item you may have noticed in viewing information from the USER_JOBS view was a column named FAILURES. If a scheduled job contains execution errors, the job scheduler attempts to process it up to 16 times. At that point, if the job execution remains unsuccessful, the system automatically marks the job as broken using the BROKEN column and the job is not processed again.

One item that is considered a shortcoming of the Oracle job scheduler is the fact that no name is assigned to a job for easier reference. Thus, many developers create their own program units to use in conjunction with the DBMS_JOB package to make it more user friendly, such as allowing job names.

Note that creating a date expression to reflect odd schedules can be a challenge. However, you can indicate specific times to execute jobs within the interval parameter by using the TRUNC function on the current date and adding hours to hit the specific time of day needed. An example is an interval of TRUNC(SYSDATE) + 22/24 or midnight plus 22 hours to indicate that the job should be executed at 10 p.m.

This section has not covered every procedure within the DBMS_JOB package; however, you now have a foundation to use the job scheduler and determine how the other programs fit in. To review all the procedures available in the DBMS_JOB package, read the documentation within the dbmsjob.sql file.

DBMS_DDL Package

The **DBMS_DDL** package allows access to two DDL statements that cannot be used directly within a PL/SQL block: ALTER_COMPILE and ANALYZE_OBJECT. The ALTER_COMPILE procedure allows the compilation of program units and the ANALYZE_OBJECT procedure creates statistics for database objects to be utilized to enhance performance.

The developers at Brewbean's want to explore this package to enable the creation of a procedure to compile any INVALID objects and another to analyze all the tables. If they can perform this functionality within procedures, these jobs can be placed into the job queue and be scheduled to run routinely during off-hours. Let's look at each of these and how the DBMS_DDL package allows their inclusion into PL/SQL code.

ALTER_COMPILE

Recall that object modifications can change the status of dependent objects to INVALID and these need to be recompiled before being executed again. That is, if we change a table and a procedure references this table, the procedure is marked as INVALID. The ALTER_COMPILE procedure of the DBMS_DDL package allows developers to programmatically compile database objects. The header of this procedure shows that it contains three IN parameters (recall if no mode is indicated, the default is IN):

```
DBMS_DDL.ALTER_COMPILE (
    type   VARCHAR2,
    schema VARCHAR2,
    name   VARCHAR2);
```

The TYPE parameter value can be one of the following: PROCEDURE, FUNCTION, TRIGGER, PACKAGE, PACKAGE BODY, or PACKAGE SPECIFICATION. Indicating just PACKAGE compiles both the specification and body. The SCHEMA is the object's owner and the NAME is the object name. The SCHEMA and NAME parameter values are case sensitive. By default, these values are stored in the database as uppercase and, therefore, should be uppercase in this statement. However, if you used double quotes to enter object names with mixed-case, then this parameter value must match that mixed-case spelling exactly. Note that all these parameters have character data types, so any literal values need to be placed in single quotes.

To successfully compile a program, you must either own the object or you must be granted the ALTER ANY PROCEDURE privilege to compile an object in another schema. If a NULL value is provided for the schema parameter, then the user's schema is used. When the compile action begins, objects that the object depends on are checked for an INVALID status. If INVALID, this object is recompiled prior to recompiling the targeted object.

Although the COMPILE command can be entered in SQL*Plus, it is an important benefit that we can accomplish the COMPILE action within a PL/SQL block, thereby allowing the creation of procedures that can be placed in the job queue. Depending on specific needs, developers will typically create a procedure that allows them to easily recompile all invalid objects or invalid objects of a specific type. In this scenario, the data dictionary ALL_OBJECTS view could be used to retrieve all the objects with a status of INVALID. These object names could be retrieved into a cursor that can be processed through a loop using the ALTER_COMPILE procedure. This is quite valuable because we can avoid the task of tracking dependencies as we complete modifications. To verify successful compilation, the procedure can scan the STATUS and LAST_DDL_TIME columns of the USER_OBJECTS view.

Typical exceptions raised by the ALTER_COMPILE procedure are listed in Table 9-10.

9

Table 9-10   ALTER_COMPILE exceptions

Oracle Error Number	Description
ORA-20000	Object does not exist or you have insufficient privileges to this object
ORA-20001	Attempted to compile remote object; only local database objects can be compiled
ORA-20002	Not a valid object type value

ANALYZE_OBJECT

The second procedure of the DBMS_DDL package that we will review is the ANALYZE_OBJECT procedure. The analyze task is typically part of the DBA's duties, but a developer needs to be aware of the options in place to complete performance-tuning tasks. The ANALYZE_OBJECT action computes statistics on a database object, typically tables and indexes, such as the range of values. The Oracle cost-based Optimizer attempts to determine the best path of execution by using these statistics. If statistics do not exist for an object, then the rule-based Optimizer is used.

Let's take a look at the header for the ANALYZE_OBJECT procedure:

```
DBMS_DDL.ANALYZE_OBJECT (
    type            VARCHAR2,
    schema          VARCHAR2,
    name            VARCHAR2,
    method          VARCHAR2,
    estimate_rows   NUMBER    DEFAULT NULL,
```

```
estimate_percent NUMBER    DEFAULT NULL,
method_opt       VARCHAR2 DEFAULT NULL,
partname         VARCHAR2 DEFAULT NULL);
```

Table 9-11 lists a description of each of the parameters available.

Table 9-11   Parameters of the ANALYZE_OBJECT procedure

Parameter Name	Description
TYPE	Value provided must be either TABLE, INDEX, or CLUSTER
SCHEMA	Owner of object to analyze; case sensitive and NULL will default to current schema
NAME	Object name, case sensitive
METHOD	Value provided must be either ESTIMATE, COMPUTE, or DELETE; if it is ESTIMATE, then one of the next two parameters must have a nonzero value
ESTIMATE_ROWS	Number of rows to be used
ESTIMATE_PERCENT	Percent of rows to be used
METHOD_OPT	Indicates which database structure associated with the object is analyzed: FOR TABLE, FOR ALL INDEXES, or FOR ALL INDEXED COLUMNS
PARTNAME	If using partitions, a specific partition can be specified

The METHOD parameter value is most significant in how the statistics for the object are calculated. The COMPUTE method uses the contents of the object in its entirety, whereas the ESTIMATE method uses only the number or percent of rows provided in the other parameters. In both cases, the statistics are stored in the data dictionary. The DELETE method simply erases the statistics that already exist in the data dictionary for an object. To verify successful analysis, check the LAST_ANALYZED column in the USER_OBJECTS view, which holds the date of the last analysis. Table 9-12 lists the typical exceptions raised by the ANALYZE_OBJECT procedure.

Table 9-12   ANALYZE_OBJECT exceptions

Oracle Error Number	Description
ORA-20000	Object does not exist or you have insufficient privileges to this object
ORA-20001	Not a valid object type value

Exploring Additional Oracle-Supplied Packages

This chapter is far from an exhaustive coverage of Oracle built-in packages and was written to simply expose you to the world of built-ins. As you journey further into PL/SQL programming, it is worthwhile to identify the available built-in packages not just to take advantage of their capabilities, but also to review their code to get ideas on coding techniques. Table 9-13 lists several more packages with a brief description that you may find useful. Again, the Oracle database documentation on the OTN Web site has a full listing and description of all Oracle built-in packages.

Table 9-13   Other built-in packages

Package Name	Description
DBMS_JAVA	Controls the behavior of the Java Virtual Machine used to run Java stored procedures
DBMS_METADATA	Retrieves information about database objects
DBMS_RANDOM	Generates random numbers
DBMS_SESSION	Allows access to session options directly from PL/SQL
DBMS_UTILITY	Contains a miscellaneous group of programs ranging from capabilities to assist in procedure management to reporting error information
DBMS_XMLGEN	Converts data from an SQL query into XML
UTL_HTTP	Accesses Web pages
UTL_INADDR	Retrieves Internet site host name or IP address

Each of these are packages with parameters just the same as those we can construct. When the database is installed, a script named catproc.sql invokes each of the scripts that create all these built-ins. It is helpful to review the individual scripts that create each package, which can be found in the rdbms\admin directory of the Oracle database software. Each script contains comments documenting the code and reviewing this not only gives you a feel for what is going on in the package, but it can also give you ideas to use in your own code. In addition, the OTN Web site includes a whole manual on Oracle-supplied packages. Scanning the list of packages gives you an appreciation for just how many built-in packages exist. This documentation shows the package header, explains each parameter, describes each program of the package, and sometimes includes an example use.

CHAPTER SUMMARY

- ❐ Oracle built-in or supplied packages refer to a group of complete packages that are available upon installation of the Oracle database.

- ❐ An Oracle-supplied package is the same type of package that you create. The package contains a set of procedures and functions.

- ❐ The DBMS_PIPE package is used to communicate between different database sessions.

- ❐ The DBMS_ALERT package enables user notification of database events.

- ❐ The UTL_SMTP package simplifies the process of generating and sending e-mail from within PL/SQL code.

- The UTL_HTTP package allows the retrieval of HTML code from specified URLs. This capability is used to scan information from Web pages.

- The UTL_TCP packages allow TCP/IP communication. The UTL_SMTP and UTL_HTTP packages use the UTL_TCP package to achieve their capabilities.

- The DBMS_OUTPUT package offers features to display data to the screen.

- The UTL_FILE package contains various features to interact with external operating system files. Features include opening a file, writing to a file, reading a file, and deleting a file.

- The DBMS_LOB package is used to manage large objects such as a BLOB or CLOB in the database. The package programs allow the LOB to be inserted into and retrieved from the database.

- The DBMS_JOB package provides features enabling job scheduling.

- The DBMS_DDL package enables execution of DDL statements to recompile and analyze database objects from within PL/SQL code.

REVIEW QUESTIONS

1. Oracle built-in packages _____.

   a. need to be created by special scripts by each user

   b. are ready for use when the database is installed

   c. can be purchased separately

   d. are the only recommended package code

2. Oracle pipes created with the DBMS_PIPE package enable _____.

   a. e-mail from user sessions

   b. access to Web pages

   c. two users on different instances to communicate

   d. two users on the same instance to communicate

3. If your application needs to send notifications to users when a data change (such as a higher bid in an auction) occurs, which built-in package would be most helpful?

   a. DBMS_PIPE

   b. DBMS_EMAIL

   c. DBMS_ALERT

   d. DBMS_DDL

4. The DBMS_DDL package allows which of the following actions to be accomplished within PL/SQL code? (Choose all that apply.)

    a. recompile a database object

    b. communicate among sessions

    c. compute statistics for a database object

    d. mark a database object as INVALID

5. The Oracle server contains a mechanism that allows the execution of programs to be scheduled. On what package is this feature based?

    a. DBMS_JOB

    b. UTL_JOB

    c. DBMS_SQL

    d. DBMS_SCHED

6. A BLOB in Oracle represents a data type that holds _____.

    a. up to 4 GB of character data

    b. a video file

    c. a graphic image

    d. foreign-language elements

7. Which parameter of the DBMS_JOB package indicates how often a job is processed?

    a. JOB

    b. WHAT

    c. NEXT_DATE

    d. INTERVAL

8. Which package is used to display values to the screen?

    a. DBMS_OUT

    b. DBMS_OUTPUT

    c. DBMS_WRITE

    d. DBMS_PIPE

9. Which package is most helpful in automatically sending e-mails from within an application?

    a. DBMS_OUTPUT

    b. DBMS_SMTP

    c. DBMS_JOB

    d. UTL_MAIL

9

10. If an application needs to write data to an external file, which package would be most helpful?

   a. UTL_FILE

   b. DBMS_FILE

   c. UTL_OUT

   d. DBMS_WRITE

11. Describe the features available with the DBMS_JOB package and why these could be useful.

12. What is an Oracle built-in package?

13. How is an Oracle built-in package different from the packages we construct ourselves?

14. Name at least two examples of specific tasks that you would use the UTL_FILE package to accomplish.

15. Do you feel the functionality offered from the UTL_SMTP package is becoming increasingly important or less important? Why?

ADVANCED REVIEW QUESTIONS

1. The order of the four basic steps of a pipe operation is _____.

   a. pack, send, receive, unpack

   b. pack, send, unpack, receive

   c. send, pack, unpack, receive

   d. send, pack, receive, unpack

2. Which data dictionary view contains information regarding jobs scheduled via the DBMS_JOB package?

   a. USER_JOBS

   b. USER_JOB

   c. USER_LIST

   d. USER_SCHED

3. The DBMS_PIPE package operates asynchronously, which means the action is _____.

   a. independent of the transaction

   b. dependent on the transaction

   c. only completed when a COMMIT is issued

   d. none of the above

4. The DBMS_ALERT actions are transaction-based, which means the action is
_____ .

   a. independent of the transaction

   b. only completed when a COMMIT is issued

   c. a DML statement

   d. unpredictable

5. The broken parameter is a BOOLEAN value in which _____ .

   a. a TRUE indicates the job should not be executed by the job queue

   b. a TRUE indicates the alert should not be sent

   c. a FALSE indicates the job should not be executed by the job queue

   d. a FALSE indicates the pipe is not secure

HANDS-ON ASSIGNMENTS

Assignment 9-1: Use the DBMS_PIPE Package

A pipe is used frequently to send data to external services such as credit card validation. Brewbean's needs to develop a pipe that will send a customer's last name, credit card number, and expiration date to a credit service.

1. Start SQL*Plus by logging on as **SYSTEM, MANAGER**.

2. Type **SET SERVEROUTPUT ON**, and then press **Enter**.

3. Start a second SQL*Plus session by logging on as **SYSTEM, MANAGER** again.

4. Type **SET SERVEROUTPUT ON**, and then press **Enter**.

5. In the first SQL*Plus session, click **File** and then **Open** from the main menu.

6. Navigate to and select the **assignment01a.sql** file from the Chapter.09 folder.

7. Click **Open** and a copy of the file contents appears in the SQL*Plus screen. Review the code.

8. To run the code and create the procedure, click **File** and then **Run** from the main menu.

9. In the second SQL*Plus session, click **File** and then **Open** from the main menu.

10. Navigate to and select the **assignment01b.sql** file from the Chapter.09 folder.

11. Click **Open** and a copy of the file contents appears in the SQL*Plus screen. Review the code.

12. To run the code and create the procedure, click **File** and then **Run** from the main menu. Your screen should resemble Figure 9-15. Close both sessions.

9

```
Oracle SQL*Plus                                    [_][□][X]
File  Edit  Search  Options  Help
SQL> DECLARE
  2      lv_status_int INTEGER;
  3      lv_msg VARCHAR2(40);
  4   BEGIN
  5      lv_status_int := DBMS_PIPE.RECEIVE_MESSAGE('Testpipe');
  6      IF lv_status_int <> 0 THEN
  7        DBMS_OUTPUT.PUT_LINE('Error');
  8      ENd IF;
  9      DBMS_PIPE.UNPACK_MESSAGE(lv_msg);
 10      DBMS_OUTPUT.PUT_LINE(lv_msg);
 11   END;
 12   /
CASTEEL-1234567890-09/05

PL/SQL procedure successfully completed.

SQL> |
```

Figure 9-15   Receiving a message via an Oracle pipe

 After you have pipe code working successfully, for an extra challenge, try to convert your anonymous blocks to two procedures: one for sending and one for receiving pipes.

Assignment 9-2: Use the DBMS_ALERT Package

Brewbean's wants to add an alert in the product management screen to advise the manager of a product stock level falling below the reorder point. The alert simply needs to advise the recipient of which product needs reordering.

1. Open one session of SQL*Plus as **Scott**, **tiger**.

2. Create a database trigger on the BB_PRODUCT table using the DBMS_ALERT package to send an alert when the stock value falls below the reorder value. Name the alert reorder and have it contain a message that states which product number needs to be reordered.

3. If an error is raised due to not having the correct permission to use the DBMS_ALERT package, you will need to log on as SYSDBA and enter the following command to grant appropriate permissions (this code assumes that you are user SCOTT):

   ```
   GRANT EXECUTE ON DBMS_ALERT TO SCOTT;
   ```

4. Open a second session of SQL*Plus as **SYSTEM**, **MANAGER**. Type **SET SERVEROUTPUT ON**, and then press **Enter**.

5. Type and execute the following block to register the alert:

   ```
   BEGIN
    DBMS_ALERT.REGISTER('reorder');
   END;
   /
   ```

6. Type and execute the following block to initiate the wait for an alert:

```
DECLARE
 lv_msg_txt VARCHAR2(25);
 lv_status_num NUMBER(1);
BEGIN
 DBMS_ALERT.WAITONE('reorder', lv_msg_txt, lv_status_num,
                                              120);
 DBMS_OUTPUT.PUT_LINE('Alert: '||lv_msg_txt);
 DBMS_OUTPUT.PUT_LINE('Status: '||lv_status_num);
END;
/
```

7. Note that a wait period of two minutes was set in the preceding code. Return to the SQL*Plus session of Scott, tiger. Type and execute the following code to cause the alert to fire:

```
UPDATE bb_product
  SET stock = stock - 2
    WHERE idproduct = 4;
COMMIT;
```

8. Return to the SQL*Plus session of SYSTEM, MANAGER and the alert message should be displayed, as shown in Figure 9-16.

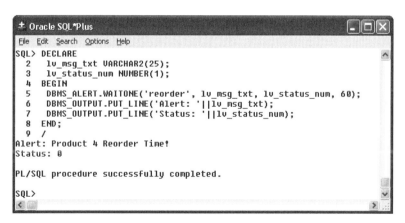

Figure 9-16   Alert message display

Assignment 9-3: Use the DBMS_DDL Package

Because Brewbean's is in the midst of constructing their application, including a good deal of database modifications, the developers will build a procedure to automatically recompile all the INVALID objects.

1. Open SQL*Plus. Type **SET SERVEROUTPUT ON**, and then press **Enter**.

2. Type and run the following code:

```
SELECT object_name, status
 FROM user_objects
 WHERE object_type = 'PROCEDURE';
```

3. If you see the BB_JOBTEST object with a VALID status on your screen, skip to Step 8. If the BB_JOBTEST procedure does not exist, then open the **assignment03a.sql** file from the Chapter.09 folder by clicking **File** and then **Open** from the main menu.

4. Navigate to and select the **assignment03a.sql** file from the Chapter.09 folder. Click **Open** and a copy of the file contents appears in the SQL*Plus screen. Review the code.

5. To run the code and create the procedure, click **File** and then **Run** from the main menu.

6. Type and run the following code:

```
SELECT object_name, status
 FROM user_objects
 WHERE object_type = 'PROCEDURE';
```

7. Verify that you see the BB_JOBTEST object with a VALID status listed from the query. This tells you that the procedure exists and that its status is VALID.

8. Type and execute the following ALTER TABLE statement to modify the BB_JOBQ table. Because the BB_JOBTEST procedure uses this table, the procedure will now change to a status of INVALID, indicating the need for recompilation.

```
ALTER TABLE bb_jobq
 MODIFY (msg VARCHAR2(30));
```

9. Type and run the following query to confirm a status of INVALID:

```
SELECT object_name, status
 FROM user_objects
 WHERE object_type = 'PROCEDURE';
```

10. Create an anonymous block that uses DBMS_DDL.ALTER_COMPILE to compile all objects that are INVALID. Retrieve all the INVALID objects in a cursor to simplify this process.

11. Execute the anonymous block.

12. Type and run the following query to confirm the status is now VALID:

```
SELECT object_name, status
 FROM user_objects
 WHERE object_type = 'PROCEDURE';
```

Assignment 9-4: Use the UTL_FILE Package to Read and Insert Data

Brewbean's has struck a deal with a new tea product supplier. The tea supplier has sent a test file of how they will submit a file containing all the product names and descriptions. We will set up a PL/SQL block using the UTL_FILE package to read the file and insert the data into the BB_PRODUCT table.

1. Start Windows Explorer.

2. Open the **tea.txt** file from the Chapter.09 folder. Read the text in the file to confirm that it contains product names and descriptions.

3. Close the text editor.

4. In Windows Explorer, right-click the filename and then click **Copy**.

5. Move to the c:\oraclass directory and place a copy of the **tea.txt** file in this folder.

6. Open SQL*Plus. Type **SET SERVEROUTPUT ON**, and then press **Enter**.

7. Create an anonymous block that will read each of the lines in the **tea.txt** file and complete an INSERT of the values into the BB_PRODUCT table. Use the UTL_FILE reading feature in a loop to read each of the lines from the text file.

8. Execute the block to accomplish the INSERTS.

9. Type and execute the following code. Your screen should include the three new tea products contained in the **tea.txt** file.

```
COLUMN description FORMAT A30
SELECT productname, description
 FROM bb_product;
```

Assignment 9-5: Use the UTL_FILE Package to Insert Data Columns

The Brewbean's manager wants to have a file extracted from the database that contains product information for inventory and cash flow analysis. The manager uses the file in spreadsheet software on a laptop computer. Create a PL/SQL block using the UTL_FILE package to place the data columns in a comma delimited text file named prod_ext.txt in the c:\oraclass directory. Comma delimited is a popular file format that contains commas between each of the values within a line. The extract file should contain one line per product and the following columns of the BB_PRODUCT table: IDPRODUCT, PRODUCTNAME, PRICE, TYPE, STOCK, ORDERED, and REORDER.

9

Assignment 9-6: Send E-mail Using UTL_SMTP

This assignment requires an SMTP server address.

The Brewbean's manager decided he wants to receive notification of product stock levels that fall below the reorder point via e-mail. Create a trigger named BB_STKALERT_TRG on the BB_PRODUCT table that will accomplish this task. The e-mail body should indicate the product id and name. Use the following UPDATE statement to set up a product to test:

```
UPDATE bb_product
   SET stock = 26
   WHERE idProduct = 4;
COMMIT;
```

Test the trigger with this code:

```
UPDATE bb_product
   SET stock = stock - 2
   WHERE idproduct = 4;
COMMIT;
```

When you complete the assignment, disable the trigger so that it does not affect other assignments.

Assignment 9-7: Use the DBMS_JOB Package

Create a procedure named BB_STK_SP that will check each of the product stock levels and display a line to the screen for each product that needs reordering. Display the product id, stock amount, and reorder point. Put the job in the job queue using the DBMS_JOB package. Schedule the job to run every day at 11 a.m. Display the information from the job queue from the data dictionary. Remove the job from the job queue. To confirm the removal, display the information from the job queue from the data dictionary. Be sure the NEXT_DATE and INTERVAL parameters reflect 11:00 a.m. the next day.

Assignment 9-8: Use DBMS_OUTPUT

Create and execute a PL/SQL block that will display lines to the screen for each product using the DBMS_OUTPUT package. Lines for every product in the BB_PRODUCT table should be displayed when this block is executed. If the product stock level is above the reorder point, then only a single line should be displayed for the product and it should appear as shown in the following code:

```
Product 5 — Sumatra does NOT need ordering
```

If the stock level is below the reorder point, then the lines displayed should appear like the following:

```
Product 5 — Sumatra needs ordering!
   Stock = 24 , reorder point = 25
```

Before executing the block, execute the following to make sure product 4 stock level is below the reorder point:

```
UPDATE bb_product
   SET stock = 24
   WHERE idProduct = 4;
COMMIT;
```

9

CASE PROJECTS

Case 9-1: Search Oracle Built-In Packages

Review the Oracle Supplied Package documentation on the OTN Web site. Identify one that is not covered in this chapter, and describe what features the package offers, how the process works, and an example of how it might be applied.

Case 9-2: The More Movies Company

The More Movies inventory clerk has requested a text file that will contain a line for each time all copies of a movie are checked out. The clerk wants to use this information in a desktop software program to perform inventory analysis. Each line in the file should contain the date, movie id, and movie stock quantity. To accomplish this, appropriate code needs to be created in a database trigger. Keep in mind that this file should maintain a continuous record and, therefore, each new line should be *appended* to the file. Name the file **checkout.txt** and place this in the C:\oraclass directory.

Test the trigger using the following two UPDATES. Both are used to be sure the data is appended in the file.

```
UPDATE mm_movie
 SET movie_qty = 0
 WHERE movie_id = 7;

UPDATE mm_movie
 SET movie_qty = 0
 WHERE movie_id = 8;
```

 To accomplish appending to a file, research the OPEN MODE parameter of the UTL_FILE.FOPEN procedure at the OTN Web site.

10

INTRODUCTION TO DYNAMIC SQL AND OBJECT TECHNOLOGY

> **In this chapter, you will:**
> ♦ Create dynamic SQL
> ♦ Use object technology

As PL/SQL developers, we need to be aware of the more advanced technologies Oracle provides and continue to embrace these technologies where appropriate to make our applications more flexible, powerful, and easy to maintain. Two of these technologies, dynamic SQL and object technology, are introduced in this chapter.

Dynamic SQL allows us to make the SQL statements in our PL/SQL code much more flexible in regards to providing values, such as column names, at runtime. Object technologies allow us to employ a more object-oriented approach to modeling our database and applications. This is a giant methodology leap in the sense of design and structure and has gained significant support over the past decade. Both of these topics, especially object technologies, are very involved and this chapter is merely an introduction to expose you to these concepts.

THE CURRENT CHALLENGE IN THE BREWBEAN'S APPLICATION

The Brewbean's lead programmer wants to add additional query functionality to the application to enhance shopper product searches and manager data analysis capabilities. The key for accomplishing this is to enable flexibility in the queries so that selection criteria can be determined at runtime. For example, shoppers should be allowed to pick if they want to search by product name, description, etc. In addition, managers need to be able to select which data columns they want to set criteria on for a particular query. To develop this capability, it has been requested that all programmers explore applying dynamic SQL use in Oracle to the Brewbean's application.

In addition, some colleagues at other companies have been working on applying object-oriented programming principles and have purported benefits of more data consistency and more data control as the program code is more closely associated with the data. Therefore, all of the programmers have also been asked to explore Oracle object types and object views to determine if the Brewbean's application design should integrate object usage.

REBUILDING YOUR DATABASE

To rebuild the Brewbean's database:

1. Locate the **c10Dbcreate.sql** file in the Chapter.10 directory to ensure it exists. This file contains the script to create the database.

2. Start SQL*Plus and log on.

3. Create a spool file so that SQL*Plus keeps a copy of all the messages received from running the file that creates the database. On the main menu in SQL*Plus, click **File**, point to **Spool**, and then click **Spool File**. A Select File dialog box appears.

4. Browse to a folder used to contain your working files and in the File name: text box, type **DB_log10**, and then click **Save**. Now, whatever text is seen in our SQL*Plus session is saved to this file for future reference.

5. Now let's create the database. In SQL*Plus, enter the following command, which runs all the statements contained in the c10Dbcreate.sql file. Messages verifying the creation and data insertion steps scroll on the SQL*Plus screen. This may take a couple minutes to complete.

    `@<pathname to PL/SQL files>\Chapter.10\c10Dbcreate`

6. Now, we need to turn the spooling off so that we can review the results in the spool file created. On the main menu, click **File**, point to **Spool**, and then click **Spool Off**.

7. Open a text editor.

8. Open the file named **DB_log10.lst** in your working folder.

9. Review the messages for any errors.

Dynamic SQL

Due to the early binding of SQL statements, only DML and query statements can be included directly within PL/SQL blocks, and these statements are only flexible in the sense of values being provided at runtime. For example, we can provide WHERE clause values at runtime via a procedure parameter. However, what if we need to make the code even more flexible, such as allowing the user to provide identifiers of column names at runtime? That is, what if we want to allow the application users to be able to choose the column on which to set criteria to perform a query?

Let's say Brewbean's is developing a product lookup screen in which they allow the user to indicate a lookup based on the product id, name, or description. In this case, the column name on which a criterion is set must be provided at runtime, because we do not know on which column the user will set criteria. To set the column name, we need to use the dynamic SQL features available with the Oracle server.

To clarify what dynamic SQL means, let's compare some procedures. First, the procedure DYN_TEST1_SP demonstrates how we use parameters to provide values within an SQL statement at runtime. The code for this procedure is shown in Figure 10-1. Notice the results of executing this procedure, which is successful in returning product information based on the parameter value (P_VALUE) provided for the product id.

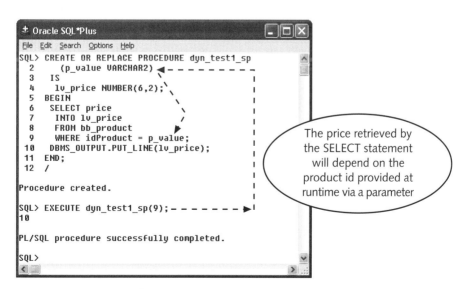

Figure 10-1 Testing the dynamic nature of SQL within PL/SQL

However, what if we attempt to expand the dynamic nature by supplying both a column name and value at runtime? Procedure DYN_TEST2_SP attempts this task, as shown in Figure 10-2. This block allows the criteria in the WHERE clause, including the column name and value, to be set at runtime.

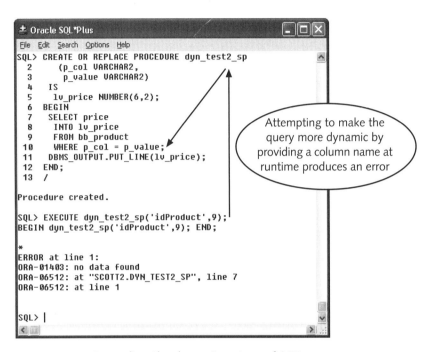

Figure 10-2 Expanding the dynamic nature of SQL

Notice that an error is raised when the DYN_TEST2_SP procedure is executed. Attempting to use a parameter to provide an identifier, such as a column name or table name, in an SQL statement within PL/SQL raises an error. In addition, the use of a DDL statement, such as ALTER TABLE or CREATE TABLE, directly within PL/SQL also produces an error.

Oracle offers two different features of dynamic SQL that provide workarounds for these situations. The DBMS_SQL package was introduced with Oracle version 7 to allow more dynamic SQL code within PL/SQL coding. In Oracle version 8, native dynamic SQL was introduced to simplify the coding of dynamic SQL and improve the performance of the code.

In this part of the chapter, both of these features are introduced and followed with a discussion of which method is more appropriate in given circumstances.

The DBMS_SQL Package

The process flow of the DBMS_SQL procedures is slightly different for each type of SQL statement, which include: DML, DDL, and queries. The general process involves the establishment of a cursor to contain and manipulate the SQL statement. This is somewhat similar to the handling of cursors we have already seen in that we must handle each process of the cursor explicitly, including opening, parsing, executing, and closing.

One distinctive element in the DBMS_SQL package is the use of placeholders as a mechanism to provide values to the SQL statement. This package allows parameters to be directly used as identifiers, such as column names and table names. However, parameters can still provide values to check criteria such as the values supplied in a WHERE clause. Therefore, placeholders are used to distinguish parameter values that are used as criteria, versus parameters that provide identifiers such as column names.

When the SQL statement is built in the PL/SQL block, placeholders with colon prefixes are used to mark where values are supplied at runtime. These placeholders are associated with a PL/SQL variable, typically a parameter, by using the BIND_VARIABLE program of the DBMS_SQL package. These placeholders are referred to as bind variables but should not be confused with the bind or host variables we have previously used, which are variables created in the host environment.

We next look at an example of deploying each type of SQL statement with the DBMS_SQL package.

10

DML Statements with DBMS_SQL

First, let's look at the steps involved with performing a DML statement via the DBMS_SQL package. In this case, Brewbean's needs a procedure to support an application screen that allows a user to set a new product price. The new price can be either a sales price or the regular product price as indicated by the user; therefore, the block needs to allow a column name—either PRICE or SALEPRICE—to be provided at runtime. Table 10-1 lists the steps and associated packaged program that needs to be used to complete a dynamic DML statement.

Table 10-1   Steps to perform DML statements

Step	Program	Description	Parameters
1	OPEN_CURSOR	Establishes a cursor or work area for the statement to be handled	N/A
2	PARSE	Checks syntax of statement	Cursor, statement to parse, and version behavior. Note: Using NATIVE for the version behavior indicates the statement should be processed based on the database version on which the program is executed

Table 10-1 Steps to perform DML statements (continued)

Step	Program	Description	Parameters
3	BIND_VARIABLE	Associates placeholders in the statement with PL/SQL variables	Cursor, placeholder name, and value to be assigned
4	EXECUTE	Runs the statement and returns the number of rows affected	Cursor
5	CLOSE_CURSOR	Frees the resources allocated to the cursor	Cursor

The following procedure accomplishes the price setting task by using the DBMS_SQL package. The procedure code includes comments to assist in highlighting the steps involved to manage a DML statement via dynamic SQL.

```
CREATE OR REPLACE PROCEDURE dyn_dml_sp
   (p_col VARCHAR2,
    p_price NUMBER,
    p_id NUMBER)
 IS
  lv_cursor INTEGER;
  lv_update VARCHAR2(150);
  lv_rows NUMBER(1);
BEGIN
  --Open Cursor
  lv_cursor := DBMS_SQL.OPEN_CURSOR;
  --Create DML statement
  lv_update := 'UPDATE bb_product
              SET ' || p_col || ' = :ph_price
              WHERE idProduct = :ph_id';
  --Parse the statement
  DBMS_SQL.PARSE(lv_cursor, lv_update, DBMS_SQL.NATIVE);
  --Associate parameters with placeholders in the statement
  DBMS_SQL.BIND_VARIABLE(lv_cursor, ':ph_price', p_price);
  DBMS_SQL.BIND_VARIABLE(lv_cursor, ':ph_id', p_id);
  --Run the DML statement
  lv_rows := DBMS_SQL.EXECUTE(lv_cursor);
  --Close the cursor
  DBMS_SQL.CLOSE_CURSOR(lv_cursor);
  --Save changes
  COMMIT;
  --Check how many rows affected
  DBMS_OUTPUT.PUT_LINE(lv_rows);
END;
```

Notice that in the construction of the DML statement, which is held in the LV_UPDATE variable, that everything, including the placeholder, is placed within single quotes as a text string. The only item not placed in single quotes is the parameter P_COL that supplies the identifier value or the column name. Figure 10-3 displays two executions of this procedure. First, the regular product price of product 3 is modified. Second, the sale price of the same product is modified.

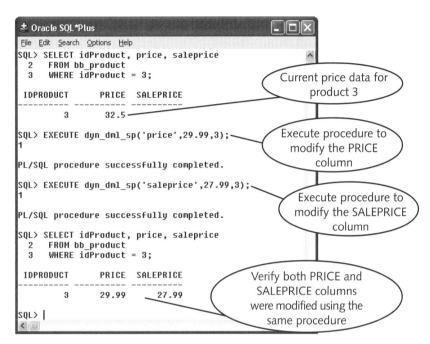

Figure 10-3   Performing DML with column names provided at runtime

The application screen for the price change allows the user to make choices via drop-down lists or radio buttons to identify which column and product should be modified. An input box allows the entry of the price value. Figure 10-4 displays the Brewbean's price change application screen.

DDL Statements with DBMS_SQL

The DBMS_SQL package not only enables the inclusion of DDL statements such as CREATE, ALTER, and DROP, but also allows these statements to operate dynamically in the sense that column and table names can be provided at runtime. This can provide a mechanism to build application screens to simplify database modification.

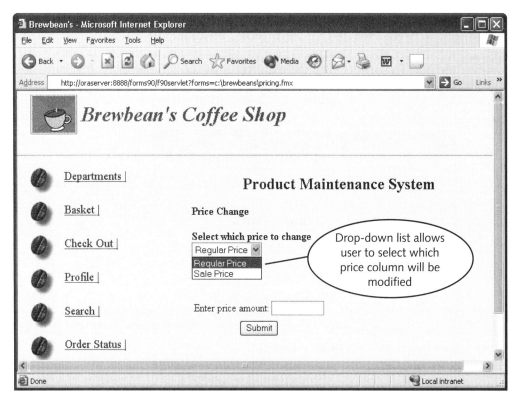

Figure 10-4   Brewbean's price change application screen

This capability could be invaluable to Brewbean's as the company does not need a full-time DBA to manage their Oracle database but desires the flexibility of making database modifications such as adding new columns on their own. The DBMS_SQL steps necessary to create dynamic DDL statements are listed in Table 10-2.

Table 10-2   Steps to perform DDL

Step	Program	Description	Parameters
1	OPEN_CURSOR	Establishes a cursor or work area for the statement to be handled	N/A
2	PARSE	Checks syntax and executes the DDL statement	Cursor, statement to parse, and version behavior
3	CLOSE_CURSOR	Frees the resources allocated to the cursor	Cursor

An EXECUTE step is not needed with DDL statements because the PARSE step automatically executes the statement upon successful parsing. The following procedure supports an application screen that allows Brewbean's employees to add a column to any table:

```
CREATE OR REPLACE PROCEDURE dyn_ddl_sp
  (p_table IN VARCHAR2,
   p_col IN VARCHAR2,
   p_type IN VARCHAR2)
 IS
  lv_cursor INTEGER;
  lv_add VARCHAR2(100);
BEGIN
  --Open the cursor
  lv_cursor := DBMS_SQL.OPEN_CURSOR;
  --Build the DDL statement
  lv_add := 'ALTER TABLE '|| p_table || ' ADD ('||
            p_col || ' ' || p_type || ')';
  --Parse and execute the statement
  DBMS_SQL.PARSE(lv_cursor, lv_add, DBMS_SQL.NATIVE);
  --Close the cursor
  DBMS_SQL.CLOSE_CURSOR(lv_cursor);
END;
```

10

Notice that parameters are provided to supply the values for the table name, column name, and the data type. The Brewbean's manager has decided to assign an employee as the leader of each product line. To do so, the manager needs to add a column to the BB_DEPARTMENT table to store the employee name for each department. Let's try completing this task.

To use dynamic DDL via the DBMS_SQL package:

1. Open or return to SQL*Plus.

2. Click **File** and then **Open** from the main menu.

3. Navigate to and select the **ddl10.sql** file from the Chapter.10 folder.

4. Click **Open** and a copy of the file contents appears in the SQL*Plus screen. Review the code.

5. To run the code and create the procedure, click **File** and then **Run** from the main menu.

6. Add a column to the BB_DEPARTMENT table by entering the following procedure invocation:

```
EXECUTE dyn_ddl_sp('bb_department','leader',
                                  'VARCHAR2(20)');
```

7. Confirm the addition of the new column, as shown in Figure 10-5.

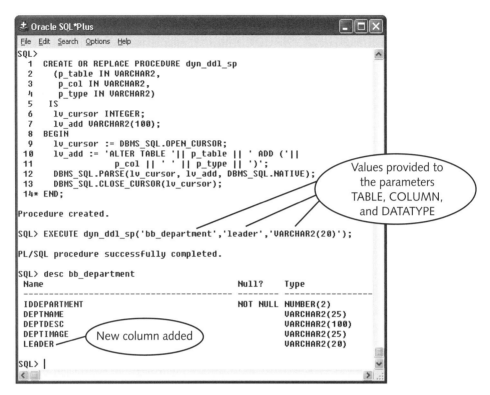

```
± Oracle SQL*Plus                                              [_][□][X]
File  Edit  Search  Options  Help
SQL>
  1   CREATE OR REPLACE PROCEDURE dyn_ddl_sp
  2     (p_table IN VARCHAR2,
  3      p_col IN VARCHAR2,
  4      p_type IN VARCHAR2)
  5   IS
  6     lv_cursor INTEGER;
  7     lv_add VARCHAR2(100);
  8   BEGIN
  9     lv_cursor := DBMS_SQL.OPEN_CURSOR;
 10     lv_add := 'ALTER TABLE '|| p_table || ' ADD ('||
 11              p_col || ' ' || p_type || ')';
 12     DBMS_SQL.PARSE(lv_cursor, lv_add, DBMS_SQL.NATIVE);
 13     DBMS_SQL.CLOSE_CURSOR(lv_cursor);
 14*  END;

Procedure created.

SQL> EXECUTE dyn_ddl_sp('bb_department','leader','VARCHAR2(20)');

PL/SQL procedure successfully completed.

SQL> desc bb_department
 Name                                        Null?    Type
 ------------------------------------------  -------- -----------------

 IDDEPARTMENT                                NOT NULL NUMBER(2)
 DEPTNAME                                             VARCHAR2(25)
 DEPTDESC                                             VARCHAR2(100)
 DEPTIMAGE                                            VARCHAR2(25)
 LEADER                                               VARCHAR2(20)

SQL> |
```

Values provided to the parameters TABLE, COLUMN, and DATATYPE

New column added

Figure 10-5   Performing DDL with table and column values provided at runtime

Queries with DBMS_SQL

Probably the most popular use of dynamic SQL is providing easy-to-use query capability with great flexibility for end users to perform data analysis and reporting. The goal is to empower users with little or no SQL skills to be able to perform a variety of data queries. Dynamic SQL allows the criteria of a query such as the column checked in the WHERE clause to be identified at runtime.

Let's say the Brewbean's application needs more flexibility in the product lookup screen so that the user can either provide a product id or name. In this case, we can build one procedure using dynamic SQL to support this application screen. First, Table 10-3 lists all the steps involved in processing a query with the DBMS_SQL package. Because we are now retrieving data, the DEFINE_COLUMN and COLUMN_VALUE programs are used to associate a PL/SQL variable to each of the column values returned in the query. Also, the FETCH_ROWS program is used to retrieve the data returned.

Table 10-3   Steps to perform queries

Step	Program	Description	Parameters
1	OPEN_CURSOR	Establishes a cursor or work area for the statement to be handled	N/A
2	PARSE	Checks syntax of statement	Cursor, statement to parse, and version behavior
3	BIND_VARIABLE	Associates placeholders in the statement with PL/SQL variables	Cursor, placeholder name, and value to be assigned
4	DEFINE_COLUMN	Identifies type and length of the PL/SQL variables that hold the column values selected when fetched	Cursor, column position in query, associated PL/SQL variable, and size. Note: Size only needed for columns of type VARCHAR2, CHAR, and RAW
5	EXECUTE	Runs the statement and returns the number of rows affected	Cursor
6	FETCH_ROWS	Retrieves the data returned by the query	Cursor
7	COLUMN_VALUE	Returns the individual values fetched to the variables indicated	Cursor, column position in query, and associated PL/SQL variable
8	CLOSE_CURSOR	Frees the resources allocated to the cursor	Cursor

10

Let's first review the following code, including comments, that accomplishes the query task and then we can execute the procedure. Note that the DEFINE_COLUMN program is now used to associate a data type to each value being returned by the query by matching a declared PL/SQL variable with each column queried. In addition, the FETCH_ROWS and COLUMN_VALUE programs are used to retrieve the query results from the cursor and pass the values returned to PL/SQL variables.

```
CREATE OR REPLACE PROCEDURE dyn_query1_sp
   (p_col IN VARCHAR2,
    p_value IN VARCHAR2)
 IS
  lv_query LONG;
  lv_status INTEGER;
  lv_cursor INTEGER;
  lv_col1 NUMBER(2);
  lv_col2 VARCHAR2(25);
  lv_col3 NUMBER(6,2);
  lv_col4 NUMBER(5,1);
BEGIN
  --Open the cursor
```

```
lv_cursor := DBMS_SQL.OPEN_CURSOR;
--Build the query
lv_query := 'SELECT idProduct, productname, price, stock
            FROM bb_product
            WHERE '|| UPPER(p_col) ||' = ' ||
                                'UPPER(:ph_value)';
--Parse the statement
DBMS_SQL.PARSE(lv_cursor, lv_query, DBMS_SQL.NATIVE);
--Identify data types for each item selected
DBMS_SQL.DEFINE_COLUMN(lv_cursor, 1, lv_col1);
DBMS_SQL.DEFINE_COLUMN(lv_cursor, 2, lv_col2, 25);
DBMS_SQL.DEFINE_COLUMN(lv_cursor, 3, lv_col3);
DBMS_SQL.DEFINE_COLUMN(lv_cursor, 4, lv_col4);
--Associate placeholder with a parameter
DBMS_SQL.BIND_VARIABLE(lv_cursor, ':ph_value', p_value);
--Execute the query
lv_status := DBMS_SQL.EXECUTE(lv_cursor);
--Fetch row returned and place into PL/SQL variables
IF (DBMS_SQL.FETCH_ROWS(lv_cursor) > 0) THEN
    DBMS_SQL.COLUMN_VALUE(lv_cursor, 1, lv_col1);
    DBMS_SQL.COLUMN_VALUE(lv_cursor, 2, lv_col2);
    DBMS_SQL.COLUMN_VALUE(lv_cursor, 3, lv_col3);
    DBMS_SQL.COLUMN_VALUE(lv_cursor, 4, lv_col4);
    DBMS_OUTPUT.PUT_LINE(lv_col1||' '||lv_col2||'
                        '||lv_col3||' '||lv_col4);
END IF;
--Close cursor
DBMS_SQL.CLOSE_CURSOR(lv_cursor);
END;
```

Now, let's create and execute the procedure to test the query activity. We will test first based on a product id and then based on a product name.

To create and test a dynamic query procedure:

1. Open or return to SQL*Plus.

2. Type **SET SERVEROUTPUT ON**.

3. Click **File** and then **Open** from the main menu.

4. Navigate to and select the **query10.sql** file from the Chapter.10 folder.

5. Click **Open** and a copy of the file contents appears in the SQL*Plus screen.

6. To run the code and create the procedure, click **File** and then **Run** from the main menu.

7. Test the procedure to query based on a product id by entering the following procedure invocation. Results are shown in Figure 10-6.

   ```
   EXECUTE dyn_query1_sp('idProduct',8);
   ```

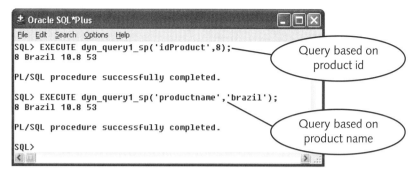

Figure 10-6   Dynamic query results

8. Now test the procedure to query based on a product name by entering the following procedure invocation. Results are also shown in Figure 10-6.

```
EXECUTE dyn_query1_sp('productname','brazil');
```

Even though this query is more dynamic than what we could accomplish directly in a PL/SQL block, it is just the tip of the iceberg in regards to the full dynamic potential available. In this procedure, we could have used parameters to provide the table name at runtime as well. In this case, the procedure could query columns from any table based on a chosen criteria.

The procedure as is handles only one row being returned from the query. What if we want the end user to be able to identify more than one criteria and return more than one row of data? This could be a powerful query application for Brewbean's to provide a product search, which can be based on the product description, price, availability, and so forth—whatever the end user is interested in! The DBMS_SQL package is particularly suited to handle array-type processing to simplify this task. Let's try it out.

To use array-like processing with DBMS_SQL:

1. Open or return to SQL*Plus.

2. Type **SET SERVEROUTPUT ON**.

3. Create a data type using the following TYPE command. The type named ARRAY is an index-by table that enables the array-like processing in this procedure.

```
CREATE TYPE array IS TABLE OF VARCHAR2(100)
/
```

4. Click **File** and then **Open** from the main menu.

5. Navigate to and select the **queryb10.sql** file from the Chapter.10 folder.

6. Click **Open** and a copy of the file contents appears in the SQL*Plus screen. Notice a FOR loop is used to construct a query containing multiple criteria.

7. To run the code and create the procedure, click **File** and then **Run** from the main menu.

8. Invoke the procedure using the following anonymous block, which provides two search criteria, one for the description to include the term "nut" and another for the stock level to be above 25. The results are displayed in Figure 10-7.

```
BEGIN
  dyn_query2_sp(array('description','stock'),
                array('LIKE','>'),
                array('%nut%','25'));
END;
/
```

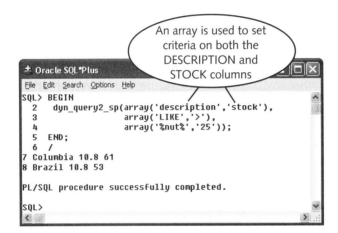

Figure 10-7 Using array-like processing in DBMS_SQL queries

Native Dynamic SQL

Native dynamic SQL was introduced with Oracle8*i* and provides a declarative method of performing dynamic SQL. The native dynamic SQL statements are much simpler to code than the DBMS_SQL statements and process more efficiently because support for native dynamic SQL statements is built into the PL/SQL interpreter. Two methods can be used to implement native dynamic SQL: EXECUTE IMMEDIATE and OPEN FOR. Let's take a look at each of these. We'll then look at the differences between the DBMS_SQL package and native dynamic SQL.

Using EXECUTE IMMEDIATE

The EXECUTE IMMEDIATE statement contains four clauses with the following syntax:

```
EXECUTE IMMEDIATE 'SQL statement'
    [ INTO ( var1, var2, … |  record ]
    [ USING [ IN | OUT | IN OUT ] bindvar1, bindvar2, …]
    [ RETURNING | RETURN INTO outvar1, outvar2, …];
```

The SQL statement is built as a text string using parameters and placeholders just as we did with the DBMS_SQL package. The INTO clause is used with queries and indicates which PL/SQL variables should hold the values returned from the SELECT statement. The USING clause associates placeholders in the SQL statement with PL/SQL variables or parameters. The RETURNING clause associates values to OUT parameters or a RETURN clause in a function.

To compare the coding of the DBMS_SQL to native dynamic SQL, let's redo the DML statement and query we already accomplished with the DBMS_SQL package. First, we redo the product price UPDATE statement, as shown in following code. Notice how compact the code is as compared to the DBMS_SQL code.

```
CREATE OR REPLACE PROCEDURE dyn_dml2_sp
   (p_col VARCHAR2,
    p_price NUMBER,
    p_id NUMBER)
 IS
BEGIN
 EXECUTE IMMEDIATE 'UPDATE bb_product
                    SET ' || p_col || ' = :ph_price
                    WHERE idProduct = :ph_id'
                    USING p_price, p_id;
END;
/
```

Figure 10-8 demonstrates the execution of this procedure.

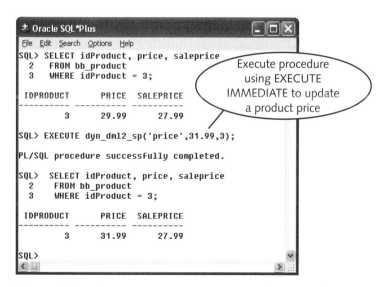

Figure 10-8 Executing a DML statement with native dynamic SQL

Native dynamic SQL provides two methods to perform queries: the EXECUTE IMMEDIATE statement for single row queries and the OPEN FOR statement for multiple row queries. We explore the OPEN FOR statement in the next section.

Let's first look at a query performed using EXECUTE IMMEDIATE. We can redo the product search query we performed earlier with the DBMS_SQL package. The following code is a procedure using EXECUTE IMMEDIATE to allow a product lookup based on either the product id or name. Review the code, noting the comments describing the activity.

```
CREATE OR REPLACE PROCEDURE dyn_query3_sp
   (p_col IN VARCHAR2,
    p_value IN VARCHAR2)
 IS
  lv_query VARCHAR2(200);
  lv_id bb_product.idProduct%TYPE;
  lv_name bb_product.productname%TYPE;
  lv_price bb_product.price%TYPE;
  lv_stock bb_product.stock%TYPE;
BEGIN
  --use a variable to hold the query construction to
  -- make it more readable
  lv_query := 'SELECT idProduct, productname, price, stock
               FROM bb_product
               WHERE UPPER(' || p_col || ') =
                                    UPPER(:ph_value)';
  --Run the dynamic query supplying variables to hold the
  -- return values in the INTO clause and associate the
  -- parameter to the placeholder with the USING clause
  EXECUTE IMMEDIATE lv_query
               INTO lv_id, lv_name, lv_price, lv_stock
               USING p_value;
  DBMS_OUTPUT.PUT_LINE(lv_id||' '||lv_name||' '
                              ||lv_price||' '||lv_stock);
END;
/
```

Figure 10-9 demonstrates invoking this procedure with both a product id search and a product name search.

Using OPEN FOR

Now, let's take a look at the OPEN FOR native dynamic SQL statement to enable multiple row queries. A REF CURSOR is used to reference the values returned by the query. Recall that a REF CURSOR is a pointer to a cursor or SQL work area. The package in the following code contains a TYPE statement to create a REF CURSOR type and one procedure, which allows the user to select all rows from the BB_PRODUCT table based on a single criterion to include whatever column and value is desired.

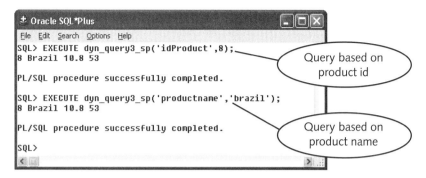

Figure 10-9   Using native dynamic SQL to query data

```
CREATE OR REPLACE PACKAGE dyn_pkg
  AS
    TYPE refcur_type IS REF CURSOR;
    PROCEDURE dyn_query4_sp
      (p_col IN VARCHAR2,
       p_value IN VARCHAR2,
       p_cursor IN OUT refcur_type);
END;

CREATE OR REPLACE PACKAGE BODY dyn_pkg
  AS
  PROCEDURE dyn_query4_sp
    (p_col IN VARCHAR2,
     p_value IN VARCHAR2,
     p_cursor IN OUT refcur_type)
   IS
    lv_query VARCHAR2(200);
    lv_bind VARCHAR2(20);
  BEGIN
   --Build query
   lv_query := 'SELECT idProduct, productname, price, stock
                 FROM bb_product
                 WHERE UPPER(' || p_col || ') LIKE
UPPER(:ph_value)';
   --Use variable to add wildcard characters to value
   lv_bind := '%'|| p_value || '%';
   --Open cursor with the query and associate a variable
   -- to the placeholder
   OPEN p_cursor FOR lv_query USING lv_bind;
  END;
END;
/
```

10

Figure 10-10 displays an execution of the DYN_QUERY4_SP procedure, checking for any product that contains the term "nut" in the DESCRIPTION column.

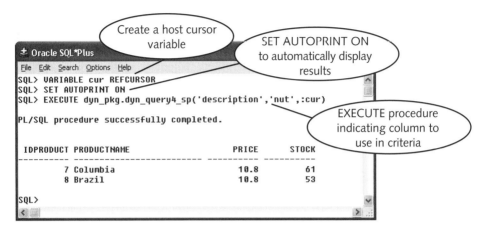

Figure 10-10   Using the OPEN FOR native dynamic SQL statement

DBMS_SQL Versus Native Dynamic SQL

Native dynamic SQL has not completely replaced the functionality provided with the DBMS_SQL package in regards to processing dynamic SQL. Native dynamic SQL is simpler to code and processes more efficiently; therefore, it should be used if possible. However, not all situations can be handled with the functionality of native dynamic SQL and, in these cases, the DBMS_SQL package must be used. The following list provides guidelines as to when it is appropriate to use native dynamic SQL:

- The number and types of columns to be used is known.

- The number and type of bind variables is known.

- You want to perform DDL.

- You are executing the statement only once or twice.

- User-defined types such as object and collections (not supported by DBMS_SQL) are used.

- You are fetching rows of data into PL/SQL records (not supported by DBMS_SQL).

- The SQL statement is less than 32 KB in size.

EXECUTE IMMEDIATE cannot reuse parsed statements; therefore, using the DBMS_SQL package offers gains in performance if a statement is to be used repeatedly, which is particularly an issue in multiuser environments. In addition, the array-like processing available in DBMS_SQL can make working with many rows easier. The DBMS_SQL package should be used if statements are repeatedly used, the number and types of columns are unknown, the number and types of bind variables are unknown, or the SQL statement is large (over 32 KB). For additional comparisons of DBMS_SQL to native dynamic SQL, reference the "Oracle9*i* Application Developer's Guide – Fundamentals Release 2 (9.2)" on the OTN Web site.

OBJECT TECHNOLOGY

Object-oriented programming has been gaining popularity over the past decade and is an approach to database and application design attempting to more closely model the "real world." This approach combines both the data types and applicable operations (program units) into the data structure or object.

Note that this section is not meant to be an introduction to object-oriented programming concepts, including topics such as inheritance, but is an introduction to the object technologies offered by Oracle, including object types, methods, and views. In other words, the object technologies or mechanisms used to build an object-oriented application are addressed.

Oracle's object-type model is similar to the class mechanism found in C++ and Java. With Oracle objects, entities such as a customer or an order are represented as an object, which contains attributes and methods. Attributes are much the same as columns and methods are similar to program units. The data structure and the programmatic actions related to the specific data are stored together as a unit called an object type. Many developers feel there are advantages to this approach, such as standardization of data and tighter control of program units as they are associated with the data with which their action is intended.

This is an extensive subject; however, as a PL/SQL developer, you should still be particularly aware of the object technology features available in Oracle. You should be aware of them because of the many purported advantages of object technology, which include ease of modeling business operations and more efficient development due to reusability of objects and simplification of code design. Because this methodology is gaining in popularity, you most likely will encounter existing code using these technologies and will need to be prepared to handle such code.

This section provides a brief introduction to object technology. A document titled "Application Developer's Guide – Object-Relational Features" is available on the OTN Web site and provides more in-depth coverage of this topic.

Creating Object Types

The Brewbean's database stores a number of addresses, such as shipping, billing, and employee addresses. We will create an object type to contain the various components of an address and use the object type as the data type of each column of a database table

10

in which address data is stored. Creating a single object to represent all address data ensures consistency of data for every address that is in the database because each uses the same object type. The following code creates an object type named ADDR_OT, which contains five attributes: STREET1, STREET2, CITY, STATE, and ZIP.

```
CREATE OR REPLACE TYPE addr_ot AS OBJECT
  (street1 VARCHAR2(25),
   street2 VARCHAR2(25),
   city VARCHAR2(25),
   state CHAR(2),
   zip NUMBER(9) );
/
```

The ADDR_OT object type is used as the data type for the billing address and shipping address columns of the BB_ORDER table. Object types used as database table columns are considered persistent because they are stored in the database. Object types can also be used as data types for PL/SQL variables, which are considered transient because they exist only for the duration of the PL/SQL block.

The following code creates the BB_ORDER table using the ADDR_OT object type for the billing and shipping address columns:

```
CREATE TABLE bb_order
 (ord_id NUMBER(4),
  cust_id NUMBER(4),
  ord_date DATE,
  total NUMBER(6,2),
  bill_addr addr_ot,
  ship_addr addr_ot );
```

Using an Object Type

Now we have individual columns in the BB_ORDER table that actually contain five elements of data. So, how do we insert data into such a structure? When an object type is created, a default constructor is automatically created. This constructor is a method that enables the creation of an object of the named type. For example, if we need to add a new order to the BB_ORDER table, the ADDR_OT constructor would be invoked for the BILL_ADDR and SHIP_ADDR columns to match provided data values to each of the attributes in the object type.

Let's go ahead and create the object type so that we can work with it to better visualize how it is used. We first create the object type and table as discussed previously. We then move on to insert data into the object type by using the constructor.

To create and use an object type:

1. Open or return to SQL*Plus.

2. Click **File** and then **Open** from the main menu.

3. Navigate to and select the **objecta10.sql** file from the Chapter.10 folder.

4. Click **Open** and a copy of the file contents appears in the SQL*Plus screen.

5. To run the code and create the object type, click **File** and then **Run** from the main menu.

6. To create the BB_ORDER table in which we use the ADDR_OT object type, repeat Steps 2 to 5 using the **objectb10.sql** file.

7. Now, we can use the constructor to insert data into the object type columns. Enter the following statement. Notice the first two assignment statements use the constructor to create a variable that contains data for each attribute of the ADDR_OT object type.

```
DECLARE
  lv_bill addr_ot;
  lv_ship addr_ot;
BEGIN
  lv_bill := addr_ot('11 Bush Dr' ,NULL, 'Savannah',
                                    'GA',346668229);
  lv_ship := addr_ot('812 Scott Lane','Apt #52',
                         'Savannah','GA',346668227);
  INSERT INTO bb_order
      VALUES (102,31,'11-NOV-03',34.50,lv_bill,lv_ship);
END;
/
```

The constructor can be invoked directly within an INSERT statement; however, creating variables to hold such values makes the INSERT statement more readable.

8. Now that we have data in an object type column, how do we retrieve data from it? First, enter a basic SELECT statement, as shown in Figure 10-11. Notice that the object type and a list of all the attribute values for that column are presented as a group.

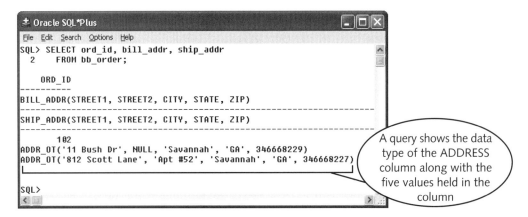

Figure 10-11   Basic SELECT of object type column

9. What if we want to retrieve an individual value from within the object type? Attempt a query using a prefix of the individual attribute name with the column name using a dot notation, as shown in Figure 10-12.

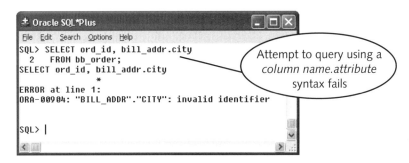

Figure 10-12   Retrieving individual attribute values incorrectly

10. Now query referencing an individual value with a table prefix and column prefix, as shown in Figure 10-13. The table name must be used even if there is only a single table in the query statement.

11. In reviewing the structure of a table, what if we want to view the attributes of the construct? Try a basic DESCRIBE command on the BB_ORDER table, as shown in Figure 10-14.

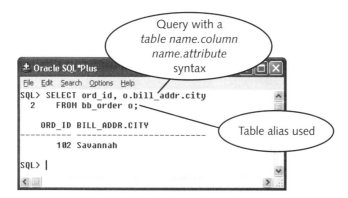

Figure 10-13   Retrieving individual attribute values correctly

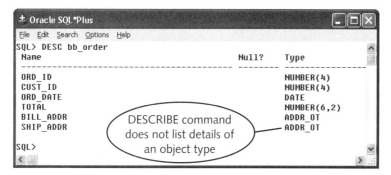

Figure 10-14   DESCRIBE command with object type columns

> 12. Now enter the DESCRIBE command using the DEPTH ALL option, as shown in Figure 10-15.

Object Methods

Beyond data standardization, another reason for using object types is that it enables procedures and functions that are to perform actions specific to this data to be stored in the object type. This forms a bond with the data and associated programmatic actions. The program units stored in an object type are called methods and are invoked by using a dot notation referencing the object and method. An example of a method Brewbean's wants to associate with the address object type is formatting the address to print in a label style.

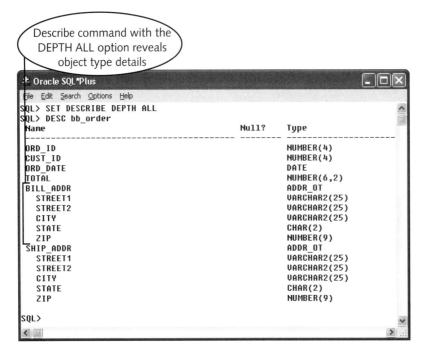

Figure 10-15 Using the DEPTH ALL option of the DESCRIBE command

Adding a program unit to an object type is very similar to setting up packages. First, we alter the object type to contain a specification for a function. Second, we create a type body that contains the full code of the function. Because the ADDR_OT object type already exists and we just want to add to it, we use the ALTER TYPE command to add the method or program unit to the object type. The key word MEMBER is used to identify each method followed by a program unit specification. Let's alter the ADDR_OT object type we created earlier to contain a method.

To add a method to an object type:

1. Open or return to SQL*Plus.

2. Enter the following ALTER TYPE statement to add a method named LBL_PRINT. Note the similarity to the code we would include in a package specification. The CASCADE option is used because the ADDR_OT type has dependent objects (the BB_ORDER table also uses this object type).

```
ALTER TYPE addr_ot
   ADD MEMBER FUNCTION lbl_print
RETURN VARCHAR2 CASCADE;
```

3. Close SQL*Plus and then open it again. Enter the DESCRIBE command, as shown in Figure 10-16, to confirm that the method information can also be viewed for an object type.

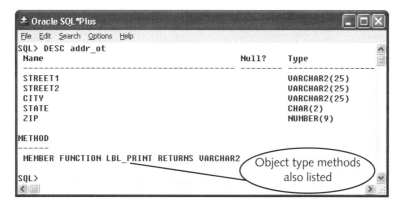

Figure 10-16 Using DESCRIBE to view methods of an object type

4. Now, to establish the function, we use the CREATE TYPE BODY command, as shown in Figure 10-17. This is quite similar to the PACKAGE BODY statement. The code is available in the body10.sql file in the Chapter.10 folder for your reference.

10

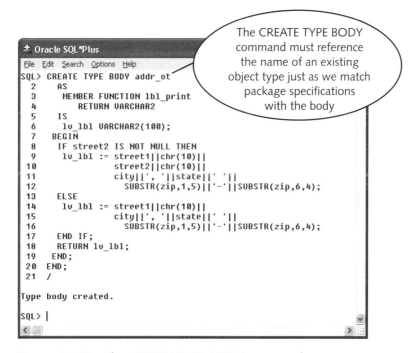

Figure 10-17 The CREATE TYPE BODY command

5. Now that the object type includes a method, how do we use it? Let's try a simple select of the SHIP_ADDR column using the LBL_PRINT method,

as shown in Figure 10-18. Notice the method name is followed by open and close parentheses that instruct the system to look for a method with this name in the object.

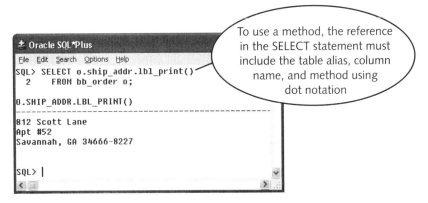

Figure 10-18   Invoking the LBL_PRINT method

> The ALTER TYPE options are fairly rigid in Oracle8*i* and many times developers are forced to rebuild tables based on an object after the object is altered. However, Oracle9*i* has added many features in this arena making object types much more flexible when making database changes. For more information on the full list of added features, reference the "Oracle9*i* Application Developer's Guide – Object-Relational Features Release 2 (9.2)" available on the OTN Web site.

Object Relations

Creating a database entirely based on an object relational model typically involves using tables of object types. Object types are created to represent all the needed data structures and then tables are built based on these object types. For example, let's approach the customer and order entities of the Brewbean's operation in this manner. First, we construct an object type to reflect the customer data and a table based on this object type. Then, we add the ORDER table in which a one-to-many relationship between the customer and ORDER table based on customer id needs to be established. In our traditional relational database tables, we would use a foreign key constraint to enforce these relationships. Object types use REF variables to establish relationships and the REF variables act more as pointers to rows in another table.

To create an object relational design:

1. Open or return to SQL*Plus.

2. Enter the following code to create an object type to represent customer data:

```
CREATE TYPE cust_ot AS OBJECT
  (cust_id NUMBER(4),
```

```
   first VARCHAR2(15),
   last VARCHAR2(20),
   email VARCHAR2(25),
   phone NUMBER(10) );
/
```

As we know, Brewbean's stores more data attributes regarding customers; however, these examples use only a few attributes in an attempt to more clearly illustrate the concepts.

3. Now create the customer table based on this object type with the following CREATE TABLE statement. Notice the PRIMARY KEY clause establishes a row uniqueness constraint based on an attribute of the object type.

```
CREATE TABLE bb_cust OF cust_ot
  (PRIMARY KEY (cust_id));
```

4. Create an object type for the order data that establishes a relationship with the customer data, as shown in the following code. The CUST_REF column is created with a REF data type, which references the CUST_OT object type. Recall the customer table was created with a primary key of CUST_ID. The REF column looks for this primary key to establish the relationship.

```
CREATE TYPE ord_ot AS OBJECT
  (ord_id NUMBER(4),
   cust_ref REF cust_ot,
   ord_date DATE,
   total NUMBER(6,2));
/
CREATE TABLE bb_ord OF ord_ot
  (PRIMARY KEY (ord_id));
```

5. Next, let's put some data into the customer and order tables using the following INSERTs to test how the REF column works. The REF operator needs a table alias that references the table for which a relationship should be established. The FROM clause identifies this table, and the WHERE clause instructs the REF operator to the row to which the relationship or correlation variable should be established.

```
INSERT INTO bb_cust
  VALUES
    (cust_ot(12,'Joe','Cool','jcool@yahoo.com',7773335555));
INSERT INTO bb_ord
  SELECT ord_ot(1,REF(c),SYSDATE,24.50)
    FROM bb_cust c
    WHERE cust_id = 12;
COMMIT;
```

6. Query the order table as shown in Figure 10-19 to see how the REF column appears. Notice the CUST_REF column contains an internally generated reference value that serves as the row pointer.

10

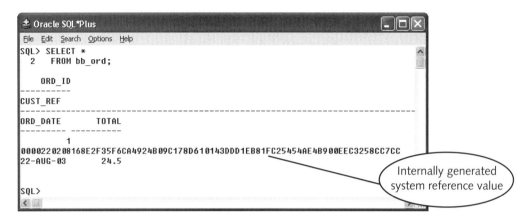

Figure 10-19 Listing the contents of a REF column

7. We can query individual attributes of the customer object type using a dot notation that provides the table, REF column name, and attribute name, as shown in Figure 10-20, which displays the order number and the associated customer last name.

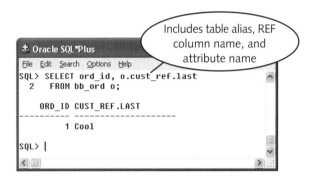

Figure 10-20 Using the REF column to retrieve values

REF Pointers Versus Foreign Keys

As we work with object types, it is critical to recognize that the behavior of REF pointers and foreign keys are quite different when it comes to managing relationships. If we create a foreign key constraint between the customer and order tables based on the customer id, an error is raised if we attempt to delete a customer record that is referenced in the order table. However, the same is not true when using a REF pointer to establish relationships. To confirm this, let's experiment with a DELETE action in the following steps.

To explore the REF pointer relationship:

1. Open or return to SQL*Plus.

2. Delete the data in the BB_CUST table with the following statements. Notice the DELETE action is accepted even though the BB_ORD table references the customer data.

```
DELETE FROM bb_cust;
COMMIT;
```

3. Query the BB_ORD table, as shown in Figure 10-21, to confirm that a NULL value is now shown for the customer last name.

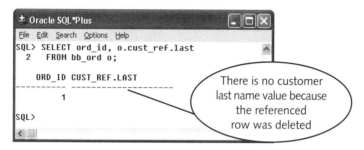

Figure 10-21   Confirmation of the deletion

4. REF pointers do provide a feature to identify when relationships are being broken. This feature is the dangling check, which identifies if the REF pointer now has no associated row. Let's check if this condition is true for the row in the order table for which we deleted the associated customer row. Enter the query in Figure 10-22 to check whether the IS DANGLING condition is true.

Figure 10-22 shows a value of 1 returned. Note that IS DANGLING returns either a 0 (zero) or a 1, which mean FALSE and TRUE, respectively.

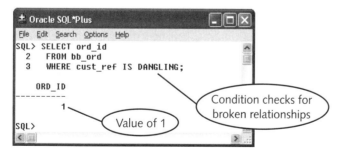

Figure 10-22   Checking the CUST_REF column for a lost REF value

Even though the REF pointer allows the deletion of a parent record, this situation is usually contained by using a BEFORE DELETE trigger to check if child records exist.

Object Views

Oracle's object technology now offers the possibility of an object-oriented approach to designing the database; however, what about the numerous shops that have the traditional Oracle relational database construction supporting their existing applications? For example, Brewbean's has already invested much time in building their database using the traditional relational design; can they take advantage of Oracle's object technology features?

Good question! They can in fact introduce object technology features into their existing databases by using object views, which basically creates a layer on top of the relational database so that the database can be viewed in terms of objects. Object views are the key to creating object-oriented applications without modifying existing relational database schemas.

Let's return to the original Brewbean's database that contains the BB_SHOPPER and BB_BASKET tables. These two tables were created with the CREATE TABLE statement, which includes the establishment of a foreign key constraint, based on the SHOPPER ID column.

Let's create an object view to represent these tables in an object structure. Figure 10-23 displays a diagram of the various components that are involved in this process. The view created in essence maps to base tables (just like views that we have created before); however, it also places the data into object form by associating object types to the data being retrieved. Therefore, we need to create the object types needed before creating the object view.

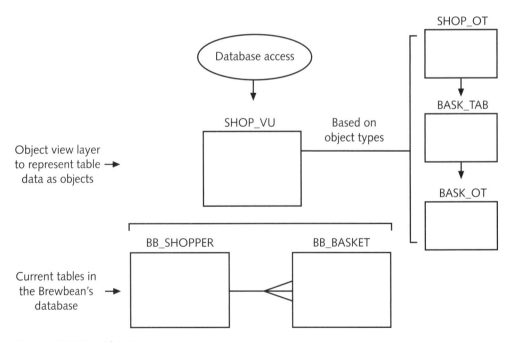

Figure 10-23 Object view components

The object view is based on the shopper and must reflect that each shopper may have many baskets as a one-to-many relationship was established between these tables with a foreign key. Let's review the new operators that we encounter in the CREATE VIEW statement:

- OBJECT OID indicates which column, typically the primary key column, of the base table provides a unique identifier for each object created.

- CAST allows a subquery to provide values for a nested table column in the view. The selected column list used in the CAST expression must match the attribute list of the object type used for this column.

- MULTISET enables the CREATE VIEW statement to handle multiple rows returned from the subquery in the CAST expression rather than a single row.

Take a look at the following code that establishes the object view for the BB_SHOPPER and BB_BASKET tables. The first statements create object types to represent the table data in the object view. A table of the BASK_OT object type is created to use as an attribute in the SHOP_OT object type to reflect that many baskets can be associated with a single shopper.

```
CREATE TYPE bask_ot AS OBJECT
  (idBasket NUMBER(5),
   total NUMBER(7,2))
/
CREATE TYPE bask_tab AS TABLE OF bask_ot
/
CREATE TYPE shop_ot AS OBJECT
  (idShopper NUMBER(4),
   last_name VARCHAR2(20),
   city VARCHAR2(20),
   idBasket bask_tab)
/
CREATE VIEW shop_vu OF shop_ot
  WITH OBJECT OID(idShopper)
 AS
   SELECT s.idShopper, s.lastname, s.city,
     CAST(MULTISET(SELECT idBasket, total
                   FROM bb_basket b
                   WHERE b.idShopper = s.idShopper)
        AS bask_tab)
     FROM bb_shopper s;
```

Note the OF SHOP_OT portion on the first line of the CREATE VIEW statement, which associates an object type to the view. In addition, the unique identifier of each object is the IDSHOPPER value, as indicated by the OBJECT OID expression. In addition, the CAST and MULTISET expressions are used to provide the values for the last attribute of the SHOP_OT object type, which is the table of baskets. All of the operations on the data can now be handled through the object view to treat the database in an object-oriented

10

fashion. Figure 10-24 shows a simple SELECT statement on the object view to demonstrate the values returned for an object. Note that shopper 21 has two baskets.

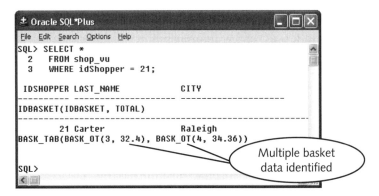

Figure 10-24   Query an object view

Sorting and Comparing Object Type Columns

Sorting and comparing simple data such as scalar data types is fairly straightforward. However, these tasks are a bit more challenging when more complex data types such as object types are introduced. Keep in mind that an object type can contain a variety of attributes. Figure 10-25 demonstrates what happens when attempting an ORDER BY clause on an object type column in a query.

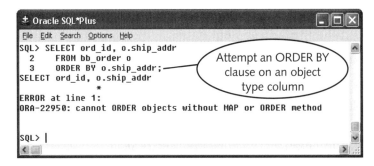

Figure 10-25   Sorting error on an ADDR_OT object type column

Two special methods of object types named ORDER and MAP are available to enable the sorting and comparison of object type columns. An object type can contain only one of these methods. (I introduce only one of these, the MAP method, just to expose you to the concept.) Let's use the BB_ORDER table in the following steps to use the MAP method.

To sort object type columns:

1. Open or return to SQL*Plus.

2. First, we need to add data to the BB_ORDER table so that we have two rows to be sorted. Click **File** and then **Open** from the main menu.

3. Navigate to and select the **map10.sql** file from the Chapter.10 folder.

4. Click **Open** and a copy of the file contents appears in the SQL*Plus screen.

5. To run the code and create the object type, click **File** and then **Run** from the main menu.

6. Verify the data in the BB_ORDER table using the query in Figure 10-26.

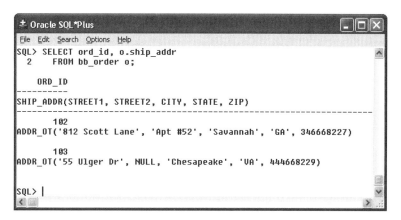

Figure 10-26   The BB_ORDER table contains two rows

7. Next, we alter the ADDR_OT object type to add a MAP method specification, as shown in the following ALTER TYPE statement:

```
ALTER TYPE addr_ot
   ADD MAP MEMBER FUNCTION mapping RETURN VARCHAR2 CASCADE;
```

8. Add the MAP method to the TYPE BODY, as shown in Figure 10-27, using a CREATE OR REPLACE TYPE BODY statement. In this case, Brewbean's decided they want to sort the address fields based on the state and city.

9. Enter the query shown in Figure 10-28 to confirm the ORDER BY clause that is now successful due to the MAP method being added to the object type.

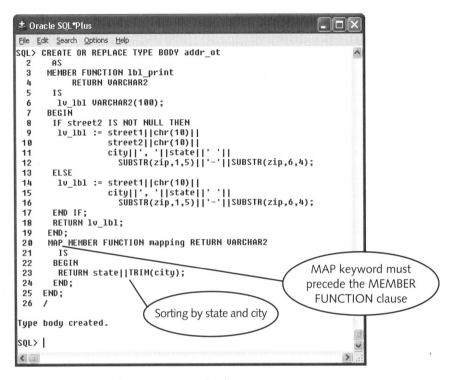

Figure 10-27   Adding a MAP method

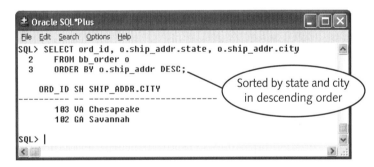

Figure 10-28   Sorting an object type column

CHAPTER SUMMARY

❏ Dynamic SQL allows identifiers, such as table names and column names, to be provided at runtime.

❏ Dynamic SQL allows DDL statements to be included in PL/SQL programs.

❐ Dynamic SQL can be implemented with the DBMS_SQL package or by using native dynamic SQL.

❐ All processing is handled manually with DBMS_SQL, including opening the cursor, parsing, executing the statement, and closing the cursor.

❐ DBMS_SQL uses placeholders to provide values into SQL statements.

❐ Multiple row queries are managed with DBMS_SQL using the FETCH_ROWS program.

❐ Native dynamic SQL is simpler to code and it processes statements more efficiently; however, it cannot accomplish all the tasks that can be done using the DBMS_SQL package.

❐ Native dynamic SQL uses the EXECUTE IMMEDIATE and OPEN FOR statements. DBMS_SQL is most appropriate when the number of columns or bind variables used in the statement is unknown until runtime.

❐ Object types knit the data structure and associated program units into a single object.

❐ Attributes are the data elements within an object type.

❐ Methods are procedures and functions within an object type and are also called members.

❐ Special methods of MAP and ORDER are available to enable sorting and comparing of object types.

❐ Relationships between objects are managed using REF pointers.

❐ Object views allow the creation of a layer on top of traditional relational databases that allow object technology features to be used.

10

REVIEW QUESTIONS

1. Dynamic SQL allows more flexibility in runtime values of an SQL statement in that _____.

   a. identifier values such as table and column names can be provided at runtime

   b. the WHERE clause of the SQL statement can affect different rows based on input at runtime

   c. users can write their own SQL statements

   d. it allows parameters that provide values in the SQL statement

2. Which SQL statement can be included in a PL/SQL block only via dynamic SQL?

   a. INSERT

   b. UPDATE

   c. DELETE

   d. CREATE TABLE

3. Dynamic SQL is implemented by using either of which two of the following features?

   a. EXECUTE IMMEDIATE

   b. DBMS_EXECUTE

   c. DBMS_SQL

   d. DYNAMIC EXECUTE

4. Which of the following is not a program in the DBMS_SQL package?

   a. OPEN_CURSOR

   b. GET_ROWS

   c. BIND_VARIABLE

   d. DEFINE_COLUMN

5. In which of the following scenarios would it be appropriate to use native dynamic SQL versus the DBMS_SQL package?

   a. to perform DDL operations

   b. when the statement is to be executed repeatedly

   c. when the number of columns used is not known until runtime

   d. when the column data types used is not known until runtime

6. Object type attributes are similar to _____.

   a. object methods

   b. table columns

   c. data types

   d. package variables

7. Object type methods can be which of the following?

   a. packages

   b. procedures

   c. functions

   d. specifications

8. An object type can be used as which of the following?

   a. function

   b. method

   c. table column

   d. PL/SQL variable

9. Which option of the DESCRIBE command must be in effect to display the attribute information about an object type table column?

    a. TREE

    b. DEPTH

    c. OBJECT

    d. SET

10. Relationships between object type tables are established using a _____.

    a. REF column

    b. foreign key constraint

    c. link

    d. bind variable

11. How would dynamic SQL assist in creating a database query system to be used by all Brewbean's employees?

12. Describe when it is appropriate to use native dynamic SQL versus the DBMS_SQL package to deploy dynamic SQL.

13. Describe what an object structure can include and how it is different than a traditional relational table structure.

14. Describe the differences in how data relationships are established and maintained when using object types versus traditional relational tables.

15. Describe an object view and what it is used for.

16. In reviewing the following CREATE statement, which items represent placeholder names?

```
CREATE OR REPLACE PROCEDURE update_prod
   (p_one VARCHAR2,
    p_two NUMBER,
    p_three NUMBER)
 IS
  lv_a INTEGER;
  lv_b VARCHAR2(150);
  lv_c NUMBER(1);
BEGIN
  lv_a := DBMS_SQL.OPEN_CURSOR;
  lv_b := 'UPDATE product
                SET ' || p_one || ' = :price
                WHERE idProd = :id';
  DBMS_SQL.PARSE(lv_a, lv_b, DBMS_SQL.NATIVE);
  DBMS_SQL.BIND_VARIABLE(lv_a, ':price', p_two);
  DBMS_SQL.BIND_VARIABLE(lv_a, ':id', p_three);
  lv_c := DBMS_SQL.EXECUTE(lv_a);
```

10

```
          DBMS_SQL.CLOSE_CURSOR(lv_a);
          COMMIT;
          DBMS_OUTPUT.PUT_LINE(lv_c);
        END;
```

    a. P_COL

    b. :PRICE

    c. IDPRODUCT

    d. :ID

17. In the code listed in Question 16, which variable contains a cursor reference?

    a. LV_A

    b. LV_B

    c. LV_C

    d. OPEN_CURSOR

18. In the code listed in Question 16, what value should be provided for the P_ONE parameter?

    a. table name

    b. column name

    c. column value

    d. placeholder

19. In reviewing the following EXECUTE IMMEDIATE statement (assume it is included as a part of a complete block), what do items two and three represent?

```
    EXECUTE IMMEDIATE 'UPDATE product
                      SET ' || one || ' = :price
                      WHERE idProd = :id'
                      USING two, three;
```

    a. placeholders

    b. parameters

    c. columns

    d. data types

20. Object types are considered transient if they are used for which of the following?

    a. database columns

    b. placeholders

    c. PL/SQL variables

    d. Object types are always transient.

HANDS-ON ASSIGNMENTS

Assignment 10-1: Use the DBMS_SQL Package

Brewbean's employees want an application screen in which they can query the customer table for id and last name based on one criterion on any customer information column. To accomplish this task, you create a procedure using DBMS_SQL to set up the dynamic query. Name the procedure DYN_CUST_SP and demonstrate that the procedure works by executing two different queries: one using a state of NC and another using an e-mail value of ratboy@msn.net. Keep in mind that any column could be used as criteria, which could result in more than one row returning from the query, such as two customers having the same last name. Therefore, this procedure needs to be able to handle multiple rows returned.

1. Open SQL*Plus and type **SET SERVEROUTPUT ON**.
2. Click **File** and then **Open** from the main menu.
3. Navigate to and select the **assignment01.sql** file from the Chapter.10 folder.
4. Click **Open** and a copy of the file contents appears in the SQL*Plus screen. Review the procedure code.
5. To run the code and create the procedure, click **File** and then **Run** from the main menu.
6. Now, let's execute the procedure using the two test cases. First, use the STATE column criteria, as shown in Figure 10-29.

Figure 10-29   Query with state criteria

7. Now check the customer e-mail address, as shown in Figure 10-30.

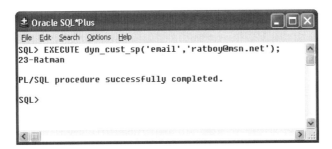

Figure 10-30 Query with e-mail address criteria

Assignment 10-2: Use Native Dynamic SQL

The Brewbean's manager wants an application screen that easily enables the addition of columns to the database by employees. Create a procedure that accepts appropriate inputs and uses native dynamic SQL to accomplish this task. Test the procedure by adding a column named MEMBER to the BB_SHOPPER table with a data type of CHAR(1).

1. Open or return to SQL*Plus and type **SET SERVEROUTPUT ON**.
2. In a text editor, open the **assignment02.txt** file from the Chapter.10 folder.
3. Review the incomplete procedure code and complete the EXECUTE IMMEDIATE statement to allow column additions.
4. Copy and paste the entire CREATE PROCEDURE statement into SQL*Plus.
5. Press **Enter** if needed to run the statement.
6. Now, let's execute the procedure using the following statement to add the MEMBER column:

```
EXECUTE dyn_addcol_sp('member','bb_shopper','CHAR(1)');
```

7. Type **DESC bb_shopper** and press **Enter** to list the table structure and confirm the addition of the MEMBER column.

Assignment 10-3: Create an Object Type

Brewbean's needs to expand the database to contain employee and vendor data. The lead programmer recognizes that contact information for a customer, employee, and vendor should all consist of the same items. To introduce more standardization to the database, an object type is requested for contact information to use as a column data type where needed. Create an object type named CONTACT_OT to contain attributes of the phone and fax numbers. In addition, include a method that formats the numbers to display with the first three digits in parentheses and a hyphen between the sixth and seventh digit. To test the object type, create a table named BB_CONTACT based on this object type. Then insert a row of data and execute a query to check the formatting method.

1. Open or return to SQL*Plus.
2. In a text editor, open the **assignment03.txt** file from the Chapter.10 folder.

3. Review the CREATE TYPE statement and complete the CREATE TYPE BODY statement needed.

4. Copy and paste all the code into SQL*Plus.

5. Press **Enter** if needed to run the statements.

6. Now, we will create a table based on this object type for testing. Type the following code to create the table:

```
CREATE TABLE bb_contact OF contact_ot;
```

7. Type the following INSERT statement to add a row of data to the table:

```
INSERT INTO bb_contact
  VALUES (contact_ot(9991114444,5552226666));
COMMIT;
```

8. Type the following query to view the data:

```
SELECT *
  FROM bb_contact;
```

9. Type the following query that invokes the formatting method. The results should match the display in Figure 10-31.

```
SELECT c.contact_fmt()
  FROM bb_contact c;
```

10

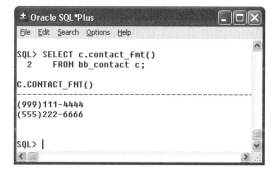

Figure 10-31   Query invoking an object type method

Assignment 10-4: Create Object Views

The Brewbean's lead programmer wants to experiment with the use of object technology on the existing database. Create an object view to represent the department and product data. The columns of the BB_PRODUCT table to be included are IDPRODUCT, PRODUCTNAME, DESCRIPTION, PRICE, and STOCK. The columns of the BB_DEPARTMENT table to be included are IDDEPARTMENT, DEPTNAME, DEPTDESC, and DEPTIMAGE.

1. Open or return to SQL*Plus.

2. Create an object type named PROD_OT to represent the product information.

3. Create a data type named PROD_TAB as a table of the product object type to use in the department object type to represent that multiple products exist per department.

4. Create an object type named DEPT_OT for the department data.

5. Create an object view name DEPT_VU to represent the department and product data using the object types created in previous steps.

6. Run the query shown in Figure 10-32 to test the view.

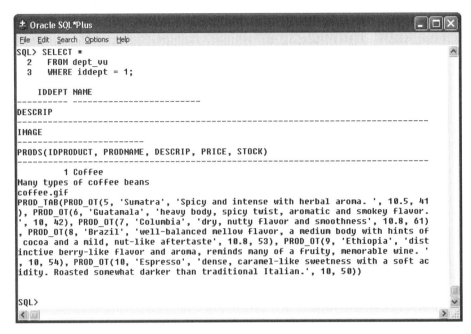

Figure 10-32   Query using an object view

Assignment 10-5: Create a Product Object Type with Sort Capability

Create an object type named ITEM_OT that contains attributes of ID, NAME, and DESCRIPTION using the same data types of the BB_PRODUCT table. Create a table named BB_COFFEE that contains three columns. One column should be named ITEM using the ITEM_OT object type as the data type. Also include columns for PRICE and TYPE in the table. Insert the three rows as listed in Table 10-4. Add a method to the object type to allow sorting by NAME. Query the table using the object type method to sort the rows displayed by the ITEM column.

Table 10-4   Rows to be added

ID	NAME	DESCRIPTION	PRICE	TYPE
12	Café Press	Steel French Coffee Press	38.00	E
76	Espresso	Roasted Italian Style	10.00	C
44	Coffee Carafe	Thermal insulated carafe	24.50	E

Assignment 10-6: Use Native Dynamic SQL

Brewbean's wants to allow customers to search either the product name or description data for particular terms. The application screen allows the user to select either product name or description and then type in a search term. Create a procedure named SEARCH_SP using native dynamic SQL that returns the product name, description, and price based on user search criteria. Keep in mind that this procedure needs to handle multiple rows returning from the query.

Assignment 10-7: Object-oriented Programming

Object-oriented programming has introduced a great deal of new terminology to the programming world. Research and then describe what is meant by the following terms used in association with object-oriented programming: abstraction, inheritance, polymorphism, and encapsulation. In addition, describe Oracle type inheritance, including a description of a type hierarchy. Give a specific example of a type hierarchy that could be used in the Brewbean's database.

Assignment 10-8: Business Intelligence

Business intelligence (BI) is a popular term used to describe providing more data power to end users. Briefly describe what business intelligence means in regards to applications, including the terms data mining, data marts, OLAP, and executive dashboards. What role might dynamic SQL play regarding BI? What products does Oracle offer in the BI area? Briefly describe each.

10

Case Projects

Case 10-1: The Brewbean's Ad Hoc Query System

Determine two queries that may be useful to customers or employees and create two program units utilizing dynamic SQL to provide the functionality to support these queries. One of the queries must allow more than one criteria to be set at runtime for the query.

Case 10-2: The More Movies Database

The More Movies lead programmer wants to begin using object technology features with their existing database. Create an object view (and the necessary object types) named MOVIE_VU that includes movie and rental data for all movies. Test the object view by executing a query against it.

11

PERFORMANCE TUNING

In this chapter, you will:

♦ Gain awareness of tuning concepts and issues

♦ Explore SQL statement tuning

♦ Explore PL/SQL statement tuning

Tuning is an ongoing effort to make our applications process efficiently to minimize the resources demanded and increase the response speed. An application typically has three areas to consider in optimizing performance: hardware, database configuration, and application source code. As a developer, you need to concentrate your tuning efforts on the source code, as this is your main area of responsibility.

In PL/SQL, SQL statements typically generate the largest resource usage because they control the database interaction. In addition, even though much of the server configuration is considered the responsibility of the DBA, developers need to be aware of the configuration settings and work with the DBA to fine-tune them.

This chapter serves as an introduction to various issues and methods regarding performance tuning. Performance tuning is such an expansive subject that this chapter is intended only to introduce the topic to create an awareness and to set the stage for your further development. We begin with a discussion of some general concepts and issues, followed by specific tuning topics for SQL and PL/SQL statements.

THE CURRENT CHALLENGE IN THE BREWBEAN'S APPLICATION

The Brewbean's application group is just beginning to test run the online application with multiple users. Brewbean's management has emphasized the importance of quick response times for their Web site shoppers. The lead programmer wants the group to be proactive in reviewing all code for potential improvements in execution efficiency; however, many are unfamiliar with performance improvement techniques. The lead programmer decided to hold a half-day seminar to introduce general performance tuning topics to all the programmers. This chapter contains the seminar contents.

REBUILDING YOUR DATABASE

To rebuild the Brewbean's database:

1. Open SQL*Plus.

2. Create a spool file so that SQL*Plus keeps a copy of all the messages received from running the file that creates the database. On the main menu in SQL*Plus, click **File**, point to **Spool**, and then click **Spool File**. A Select File dialog box appears.

3. Browse to a folder used to contain your working files and in the File name: text box, type **DB_log11**, and then click **Save**. Now, whatever text is seen in our SQL*Plus session is saved to this file for future reference.

4. Now, let's create the database. In SQL*Plus, enter the following command, which runs all the statements contained in the c11Dbcreate.sql file. Messages verifying the creation and data insertion steps scroll on the SQL*Plus screen. This may take a couple minutes to complete.

   `@<pathname to PL/SQL files>\Chapter.11\c11Dbcreate`

5. Now, we turn the spooling off so that we can review the results in the spool file created. On the main menu, click **File**, point to **Spool**, and then click **Spool Off**.

6. Open a text editor.

7. Open the file named **DB_log11.lst** in your working folder.

8. Review the messages for any errors.

TUNING CONCEPTS AND ISSUES

To begin tuning code, we must first become familiar with methods that assist us in identifying coding problems. First, we need to be able to identify which statements are causing lengthy execution times. In addition, not only do we need to know the execution time, but also we need to understand how the Oracle server is processing the statement in order to determine what potential improvements could be applied. After we attempt a modification, we need to be able to determine if it succeeded in improving performance.

In this part of the chapter, we first explore some methods of identifying statements that are resource intensive. Second, we examine how statement processing and options are managed. Third, we use database features to obtain processing statistics helpful in reviewing performance and learn about the explain plan and AUTOTRACE. Finally, we use the timing feature to measure and compare processing time.

Identifying Problem Areas in Coding

To begin tuning efforts, we need to first identify the source code areas that are likely candidates for tuning. There are two basic methods available to assist in this identification:

- First, the application-testing phase should simulate actual operations and include end users. End user feedback often pinpoints problem areas, especially regarding slow response. Set up a testing procedure that makes it easy for the testing group to document in which part of the application and what specific actions are problematic so that this information can lead you directly to the coding that requires review.

- Second, the V_$SQLAREA view provides execution details, such as disk and memory reads for all statements processed since the database startup and as memory allows. Use the DESCRIBE command on the view to list all the available columns of data.

DBA accounts (such as SYSTEM) have access to the V_$SQLAREA view; however, general users are not typically granted access to this view by default. Typically, a public synonym named V$SQLAREA is created by the DBA. The DBA then grants users access via either the view or synonym.

A few particular performance tuning statistics include number of executions, disk reads, and buffer gets. The V_$SQLAREA view and the SQL TRACE facility both assist in identifying these statistics. We discuss each in the following sections.

The V_$SQLAREA View

The number of executions indicates how many times a statement has been processed, which helps identify the more heavily used statements. The number of disk reads reflects the total amount of physical reads that were accomplished for this statement. The disk reads divided by the number of executions determine the number of reads per execution of the statement, which is a number to focus on when deciding which statements need tuning. A high number of reads indicates we may be able to improve performance by making statement modifications.

Let's look at an example query of V_$SQLAREA in Figure 11-1. Note that only partial output of one statement is shown.

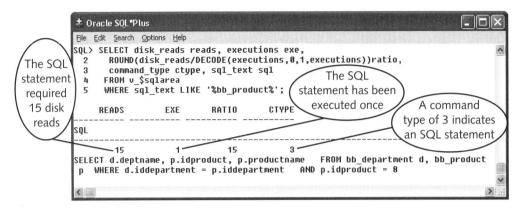

Figure 11-1   Query of V_$SQLAREA

A query on the V_$SQLAREA view should contain a WHERE clause to provide some criteria or threshold such as the number of disk reads above some amount. In addition, you should sort the results to place your most likely candidates for review at the top. If you are issuing this query from SQL*Plus, you may want to turn the output spool on because the results can become lengthy depending on the amount of activity on your server.

Table 11-1 shows a list of command types, which can be helpful in interpreting the values in the COMMAND TYPE column of the V_$SQLAREA view.

Table 11-1   Command types in the V_$SQLAREA view

Type	Statement
3	SELECT
2	INSERT
6	UPDATE
7	DELETE
47	PL/SQL program unit

Another important statistic in the V_$SQLAREA view is the number of buffer gets, which is the number of memory reads performed for the statement. The BUFFER_GETS column can be queried to show the number per execution. A high number of buffer reads can indicate an index is needed or a join could be improved.

The SQL TRACE Facility

The SQL TRACE facility can be turned on for a session, and statistics regarding SQL statement execution during that session can be stored for review. The statistics are saved in an operating system file that must be converted to a readable format using the TKPROF executable. The converted file can then be viewed using simple word processing software such as WordPad.

Setting Server Parameters

Before using the TRACE facility, several database server parameters must be set. Recall that database parameters are set in the init.ora file, which is typically managed by the DBA.

Examine the following code:

```
timed_statistics = TRUE
user_dump_dest = \oracle\admin\ora9\udump
```

The TIMED_STATISTICS setting turns statistic tracking on and the USER_DUMP_DEST setting indicates in which directory the statistics file should be saved. In addition to these settings, MAX_DUMP_FILE_SIZE indicates the maximum size of the resulting file that contains the statistics. If you discover the trace file is truncated or does not contain statistics for all the statements from your session, increase the size of this parameter.

Let's walk through an example session using the TRACE facility.

To use the TRACE facility:

1. Open or return to SQL*Plus.

2. Start tracing statistics by setting the SQL_TRACE session parameter to TRUE using an ALTER SESSION statement, as shown in the following code:

   ALTER SESSION SET SQL_TRACE = TRUE;

3. Enter the SQL statement displayed in Figure 11-2.

4. Turn the statistic tracing off using the following statement:

   ALTER SESSION SET SQL_TRACE = FALSE;

5. Now, we can open the resulting trace file to review the statistics saved for the session. Go to the directory indicated in USER_DUMP_DEST and locate the file with the name layout listed in the following code. Note the actual name of your file is dependent on the name of the database instance. For example, the database instance name in this example is oraper; therefore, there is a file named oraper_ora_###.trc. The number in the name is assigned by Oracle. If more than one .trc file exists, then review the file dates to identify the one most recently created.

   *database-instance-name*_ora_###.trc

11

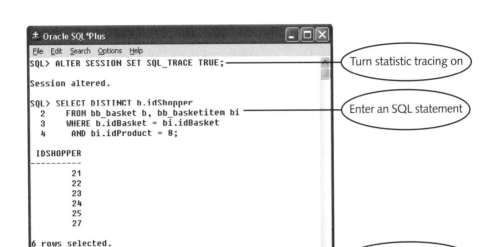

Figure 11-2 Using the TRACE facility

      6. We use TKPROF to covert this file to a readable file format. TKPROF is an executable file at the operating system level, so we execute this step at the command prompt, as shown in Figure 11-3.

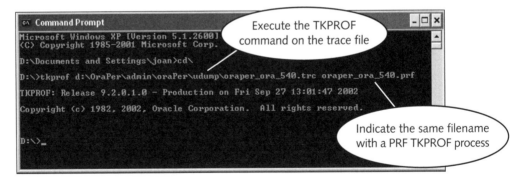

Figure 11-3 Use TKPROF to convert the trace file to a readable form

      7. In WordPad, open the PRF file created in the previous step. Review the contents, as shown in Figure 11-4.

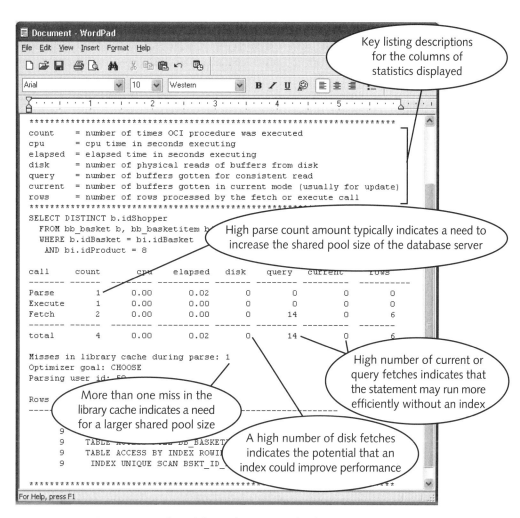

Figure 11-4 Statistics produced from the TRACE facility

SQL_TRACE can be set at the database system level using the SQL_TRACE parameter in the init.ora file or the DBMS_SYSTEM package. If SQL_TRACE is turned on at the database level, then statistics are logged for statements processed by every user. This does noticeably impact the performance of the database.

If statistics are being collected for a number of statements, the SORT option of the TKPROF program may be used to make the output file more useful. The option is listed at the end of the TKPROF statement, as shown in the following code:

```
tkprof oraper_ora_540.trc oraper_ora_540.prf sort=PRSCPU
```

The PRSCPU option in the preceding code sorts the statements in descending order by the CPU time spent parsing. More sort options are listed in Table 11-2. This list is not comprehensive; to review all options, refer to Oracle documentation available at *http://otn.oracle.com*.

Table 11-2   TKPROF sort options

Sort Option	Description
PRSELA	Time elapsed parsing
EXECPU	CPU time for execution
EXEDSK	Number of physical reads from disk
EXEROW	Number of rows processed
EXEMIS	Number of library cache misses during execution

Processing and the Optimizer

Before tackling performance tuning, we need to establish an understanding of the SQL processing architecture in order to be able to determine how statements can be modified to execute more efficiently. SQL processing contains the components pictured in Figure 11-5: parser, Optimizer, row source generator, and execution engine.

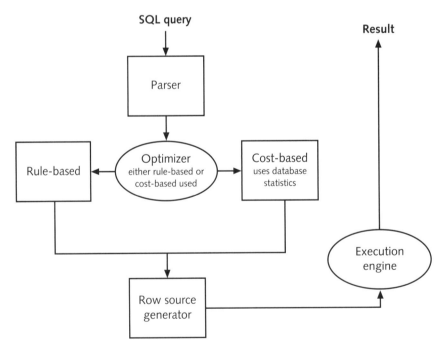

Figure 11-5   SQL processing components

The parser checks for correct statement syntax and that all objects referenced exist. The Optimizer determines the most efficient way to process the statement creating an execution plan that is followed. For example, the Optimizer determines items such as whether an index will be used or the order tables will be joined. These factors for the most part determine how efficient your statements process. The row source generator sends the execution plan and the row source for each step in the plan to the execution engine. A row source returns a set of rows for the applicable step. The execution engine processes each row source and completes the execution plan to produce the final results.

The Oracle database server contains two statement Optimizers that follow different methodologies to determine how a statement is processed. The rule-based Optimizer is the older method, which uses a list of rules to determine processing. For example, the rule-based engine typically uses an index if one is available, even if it may not be beneficial.

On the other hand, the cost-based Optimizer is a newer methodology that uses database object statistics such as distribution of data to determine how best to process statements. If particular rows are being retrieved from a small table, the cost-based engine may decide not to use an index, as it would not increase the performance of the query.

The Optimizer is one of the many settings in the init.ora file and the parameter is named OPTIMIZER_MODE. The default is CHOOSE, which uses the cost-based Optimizer if table and index statistics are available; otherwise, it uses the rule-based Optimizer. Table and index statistics are created by issuing an ANALYZE command or using the DBMS_ STATS built-in package.

 Oracle recommends the use of the cost-based Optimizer as the effectiveness of the rule-based Optimizer is reduced in future releases.

Table 11-3 lists available Optimizer mode parameter values.

Table 11-3   Optimizer mode parameter values

Value	Description
CHOOSE	If statistics exist for any of the tables accessed, the cost-basis is used with a goal of best throughput. If some of the tables do not have statistics, then estimation is used. If no statistics are available, the rule-basis is used.
ALL_ROWS	Always uses cost-basis with a goal of best throughput.
FIRST_ROWS_*n*	Always uses cost-basis with a goal of best response time to return the first *n* number of rows. *n* can be 1, 10, 100, or 1000.
FIRST_ROWS	Available for backward compatibility. Uses a mix of cost and heuristics for fast delivery of the first few rows.
RULE	Always uses rule-basis.

11

One problem may have occurred to you in regards to setting the Optimizer mode. If the mode is set in the database init.ora file, then it is effective for all database users. What if one setting is not the ideal setting for all users or applications? This is usually the case, as we tend to produce a variety of applications. Fortunately, however, the Optimizer mode can also be set by session. The OPTIMIZER_GOAL parameter of the ALTER SESSION command can override the init.ora mode setting. The OPTIMIZER_GOAL parameter can be set to any of the Optimizer mode values listed in Table 11-3.

The Cost-Based Optimizer

The cost-based Optimizer (CBO) uses a goal of best throughput by default. Best throughput means it chooses the path of execution that uses the least amount of resources needed to process all the rows in the statement. However, the CBO can run with a goal of optimizing the response time. With this goal, it generates an execution path that uses the least amount of resources to process the first row accessed by the statement.

Note that applications that center around large or batch requests such as with Oracle Reports typically optimize for best throughput. Interactive or operational applications such as those created with Oracle Forms should be optimized for best response time because users are waiting to view feedback. If many rows are returned, the end user is still interested in getting the first rows of feedback fast to begin analyzing the results.

The Explain Plan and AUTOTRACE

One of the most important items to review regarding the performance of a statement is the execution or explain plan for that statement. Recall that the Optimizer develops an execution plan that outlines the specific steps that are taken to process a statement. The Oracle AUTOTRACE tool enables the display of both the execution plan and the execution statistics.

 The terms "execution plan" and "explain plan" are used interchangeably in this text.

The AUTOTRACE tool is started by issuing the SET AUTOTRACE ON command in SQL*Plus. A number of options can be used when starting this tool, as listed in Table 11-4.

Table 11-4   AUTOTRACE tool options

SQL*Plus Command	Description
SET AUTOTRACE ON	Displays explain plan, statistics, and result set
SET AUTOTRACE ON EXPLAIN	Displays explain plan and result set
SET AUTOTRACE ON STATISTICS	Displays statistics and result set
SET AUTOTRACE TRACEONLY	Displays explain plan and statistics

Let's work through steps to set up and use the AUTOTRACE tool.

To use the AUTOTRACE tool:

1. Open or return to SQL*Plus, logging on as the SYSTEM user. Type **connect as sysdba**, and then enter the user id and password.

2. To obtain the access privileges to the needed views, a role named PLUSTRACE will be created. A script file named plustrce.sql that creates this role is located in the Oracle database directory *<your_oracle_home>*\sqlplus\admin\. A copy of this file is also located in the Chapter.11 folder for your reference. Execute the script with the following command.

   &*<your_oracle_home>*\sqlplus\admin\plustrce.sql

Recall your Oracle home is the directory in which Oracle9*i* was installed.

3. Grant the privileges to your regular user id (such as SCOTT) with the following command:

   GRANT plustrace TO *<user id>*;

4. Close SQL*Plus.

11

5. Open SQL*Plus logging on with your regular user id. This needs to be a new session.

6. Create a table named PLAN_TABLE that holds the explain plan information. A script file named utlxplan.sql creates this table and is located on the Oracle database at *<your_oracle_home>*\rdbms\admin\. A copy of this file is also available in the Chapter.11 folder. Execute the script with the following command:

   &*<your_oracle_home>*\rdbms\admin\utlxplan.sql

If an error is received that states the object already exists upon execution, this indicates the plan tables have already been created in this schema, and you are ready to AUTOTRACE.

7. Enter the following command to start the AUTOTRACE tool:

   SET AUTOTRACE TRACEONLY

8. Now, we enter an SQL command to view the execution plan and statistics that the AUTOTRACE tool produces. Enter the SQL command shown in Figure 11-6.

9. Turn AUTOTRACE off with the following command:

   SET AUTOTRACE OFF

The explain plan is read from top-down, most-indented lines first. The table access lines can include additional lines indented below it that indicate the use of an index to accomplish the table access. The two lines are both considered as part of the table access step. For example, the steps in the explain plan from our sample query in Figure 11-6 are listed in Table 11-5. The column on the left indicates the order in which these steps are executed.

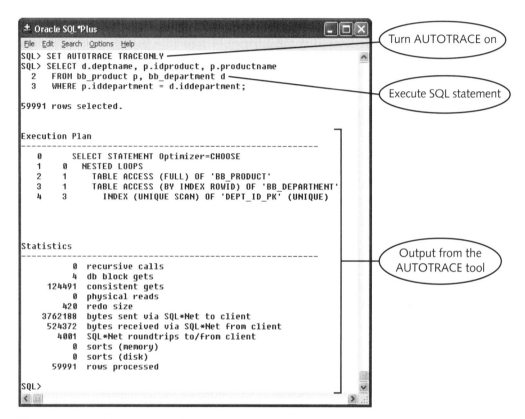

Figure 11-6   SQL command to enter

Table 11-5   Explain plan execution order

Order	Statement Execution Steps
	SELECT STATEMENT Optimizer=CHOOSE
4	NESTED LOOPS
1	TABLE ACCESS (FULL) OF 'BB_PRODUCT'
3	TABLE ACCESS (BY INDEX ROWID) OF 'BB_DEPARTMENT'
2	INDEX (UNIQUE SCAN) OF 'DEPT_ID_PK' (UNIQUE)

This indicates a sequence of processing that includes a full scan of the BB_PRODUCT, a scan of the BB_DEPARTMENT table by using the index on the IDDEPARTMENT column, and nested looping to join these results. The top line in the explain plan shows the type of statement being processed and which Optimizer mode is in effect, which is CHOOSE in our case. The fact that no cost statistics are listed on this top line indicates that the rule-based Optimizer was used. Recall that we must analyze a table for the cost-based Optimizer to be used and none of these tables have been analyzed.

 The two columns of numbers output in the execution plan indicate the parent and dependent execution steps. These do not indicate the sequence of execution.

Table 11-6 lists a description for each of the statistics displayed by the AUTOTRACE tool. Notice the SQL*Net statistics, which are helpful in identifying network traffic. In addition, the consistent reads are the same as the buffer gets we covered earlier, and the physical reads are the same as the disk reads.

Table 11-6   Statistic definitions

Statistic	Description
Recursive calls	Number of recursive calls generated at both the user and system level. Oracle maintains tables used for internal processing. When Oracle needs to make a change to these tables, it internally generates an SQL statement, which in turn generates a recursive call.
Db block gets	Number of times a CURRENT block was requested.
Consistent gets	Number of times a consistent read was requested for a block.
Physical reads	Total number of data blocks read from disk. This number equals the value of "physical reads direct" plus all reads into buffer cache.
Redo size	Total amount of redo generated in bytes.
Bytes sent via SQL*Net to client	Total number of bytes sent to the client from the foreground processes.
Bytes received via SQL*Net from client	Total number of bytes received from the client over Oracle Net.
SQL*Net roundtrips to/from client	Total number of Oracle Net messages sent to and received from the client.
Sorts (memory)	Number of sort operations that were performed completely in memory and did not require any disk writes.
Sorts (disk)	Number of sort operations that required at least one disk write.
Rows processed	Number of rows processed during the operation.

11

Timing Feature

Before we jump into some SQL statement tuning examples, it is worth looking at another basic tuning tool that can be used in SQL*Plus. The timing feature allows a developer to measure execution time of a statement.

To use the timing feature:

1. Return to or open SQL*Plus.

2. Turn timing on with the following command:

   SET TIMING ON

3. Enter the query shown in Figure 11-7. The time elapsed is displayed immediately after the SQL statement output. The timing information is displayed in terms of hours, minutes, and seconds.

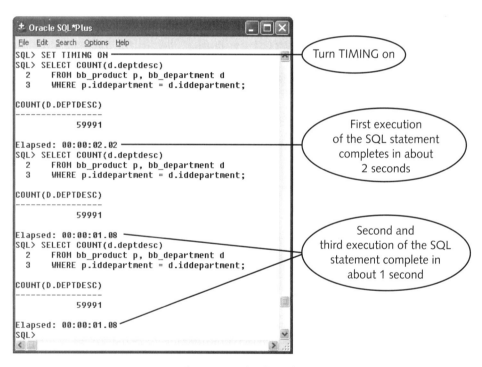

Figure 11-7   Execution time information displayed

In this query, the first execution shows a time of 2.02 seconds. Note that the query is run two more times, as shown in Figure 11-7. The second and third queries display an execution time of 1.08 seconds. Note that the execution time will be dependent on the computer sytem being used. Why the difference? The first run of a query is cached into memory of the Oracle server and, therefore, successive runs of the same query can skip the parsing, creating the execution plan, loading into the SQL area, and storing the result set in memory. Before modifying statements in a tuning effort, make sure to use the timing from a second run of the statement as your base time to improve. Otherwise, you are led to believe you improved performance when, in fact, your modification may not have done this at all.

SQL STATEMENT TUNING

In this section, we look at examples concentrating on the explain plan and identifying how statement modifications affect the plan and potentially performance. The AUTOTRACE and timing features are both turned on for the execution of these statements. It is important to test statements on a test database that resembles the production database in design and size. The database tables used in the following examples are rather small and most statements are fairly simple. This is done purposely to clearly demonstrate the explain plan and specific effects on them.

Avoiding Unnecessary Column Selection

Including columns that are not actually needed in a select list can have quite a detrimental effect on performance. Let's look at the explain plan for a query on the BB_PRODUCT table. Note that the BB_PRODUCT table has been increased to almost 3000 rows for demonstration purposes; therefore, the TRACEONLY option of AUTOTRACE is used to suppress the display of output rows. Figure 11-8 displays a query on the BB_PRODUCT table and the associated explain plan.

11

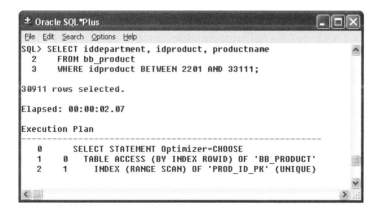

Figure 11-8   Query on BB_PRODUCT

Notice that an index scan is first performed using the index created on the IDPRODUCT column as the primary key. Then, the BB_PRODUCT table is scanned by ROWID to return the requested rows. However, what if only the IDPRODUCT column was actually needed from the query? Could this affect performance? Let's run the same statement selecting only the IDPRODUCT column, as shown in Figure 11-9.

Now, the only step that needs to be processed is an index scan. A table scan is not necessary as the index can return the IDPRODUCT value. Keep in mind that an index contains the value of the column(s) indexed and the ROWID for the associated row. The ROWID is a physical address for a table row and provides the fastest method for retrieving rows.

Cost Versus Rule Basis

As outlined earlier, the cost-based Optimizer much of the time will create a different execution plan than the rule-based Optimizer. Let's look at a query on the BB_PRODUCT table that sets criteria on the IDDEPARTMENT column. An index is created on the IDDEPARTMENT column to illustrate the differences in the two optimizers. Figure 11-10 displays the results of a query and verifies that most rows are selected by this query.

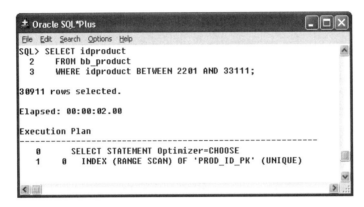

Figure 11-9 Query on BB_PRODUCT needing only an index scan

The explain plan shows that the index created on the IDDEPARTMENT column is scanned and then the table is accessed by ROWID. However, when a query retrieves a large percent of rows from the table, using an index is not always most efficient. Now, we will analyze the BB_PRODUCT table to enable the cost-based Optimizer to be used. The following ANALYZE command computes the statistics on the BB_PRODUCT table:

```
ANALYZE TABLE bb_product COMPUTE STATISTICS;
```

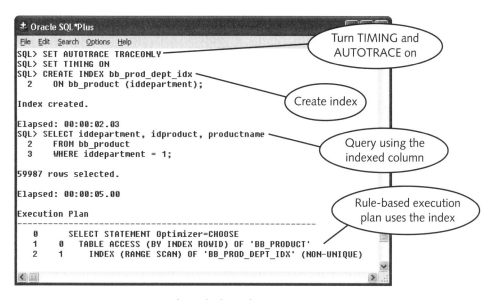

Figure 11-10   Query using the rule-based Optimizer

Now, let's run the same exact query as we did earlier. Figure 11-11 shows the results.

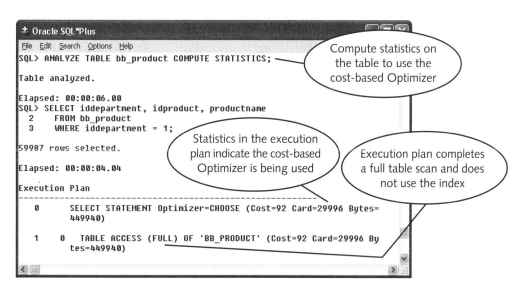

Figure 11-11   Query using the cost-based Optimizer

Notice the cost statistics that now appear on the top line of the explain plan, which indicate that the cost-based Optimizer was used. The explain plan now does not include the use of the IDDEPARTMENT column index. The only step in the execution of this statement is a full table scan on the BB_PRODUCT table. The cost-based Optimizer determined the IDDEPARTMENT column has a small distribution and many rows would be returned. Therefore, it is more efficient to skip the index and complete a full table scan.

Index Suppression

As indexes are one of the central topics within performance tuning, it is important to recognize when SQL statements suppress the Optimizer from using an index. Index suppression can occur when a WHERE clause uses a function on a column or compares different data type values.

Let's work through an example to witness how index suppression initiates.

To verify index suppression:

1. Open or return to SQL*Plus.

2. Compute statistics on the BB_PRODUCT table with the following ANALYZE command:

   ANALYZE TABLE bb_product COMPUTE STATISTICS;

3. Set AUTOTRACE and TIMING on with the following code, if necessary:

   SET AUTOTRACE TRACEONLY
   SET TIMING ON

4. Enter the query shown in Figure 11-12. Note that the execution plan indicates the index on the IDPRODUCT column is used.

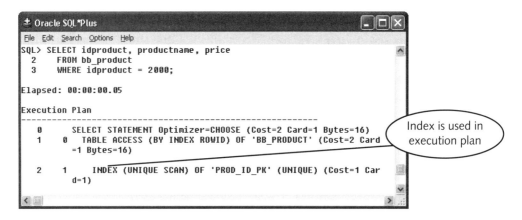

Figure 11-12   Query using index

5. Now enter a different query using the same column, as shown in Figure 11-13. Notice that the WHERE clause in the query utilizes a SUBSTR function on the IDPRODUCT column. Now, the explain plan indicates the index is not used; a full table scan is used to resolve the query.

6. Enter the query again using a BETWEEN operator in the WHERE clause, as shown in Figure 11-14. Notice the execution plan now uses the index.

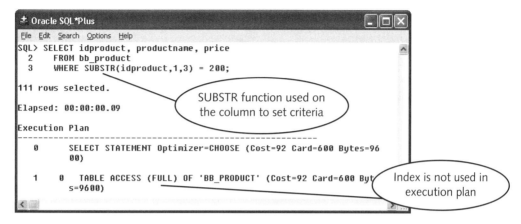

Figure 11-13   Query with function suppressing index use

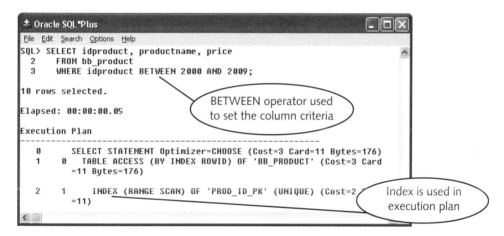

Figure 11-14   Query using BETWEEN operator uses index

Comparison of different data types can also cause index suppression in statement execution. Let's take a look at a WHERE clause that compares a numeric column to a character string value. Figure 11-15 shows a simple query on the BB_PRODUCT table comparing the numeric column of IDPRODUCT to a string value of '2111.'

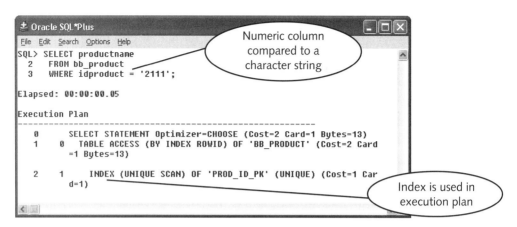

Figure 11-15   Numeric column to string value comparison

Notice the explain plan indicates that the IDPRODUCT column index is still used for execution. This is true because in this case, Oracle internally converts the string value to a number for comparison to the column and no conversion is done to the column value. However, if we reverse this comparison to have a character column compared to a numeric value, the index suppression then returns. In that case, the column value is internally converted to a numeric value and, therefore, the Optimizer considers this activity the same as when we use a function on the column value. Figure 11-16 demonstrates this index suppression. The IDPROD2 column is an exact copy of the IDPRODUCT column values but it is a character column.

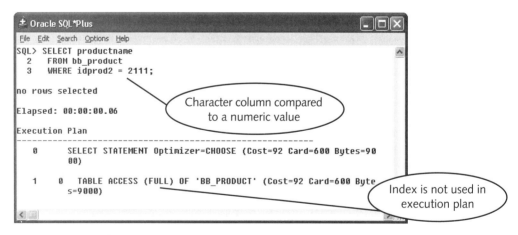

Figure 11-16   Character column-to-numeric value index suppression

Concatenated Indexes

At times, we create concatenated indexes or indexes that involve more than one column. The Optimizer uses these indexes only when the leading column indexed is included in the criteria of the SQL statement. For example, a concatenated index named BB_SHOPNAME_IDX that indexes the LASTNAME and FIRSTNAME columns of the BB_SHOPPER table would be used only if the query uses the LASTNAME column in the criteria, and if the index is created with the LASTNAME column first.

Let's try a few queries to see how the index is used or suppressed.

To test concatenated index suppression:

1. Open or return to SQL*Plus.

2. Create the concatenated index previously described by using the following statement:

   ```
   CREATE INDEX bb_shopname_idx
     ON bb_shopper (lastname, firstname);
   ```

3. Enter the query shown in Figure 11-17 that has criteria on both of the columns in the index. The execution plan shows that the index is used to execute the query.

4. Enter the query shown in Figure 11-18 that has criteria on only the LASTNAME column. Because the leading column of the index is still being used in the criteria, the execution plan shows that the index is still used to execute the query.

11

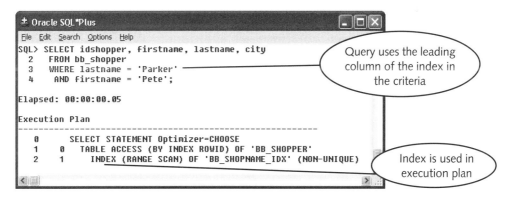

Figure 11-17   Using a concatenated index with criteria on two columns

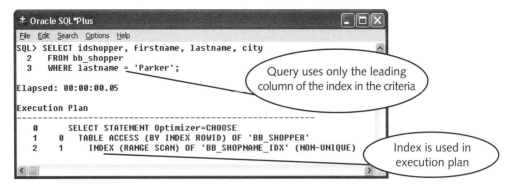

Figure 11-18 Using a concatenated index with criteria on one column

5. Now, let's see what happens when criteria is set only on the FIRSTNAME column. Enter the query shown in Figure 11-19 and note that the execution plan indicates that the index is not used. Recall that the FIRSTNAME column is the second column in the index and the leading column must be used for the Optimizer to consider the use of the index.

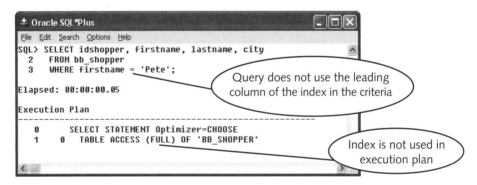

Figure 11-19 Query not using leading column of concatenated index

Subqueries

Another area worthwhile to explore is the type of subqueries used. Correlated subqueries are typically considered more efficient; however, this is not always the case. In addition, the type of optimizer in effect has an impact on execution plans of subqueries. When a subquery is noncorrelated, the inner query executes first and returns a result set that is treated like an IN list. Then, the outer query is executed and compared to the result set. In a correlated subquery, the outer query executes first and then the inner query executes for each record returned from the outer query. Also, using the EXISTS operator terminates the inner query of a subquery when a match is found, whereas the IN operator continues until fetching all the rows in the inner query.

Let's run examples of a subquery written as correlated and noncorrelated to see the statement differences and compare the execution plans.

To test correlated versus noncorrelated subqueries:

1. Open or return to SQL*Plus.

2. Enter the following statement to eliminate statistics on the BB_PRODUCT table to test with the rule-based Optimizer first:

   ANALYZE TABLE bb_product DELETE STATISTICS;

3. Enter the noncorrelated subquery in Figure 11-20, which determines the matches between the BB_PRODUCT table and the BB_PROD_LIST table based on the PRODUCTNAME column.

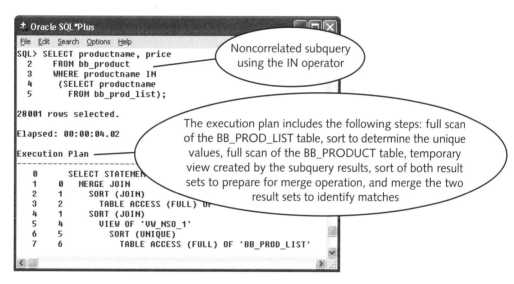

Figure 11-20   Noncorrelated subquery

4. Let's try the same statement using a correlated subquery, as shown in Figure 11-21. Notice the execution plan for the correlated subquery is much simpler; however, the execution time is slightly higher than the noncorrelated subquery.

5. Now, let's analyze both tables used in the queries using the following statement so that the cost-based Optimizer is used:

   ANALYZE TABLE bb_product COMPUTE STATISTICS;
   ANALYZE TABLE bb_prod_list COMPUTE STATISTICS;

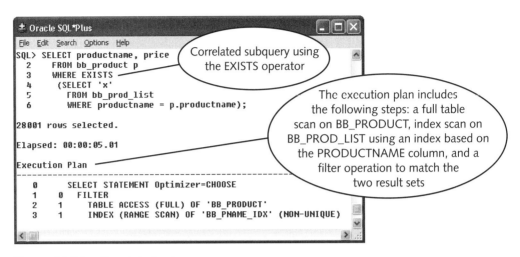

Figure 11-21   Correlated subquery

6. Enter both the noncorrelated and correlated query used in Steps 3 and 4 and compare their execution plans. Notice that both queries now result in the same execution plan using the cost-based Optimizer, as shown in Figure 11-22.

Again, it is important to consider the Optimizer mode in effect to determine the execution plan being followed and the different impact on statements.

Joins

The driving table in a join is considered the one that the execution uses first to perform the join operation. The goal is to have the smallest table involved in the join operation to be the driving table to ascertain the best performance. The rule-based Optimizer uses the last table listed in the FROM clause as the driving table. In contrast, the cost-based Optimizer uses the first table listed in the FROM clause as the driving table. Therefore, be careful in ordering the tables in your FROM clauses. The order of tables in the FROM clause in all queries involving joins should be reviewed with this consideration. Large differences in table sizes can have a significant impact on performance if not listed in the most advantageous order.

Optimizer Hints

Based on knowledge of the data and review of the execution plan, we may determine the need to alter the execution plan to improve performance. We can alter the execution plan of SELECT, INSERT, UPDATE, and DELETE statements by using hints within the statement. All hints invoke the cost-based Optimizer regardless if statistics have been generated. One exception is the hint that forces the use of the rule-based Optimizer. Hints are included in statements as comments including a plus sign.

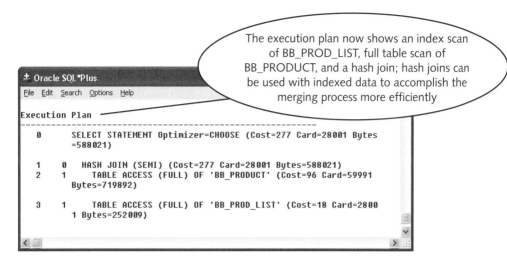

The execution plan now shows an index scan of BB_PROD_LIST, full table scan of BB_PRODUCT, and a hash join; hash joins can be used with indexed data to accomplish the merging process more efficiently

```
± Oracle SQL*Plus
File  Edit  Search  Options  Help

Execution Plan
-------------------------------------------------------
   0       SELECT STATEMENT Optimizer=CHOOSE (Cost=277 Card=28001 Bytes
           =588021)

   1    0    HASH JOIN (SEMI) (Cost=277 Card=28001 Bytes=588021)
   2    1      TABLE ACCESS (FULL) OF 'BB_PRODUCT' (Cost=96 Card=59991
           Bytes=719892)

   3    1      TABLE ACCESS (FULL) OF 'BB_PROD_LIST' (Cost=18 Card=2800
           1 Bytes=252009)
```

Figure 11-22   Using a correlated and noncorrelated subquery with the cost-based Optimizer

Numerous hints are available and are grouped in the following categories:

- The optimization approach for an SQL statement
- The goal of the cost-based Optimizer for an SQL statement
- The access path for a table accessed by the statement
- The join order for a join statement
- A join operation in a join statement

We will look at a couple of hints to give you a feel for how they work. Reference the *Oracle9i Database Performance Tuning Guide and Reference* on the OTN Web site for in-depth coverage.

Let's first consider the optimization approach category of hints. One hint available is named RULE and forces the use of the rule-based Optimizer for the statement even if statistics exist for the tables. An example of including this hint in a SELECT statement is shown in the following code:

```
SELECT /*+RULE*/ firstname, lastname
  FROM bb_shopper
  WHERE lastname = 'Parker';
```

As seen in the example, hints are embedded into statements using comments. Notice the opening comment symbol of /* must be immediately followed by a + sign. Hints can also be included by using the two hyphens comment marker. In this case, nothing else can be included in the SELECT clause line or the hint is ignored. All hints must be included in the statement immediately following the key word of SELECT, INSERT, UPDATE, or DELETE. If a hint is incorrectly added to a statement, it typically will not produce an error yet will be ignored. Therefore, as always, it is important to review the execution plan after adding a hint to not only determine if an improvement resulted, but also to confirm if the hint is being used.

Table 11-7 lists the available hint options regarding the optimization approach and goal.

Table 11-7   Available hints for the optimization approach and goal

Hint	Description	Syntax
FIRST_ROWS	Forces the Optimizer to choose a plan that returns the first row the fastest	`SELECT /*+FIRST_ROWS*/`
ALL_ROWS	Forces the Optimizer to choose a plan that returns all rows the fastest	`SELECT /*+ALL_ROWS*/`
CHOOSE	Forces the Optimizer to choose between the rule-based and cost-based approaches based on the presence of statistics	`SELECT /*+CHOOSE*/`
RULE	Forces the use of the rule-based Optimizer	`SELECT /*+RULE*/`

Figure 11-23 demonstrates the application of the RULE hint. Two executions of the same SELECT statement are displayed. The first execution uses the cost-based Optimizer by default because statistics exist for the BB_PRODUCT table. The second execution includes the RULE hint. Notice that this provides an easy way to compare execution plans and times of each Optimizer.

Another category of hints involves controlling the access path to a table. One hint in this category is FULL, which forces a full table scan, regardless if an index exists that could be used. The following statement shows the FULL hint included in a SELECT statement:

```
SELECT /*+FULL(b)*/ idshopper, firstname, lastname
 FROM bb_shopper b
 WHERE lastname = 'Parker';
```

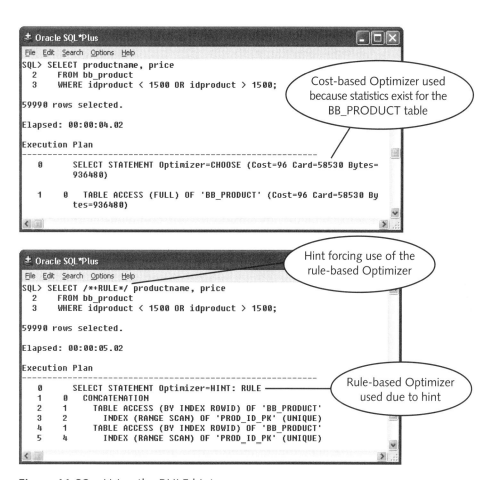

Figure 11-23   Using the RULE hint

Notice that the table name must be indicated in the hint. In addition, if a table alias is included in the statement, then the alias must be referenced in the hint.

This section provides only an introduction to hints; however, the bottom line is to recognize that you can control the Optimizer by using hints. Use hints to modify execution plans of statements and compare the execution efficiency of each plan. This is a great way to become more familiar with the Optimizer and determine which processing modifications improve efficiency.

 Multiple hints can be used in a single statement. This is accomplished by listing all the hints separated by a space within a single comment area.

PL/SQL STATEMENT TUNING

Any programming language including PL/SQL includes many options to perform needed tasks. We must consider a variety of issues that can improve the performance of PL/SQL statements. This section explores some of these issues, including minimizing statement iteration, using ROWID, performing variable comparisons, sorting conditional statements, using numeric data types, and pinning objects to memory.

Program Unit Iterations

Looping actions are always subject to review, not just to be sure to minimize the number of iterations, but also to minimize the processing time of each iteration. Minimizing the iteration processing time is particularly important with loops that have a large number of iterations. For example, let's say Brewbean's developers have created a program that allows the manager to project the cost of a rebate program being considered. Customers who purchase at least fifteen one-pound bags of coffee during a calendar year receive a $6 coupon for their next pound purchase. Customers who purchase at least twenty-five one-pound bags receive a $10 coupon. The program displays the number of each type of coupon that would be distributed and the total dollar cost of the program. The following is the PL/SQL block developed to accomplish this task. Take a close look at the cursor and the activity in the loop:

```
1 DECLARE
2  CURSOR rebate_cur IS
3    SELECT b.idshopper, SUM(bi.quantity) qty
4      FROM bb_basket b, bb_basketitem bi, bb_product p
5      WHERE b.idbasket = bi.idbasket
6        AND bi.idproduct = p.idproduct
7        AND p.type = 'C'
8        AND p.option1 = 2
9      GROUP BY idshopper;
10   lv_coupa NUMBER(4) := 0;
11   lv_coupb NUMBER(4) := 0;
12   lv_total NUMBER(5) := 0;
13 BEGIN
14  FOR rebate_rec in rebate_cur LOOP
15    IF rebate_rec.qty BETWEEN 15 AND 24 THEN
16      lv_coupa := lv_coupa + 1;
17      lv_total := lv_total + 6;
18    END IF;
19    IF rebate_rec.qty > 25 THEN
20      lv_coupb := lv_coupb + 1;
21      lv_total := lv_total + 10;
22    END IF;
23  END LOOP;
24  DBMS_OUTPUT.PUT_LINE('A= ' || lvcoupa ||' B= '||
25                  lv_coupb || ' Total $ = ' || lv_total);
26 END;
```

The block does not check for the calendar year; this is done for simplicity and to avoid further cluttering the code.

First of all, review the cursor in line 2, which sums the number of pound bags of coffee for each customer. Could we modify the cursor to further decrease the number of rows that must be processed by the loop? Yes! We can add a HAVING clause to the cursor to include only the customers with a total of at least fifteen. It is critical to retrieve only needed rows into a cursor to avoid unnecessary processing. Could we reduce the logic processing in the loop that begins on line 14? Yes! Notice that the two IF statements must be checked for each row processed. Also, the total program cost (variable LV_TOTAL) is being updated every iteration of the loop. Let's take a look at the following modified version to see the improvement.

```
 1 DECLARE
 2  CURSOR rebate_cur IS
 3    SELECT b.idshopper, SUM(bi.quantity) qty
 4      FROM bb_basket b, bb_basketitem bi, bb_product p
 5      WHERE b.idbasket = bi.idbasket
 6        AND bi.idproduct = p.idproduct
 7        AND p.type = 'C'
 8        AND p.option1 = 2
 9        HAVING SUM(bi.quantity) >= 15
10         GROUP BY idshopper;
11  lv_coupa NUMBER(4) := 0;
12  lv_coupb NUMBER(4) := 0;
13  lv_total NUMBER(5) := 0;
14 BEGIN
15  FOR rebate_rec in rebate_cur LOOP
16    IF rebate_rec.qty > 25 THEN
17      lv_coupb := lv_coupb + 1;
18    ELSE
19      lv_coupa := lv_coupa + 1;
20    END IF;
21  END LOOP;
22  lv_total := (lv_coupa * 6) + (lv_coupb * 10);
23  DBMS_OUTPUT.PUT_LINE('A= ' || lvcoupa ||' B= '||
24          lv_coupb || ' Total $ = ' || lv_total);
25 END;
```

Note that a HAVING clause is added to the cursor at line 9, which reduces the number of loop iterations by eliminating any customers with less than 15 pounds purchased in the cursor. We also reduced the processing time for each loop by only using one IF conditional check per row and moving the total calculation outside the loop (line 22). Imagine if Brewbean's has 15,000 customers to process for a rebate. If we reduce the loop processing time by just a quarter of a second, it could reduce total processing time for the program by about an hour!

11

Using ROWID When Updating

The ROWID value is the physical address of a row in a database table and provides the fastest path to retrieving records. The ROWID column is created and maintained automatically by the Oracle server. When we select data for manipulation and then update that data, ROWID should be used in the UPDATE to enhance performance.

Let's look at the following PL/SQL block that updates all product prices by a percent value provided by the user. The ROWID value is selected within the cursor and is then referenced to instruct the update as to which row should be affected.

```
1 DECLARE
2  CURSOR prod_cur IS
3    SELECT idproduct, price, ROWID
4      FROM bb_product;
5  lv_new NUMBER(7,2);
6 BEGIN
7    FOR prod_rec IN prod_cur LOOP
8      lv_new := ROUND(prod_rec.price * :g_pct,2);
9      UPDATE bb_product
10        SET price = lv_new
11        WHERE rowid = prod_rec.ROWID;
12   END LOOP;
13   COMMIT;
14 END;
```

Notice the WHERE clause of the UPDATE statement (line 11) references the ROWID to point to the correct record. Typically, we might use a primary key here, which would first perform an index lookup and then table search. The ROWID reference is quicker because it goes directly to the table with the ROWID and does not need to accomplish the extra step of an index lookup.

Variable Comparisons with the Same Data Type

We know we can compare variables that have different data types and Oracle will handle the task. For example, the IF statement in the following loop compares the numeric value of i to an IN list of values that are entered as character strings:

```
FOR i IN 1..100000 LOOP
  IF i IN('5','10','15','20') THEN
    lv_cnt := lv_cnt + 1;
  END IF;
END LOOP;
```

Note that to successfully process the comparison, Oracle implicitly converts the string values to numbers before doing the comparison. This conversion has a cost of processing time and if it occurs many times, the loss of performance adds up. Let's run an example just to verify that this causes a reduction in performance.

To check the performance cost of a different data type comparison:

1. Open or return to SQL*Plus.

2. Type **SET TIMING ON** and then press **Enter**, if necessary.

3. Type **SET SERVEROUTPUT ON** and then press **Enter**.

4. In a text editor, open the **compare11.txt** file from the Chapter.11 folder.

5. Highlight and copy the first PL/SQL block in the file. Note that this block contains the loop we just reviewed that includes a numeric to character data type comparison in the IF statement.

6. Copy and paste the block into SQL*Plus.

7. Press **Enter** if needed to make it run. Note the elapsed time.

8. Return to the **compare11.txt** file opened in the text editor.

9. Highlight and copy the second PL/SQL block in the file. Note that the comparison has been modified to the same data types.

10. Paste the second block into SQL*Plus.

11. Press **Enter** if needed to make it run. Note the elapsed time.

12. Compare the two elapsed times. The second run comparing the same data types should run noticeably faster.

11

Ordering Conditions by Frequency

When we are programming, we tend to place items in sequential or logical order, which may not be most efficient. For example, imagine an IF statement containing ten ELSIF clauses to check various conditions. If 90% of the values match the condition in the last ELSIF clause, then all of the conditions need to be checked 90% of the times the statement executes. When we know about the nature of the data values, we need to sort the condition checking to list the most frequent matching conditions at the beginning of the statement. It may look strange being out of sequential order; however, depending on the number of times the statement is processed, significant processing savings could be reaped.

Using the PLS_INTEGER Data Type

Much of the time, numeric variables are declared with the NUMBER data type. However, the Oracle server can process the PLS_INTEGER data type more efficiently. The PLS_INTEGER data type can hold whole number values with a maximum positive or negative value of 2,147,483,647. If this numeric range will suffice, the PLS_INTEGER data type should be used.

Pinning Stored Program Units

The shared pool is a part of the memory structure of the Oracle server. All users connected to the server can use items that are cached into the shared pool. PL/SQL program units that are invoked many times by users are candidates to be pinned or cached into shared memory for quick retrieval. The V$DB_OBJECT_CACHE view displays all the PL/SQL objects currently loaded into the shared pool and the size of memory required by each. The Oracle supplied package DBMS_SHARED_POOL provides capabilities to pin objects to the shared pool.

CHAPTER SUMMARY

- ❐ Performance tuning involves optimizing the hardware configuration, database configuration, and application source code.

- ❐ The goal of tuning application code is to minimize resources needed and improve response time.

- ❐ In PL/SQL, SQL statements typically generate the most resource intensive statements.

- ❐ Identifying problem coding includes the review of user feedback and system statistics regarding statement processing.

- ❐ The V_$SQLAREA view and the SQL TRACE facility allow developers to review statement processing statistics, such as disk reads and library cache misses.

- ❐ The statement Optimizer attempts to determine the most efficient way to process a statement. Two Optimizers exist: rule-based and cost-based.

- ❐ The rule-based Optimizer uses a set list of rules to determine the path of execution, whereas the cost-based Optimizer uses database statistics to determine the execution path. A table must be analyzed in order for the cost-based Optimizer to be used.

- ❐ The execution plan generated by the Optimizer is referred to as the explain plan.

- ❐ The AUTOTRACE facility enables the display of the statement explain plan and processing statistics.

- ❐ The timing feature displays the processing time required for a statement.

- ❐ In tuning SQL statements, beware of selecting unneeded columns, index suppression, and noncorrelated subqueries. Note that Optimizer hints can be used in SQL statements to force particular processing to occur.

- ❐ PL/SQL tuning includes minimizing loop iterations, using ROWID in DML, checking variable data types in comparison operations, ordering conditional checking by frequency, using the PLS_INTEGER data type, and pinning program units to the shared pool.

REVIEW QUESTIONS

1. Performance tuning involves all of the following except _____.

   a. hardware configuration

   b. database configuration

   c. user training

   d. application source code

2. Select the items that can be helpful in determining where to focus performance tuning efforts regarding application code. (Choose all that apply.)

   a. user feedback

   b. init.ora file

   c. V_$SQLAREA view

   d. SQL TRACE facility

3. Within PL/SQL, which type of statements typically generates the highest resource demand?

   a. SQL

   b. PL/SQL

   c. SESSION

   d. DDL

4. Which statement Optimizer is used if a table has been analyzed?

   a. rule

   b. cost

   c. shared

   d. cannot be determined

5. If AUTOTRACE is set to ON in SQL*Plus, what happens when a statement is executed?

   a. Processing statistics are displayed.

   b. The explain plan is displayed.

   c. The statement's results are displayed.

   d. all of the above

6. By default, the cost-based Optimizer has a goal of best _____.

   a. response time

   b. throughput

   c. response time to return one row

   d. CPU usage

7. The Optimizer uses a concatenated index only if _____.

   a. the leading column indexed is included in the criteria of the SQL statement

   b. all columns indexed are included in the criteria of the SQL statement

   c. any columns indexed are included in the criteria of the SQL statement

   d. it never uses a concatenated index

8. In the following code, what is the /*+FIRST_ROWS*/ section called?

   ```
   SELECT /*+FIRST_ROWS*/
       Last, first
       FROM bb_shopper;
   ```

   a. Optimizer force

   b. Optimizer choice

   c. Optimizer hint

   d. Optimizer input

9. The use of an index can be suppressed by a WHERE clause if the WHERE clause contains a(n) _____. (Choose all that apply.)

   a. function on a column

   b. Optimizer hint

   c. comparison of different data types

   d. calculation

10. What value is the physical address of a row in a database table that provides the fastest path to retrieving records?

    a. ROW

    b. ROWID

    c. IDROW

    d. ID

11. When would you want to pin a procedure to the shared memory?

    a. The procedure is large.

    b. The procedure involves the use of Oracle-supplied packages.

    c. The procedure is used repeatedly in applications.

    d. A procedure should never be pinned to shared memory.

12. Which numeric data type can be processed most efficiently?

    a. INTEGER

    b. PLS_INTEGER

    c. NUMBER

    d. SHORT_INTEGER

13. If the tables used in an SQL statement are not analyzed, which Optimizer is used?

    a. cost

    b. rule

    c. choose

    d. cannot determine

14. When constructing code to support an application screen in which the priority is to get the quickest feedback even if only partial results are shown initially, which Optimizer hint should be utilized?

    a. FIRST_ROWS

    b. ALL_ROWS

    c. CHOOSE

    d. RULE

15. If SHOP_ID is a numeric column, which of the following WHERE clauses would execute the slowest?

    a. WHERE shop_id IN(10)

    b. WHERE shop_id = '10'

    c. WHERE shop_id = 10

    d. All of the above execute the same.

16. What is the job of the Optimizer?

17. Assume you have a PL/SQL block that contains a CASE or IF statement that checks twelve different conditions. Does the order in which the conditions are listed have any impact on performance? Explain.

18. Does the order of the tables listed in a FROM clause have any effect on performance? Explain.

19. If the Optimizer parameter is set to CHOOSE, when would the cost-based Optimizer be used?

20. A concatenated index will be used as long as one of the columns indexed is involved in the query. Explain why this statement is true.

11

HANDS-ON ASSIGNMENTS

Assignment 11-1: Review Statement Execution Plans

The explain plan outlines how the statement is processed. You can use the AUTOTRACE feature in SQL*Plus and compare explain plans of statement modifications.

1. Open SQL*Plus.

2. If you did not create a plan table using the code in the utlxplan11.txt file in the chapter, then do so now. Find the **utlxplan.sql** file in the Chapter.11 folder and open it with a text editor. Copy and paste the code into SQL*Plus.

3. First we need to make sure no statistics currently exist for the BB_PRODUCT table using the following command:

```
ANALYZE TABLE bb_product DELETE STATISTICS;
```

4. Issue the following command to cause the display of the explain plan for SQL statements:

```
SET AUTOTRACE TRACEONLY
```

5. Run the following SQL statement. Notice the explain plan shows a full table scan is executed.

```
SELECT idproduct, productname
  FROM bb_product
  WHERE productname LIKE 'E%';
```

6. Create an index on the PRODUCTNAME column with the following statement:

```
CREATE INDEX prodname_idx
  ON bb_product(productname);
```

7. Run the same SQL query executed in Step 3 by cutting and pasting the statement within SQL*Plus. Notice the explain plan shows the index is now used to execute the statement. Also note that no statistics display at the top of the explain plan, which indicates the rule-based Optimizer is being used.

8. Compute statistics on the table using the following command. Statistics allow the cost-based Optimizer to be used.

```
ANALYZE TABLE bb_product COMPUTE STATISTICS;
```

9. Copy and paste the SQL query statement to execute again. Notice the cost-based Optimizer does not use the PRODUCTNAME index. Because we now have statistics, the cost-based Optimizer has determined the query can perform more efficiently without the index.

10. Issue the following statement to turn the AUTOTRACE feature off:

```
SET AUTOTRACE OFF
```

Assignment 11-2: Use the Timing Feature in SQL*Plus

The timing feature allows a developer to measure performance in terms of processing time.

1. Open SQL*Plus.

2. Issue the following statement to enable the display of execution time:

```
SET TIMING ON
```

3. Execute the following SQL query. Note that it may take a while to complete. Notice the time elapsed displayed after the query result set. How much time was used?

```
SELECT idProduct, price
  FROM bb_product
  WHERE idProduct LIKE '%20%'
    AND iddepartment = 1;
```

4. When comparing processing time of SQL statements, the elapsed time on the second execution of each SQL statement should be used rather than the first run. Note that the execution time will be dependent on the computer system being used. Why? Issue the following statement to turn the timing off:

```
SET TIMING OFF
```

Assignment 11-3: Compare Explain Plans

Minor SQL statement modifications can have a dramatic effect on performance; therefore, it is important to review the explain plan. Use the following steps to identify the change in execution generated by a change in the SELECT clause of an SQL query.

1. Open SQL*Plus.

2. Issue the following command to turn AUTOTRACE on:

```
SET AUTOTRACE TRACEONLY
```

3. Run the following SQL statement and review the explain plan:

```
SELECT idProduct
  FROM bb_product
  WHERE idProduct BETWEEN 1100 AND 1300;
```

4. Now run the following SQL statement, which has a modification in the SELECT clause from the previous query. Compare this explain plan to the one generated in Step 3.

```
SELECT idProduct, price
  FROM bb_product
  WHERE idProduct BETWEEN 1100 AND 1300;
```

5. Identify the difference in the execution paths and what has caused this to occur.

6. Issue the following command to turn AUTOTRACE off:

```
SET AUTOTRACE OFF
```

Assignment 11-4: Use ROWID to Improve Updates

The ROWID represents the physical address of a database row and is the fastest way to retrieve a row. Compare timing performance with and without ROWID in the following steps.

1. Open SQL*Plus.

2. Set timing to display with the following command:

```
SET TIMING ON
```

11

3. In a text editor, open the **assignment04.txt** file from the Chapter.11 folder.

4. Copy the first block in the text file into SQL*Plus.

5. Press **Enter** if needed to make the statement execute.

6. Click **Edit** on the menu bar, and then click **Paste** to run the statement a second time to get an accurate time reading.

7. Return to the assignment04.txt file that is open in the text editor.

8. Copy the second block in the text file into SQL*Plus.

9. Press **Enter** if needed to make the statement execute.

10. Click **Edit** on the menu bar, and then click **Paste** to run the statement a second time to get an accurate time reading.

11. Review the coding of the two blocks. Why does the second block process faster?

12. Set the timing off with the following command:

```
SET TIMING OFF
```

Assignment 11-5: Index Suppression

A variety of coding techniques can lead to index suppression during execution. Create two SQL queries using the SHIPZIPCODE column of the BB_BASKET table in a WHERE clause to identify two coding techniques that cause index suppression. Prior to executing these statements, create an index on the SHIPZIPCODE column. Display the execution plans of the queries executed.

Assignment 11-6: Optimizer Hints

Optimizer hints can be used to control the access path used in the execution plan of an SQL statement. Execute the following statement displaying the execution path for the following four scenarios: using the cost-based Optimizer, using the rule-based Optimizer, forcing the Optimizer to choose a plan that returns the first row the fastest, and forcing the Optimizer to choose a plan that returns all rows the fastest. Compare the execution plans of all the scenarios.

```
SELECT p.idproduct, p.productname, p.price, d.deptname
  FROM bb_product p JOIN bb_department d
       USING (iddepartment)
 WHERE SUBSTR(productname,1,4) = 'prod';
```

Assignment 11-7: Execution Plan

An SQL statement and its execution plan are listed in the following code. Identify the Optimizer mode in effect and explain the steps in executing this statement by interpreting the execution plan.

```
SQL> SELECT idProduct, price
  2    FROM bb_product
  3    WHERE idProduct BETWEEN 1100 AND 1300;
```

```
201 rows selected.

Execution Plan
-----------------------------------------------------------
   0        SELECT STATEMENT Optimizer=CHOOSE (Cost=4
                Card=202 Bytes=1212)
   1    0    TABLE ACCESS (BY INDEX ROWID) OF 'BB_PRODUCT'
                (Cost=4 Card=202 Bytes=1212)
   2    1      INDEX (RANGE SCAN) OF 'PROD_ID_PK' (UNIQUE)
                (Cost=2 Card=202)
```

Assignment 11-8: Focus Tuning Efforts

Typically, you have limited available time to dedicate to performance tuning; therefore, identifying the statements that could have the most potential for improvement is critical. Name and describe at least two methods that can be used to accomplish this task.

CASE PROJECTS

Case 11-1: Brewbean's Professional Development

Perform a Web search for articles or information concerning the Oracle explain plan. Record the Web address and describe an example in the article that clarifies a part of the execution plan or identifies new information regarding the execution plan that has not been specifically addressed in this chapter.

11

Case 11-2: More Movies Performance Tuning

The More Movies' developers are attempting to improve the performance of the movie rental system. The lead programmer gathered the SELECT statement statistics from the V_$SQLAREA view from the past hour of application testing. Review the mm_log.txt file in the Chapter.11 folder containing the information gathered from the view. Describe what each statistic identifies and explain which SQL statement the team should focus their performance tuning efforts on at this time.

TABLES FOR THE BREWBEAN'S DATABASE

The structure and data contents of each table presented in the database ERD in Chapter 1 are listed in this appendix. Note that columns containing no initial data are excluded from the data listing where space was limited. Note also that some column names are truncated in data listings because the column display width is determined by the width of the actual data. Column widths have not been increased; this is due to space limitations within this book. In addition, long text values wrap due to the width limit. Text wrapping does not occur in a uniform manner within SQL*Plus.

BB_SHOPPER

```
Name                            Null?     Type
------------------------------- --------  ------------------------
IDSHOPPER                       NOT NULL  NUMBER(4)
FIRSTNAME                                 VARCHAR2(15)
LASTNAME                                  VARCHAR2(20)
ADDRESS                                   VARCHAR2(40)
CITY                                      VARCHAR2(20)
STATE                                     CHAR(2)
ZIPCODE                                   VARCHAR2(15)
PHONE                                     VARCHAR2(10)
FAX                                       VARCHAR2(10)
EMAIL                                     VARCHAR2(25)
USERNAME                                  VARCHAR2(8)
PASSWORD                                  VARCHAR2(8)
COOKIE                                    NUMBER(4)
DTENTERED                                 DATE
PROVINCE                                  VARCHAR2(15)
COUNTRY                                   VARCHAR2(15)
PROMO                                     CHAR(1)
```

Figure A-1   Structure of BB_SHOPPER

```
IDSHOPPER FIRSTNAME  LASTNAME   ADDRESS            CITY          ST ZIPCODE  PHONE       FAX
--------- ---------- ---------- ------------------ ------------- -- -------- ----------- -----
       21 John       Carter     21 Front St.       Raleigh       NC 54822   9014317701
       22 Margaret   Somner     287 Walnut Drive   Cheasapeake   VA 23321   7574216559
       23 Kenny      Ratman     1 Fun Lane         South Park    NC 54674   9015680902
       24 Camryn     Sonnie     40162 Talamore     South Riding  VA 20152   7035556868
       25 Scott      Savid      11 Pine Grove      Hickory       VA 22954   7578221010
       26 Monica     Cast       112 W. 4th         Greensburg    VA 27754   7573217384
       27 Pete       Parker     1 Queens           New York      NY 67233   1013217384
```

```
IDSHOPPER EMAIL             USERNAME PASSWORD COOKIE DTENTERED PROVINCE COUNTRY  P
--------- ----------------- -------- -------- ------ --------- -------- -------  -
       21 Crackjack@aol.com Crackj   flyby         1 13-JAN-03          USA
       22 MargS@infi.net    MaryS    pupper        1 03-FEB-03          USA
       23 ratboy@msn.net    rat55    kile          0 26-JAN-03          USA
       24 kids2@xis.net     kids2    steel         1 19-MAR-03          USA
       25 scott1@odu.edu    fdwell   tweak         1 19-FEB-03          USA
       26 gma@earth.net     gma1     goofy         1 09-FEB-03          USA
       27 spider@web.net                           0 14-FEB-03          USA
```

Figure A-2   Data contained in BB_SHOPPER

BB_BASKET

```
Name                                              Null?       Type
-----------------------------------------------   --------    -------------
IDBASKET                                          NOT NULL    NUMBER(5)
QUANTITY                                                      NUMBER(2)
IDSHOPPER                                                     NUMBER(4)
ORDERPLACED                                                   NUMBER(1)
SUBTOTAL                                                      NUMBER(7,2)
TOTAL                                                         NUMBER(7,2)
SHIPPING                                                      NUMBER(5,2)
TAX                                                           NUMBER(5,2)
DTCREATED                                                     DATE
PROMO                                                         NUMBER(2)
SHIPFIRSTNAME                                                 VARCHAR2(10)
SHIPLASTNAME                                                  VARCHAR2(20)
SHIPADDRESS                                                   VARCHAR2(40)
SHIPCITY                                                      VARCHAR2(20)
SHIPSTATE                                                     VARCHAR2(2)
SHIPZIPCODE                                                   VARCHAR2(15)
SHIPPHONE                                                     VARCHAR2(10)
SHIPFAX                                                       VARCHAR2(10)
SHIPEMAIL                                                     VARCHAR2(25)
BILLFIRSTNAME                                                 VARCHAR2(10)
BILLLASTNAME                                                  VARCHAR2(20)
BILLADDRESS                                                   VARCHAR2(40)
BILLCITY                                                      VARCHAR2(20)
BILLSTATE                                                     VARCHAR2(2)
BILLZIPCODE                                                   VARCHAR2(15)
BILLPHONE                                                     VARCHAR2(10)
BILLFAX                                                       VARCHAR2(10)
BILLEMAIL                                                     VARCHAR2(25)
DTORDERED                                                     DATE
SHIPPROVINCE                                                  VARCHAR2(20)
SHIPCOUNTRY                                                   VARCHAR2(20)
BILLPROVINCE                                                  VARCHAR2(20)
BILLCOUNTRY                                                   VARCHAR2(20)
CARDTYPE                                                      CHAR(1)
CARDNUMBER                                                    VARCHAR2(20)
EXPMONTH                                                      CHAR(2)
EXPYEAR                                                       CHAR(4)
CARDNAME                                                      VARCHAR2(25)
SHIPBILL                                                      CHAR(1)
SHIPFLAG                                                      CHAR(1)
```

Figure A-3   Structure of BB_BASKET

```
IDBASKET   QUANTITY  IDSHOPPER ORDERPLACED   SUBTOTAL     TOTAL   SHIPPING         TAX DTCREATED     PROMO DTORDERED S S
---------- --------- --------- ----------- ---------- ---------- ---------- ---------- --------- ---------- --------- - -
       3         3        21           1       26.6       32.4          5         .8 23-JAN-03         0 23-JAN-03 N N
       4         1        21           1       28.5      34.36          5        .86 12-FEB-03         0 12-FEB-03 N N
       5         4        22           1       41.6      48.47          5       1.87 19-FEB-03         0 19-FEB-03 N N
       6         3        22           1     149.99     161.74          5       6.75 01-MAR-03         0 01-MAR-03 N N
       7         2        23           1       21.6      27.25          5        .65 26-JAN-03         0 26-JAN-03 N N
       8         2        23           1       21.6      27.25          5        .65 16-FEB-03         0 16-FEB-03 N N
       9         2        23           1       21.6      27.25          5        .65 02-MAR-03         0 02-MAR-03 N N
      10         3        24           1       38.9      45.65          5       1.75 07-FEB-03         0 07-FEB-03 N N
      11         1        24           1         10      15.45          5        .45 27-FEB-03         0 27-FEB-03 N N
      12         7        25           0       72.4      83.66          8       3.26 19-FEB-03         0 19-FEB-03 N N
      13         2        26           0         20          0          0          0 09-FEB-03         0           N N
      14         0        26           0          0          0          0          0 10-FEB-03         0           N N
      15         2        27           0       16.2      21.69          5        .49 14-FEB-03         0 19-JAN-03 N N
      16         2        27           0       16.2      21.69          5        .49 24-FEB-03         0 19-JAN-03 N N
```

Figure A-4   Data contained in BB_BASKET

BB_BASKETITEM

```
Name                          Null?     Type
----------------------------  --------  --------------------------
IDBASKETITEM                  NOT NULL  NUMBER(2)
IDPRODUCT                               NUMBER(2)
PRICE                                   NUMBER(6,2)
QUANTITY                                NUMBER(2)
IDBASKET                                NUMBER(5)
OPTION1                                 NUMBER(2)
OPTION2                                 NUMBER(2)
```

Figure A-5   Structure of BB_BASKETITEM

IDBASKETITEM	IDPRODUCT	PRICE	QUANTITY	IDBASKET	OPTION1	OPTION2
15	6	5	1	3	1	4
16	8	10.8	2	3	2	4
17	4	28.5	1	4		
18	7	10.8	1	5	2	3
19	8	10.8	1	5	2	3
20	9	10	1	5	2	3
21	10	10	1	5	2	3
22	10	10	2	6	2	4
23	2	129.99	1	6		
24	7	10.8	1	7	2	3
25	8	10.8	1	7	2	3
26	7	10.8	1	8	2	3
27	8	10.8	1	8	2	3
28	7	10.8	1	9	2	3
29	8	10.8	1	9	2	3
30	6	5	1	10	1	3
31	8	5.4	1	10	1	3
32	4	28.5	1	10		
33	9	10	1	11	2	3
34	8	10.8	2	12	2	3
35	9	10	2	12	2	3
36	6	10	2	12	2	3
37	7	10.8	1	12	2	3
38	9	10	2	13	2	3
40	8	10.8	1	15	2	3
41	7	5.4	1	15	1	3
42	8	10.8	1	16	2	3
43	7	5.4	1	16	1	3

Figure A-6   Data contained in BB_BASKETITEM

BB_PRODUCT

```
Name                              Null?     Type
------------------------------    --------  ------------------------------
IDPRODUCT                         NOT NULL  NUMBER(2)
PRODUCTNAME                                 VARCHAR2(25)
DESCRIPTION                                 VARCHAR2(100)
PRODUCTIMAGE                                VARCHAR2(25)
PRICE                                       NUMBER(6,2)
SALESTART                                   DATE
SALEEND                                     DATE
SALEPRICE                                   NUMBER(6,2)
ACTIVE                                      NUMBER(1)
FEATURED                                    NUMBER(1)
FEATURESTART                                DATE
FEATUREEND                                  DATE
TYPE                                        CHAR(1)
IDDEPARTMENT                                NUMBER(2)
STOCK                                       NUMBER(5,1)
ORDERED                                     NUMBER(3)
REORDER                                     NUMBER(3)

SQL>
```

Figure A-7   Structure of BB_PRODUCT

```
IDPRODUCT PRODUCTNAME              DESCRIPTION          PRODUCTIMAGE      PRICE SALESTART SALEEND    SALEPRICE
--------- ------------------------ -------------------- --------------- ------- --------- ---------- ----------
        1 CapressoBar Model #351   A fully programmable capresso.gif     99.99
                                   pump espresso machi
                                   ne and 10-cup coffee
                                   maker complete with
                                   GoldTone filter
        2 Capresso Ultima          Coffee and Espresso  capresso2.gif   129.99
                                   and Cappuccino Machi
                                   ne. Brews from one e
                                   spresso to two six o
                                   unce cups of coffee
        3 Eileen 4-cup French Press A unique coffeemaker frepress.gif     32.5
                                   from those proud cr
                                   aftsmen in windy Nor
                                   mandy.
        4 Coffee Grinder           Avoid blade grinders grind.gif        28.5
                                   ! This mill grinder
                                   allows you to choose
                                   a fine grind to a c
                                   oarse grind.
        5 Sumatra                  Spicy and intense wi sumatra.jpg      10.5
                                   th herbal aroma.
        6 Guatamala                heavy body, spicy tw Guatamala.jpg      10 01-JUN-04 15-JUN-04          8
                                   ist, aromatic and sm
                                   okey flavor.
        7 Columbia                 dry, nutty flavor an columbia.jpg     10.8
                                   d smoothness
        8 Brazil                   well-balanced mellow brazil.jpg       10.8
                                   flavor, a medium bo
                                   dy with hints of coc
                                   oa and a mild, nut-l
                                   ike aftertaste
        9 Ethiopia                 distinctive berry-li ethiopia.jpg       10
                                   ke flavor and aroma,
                                   reminds many of a f
                                   ruity, memorable win
                                   e.
       10 Espresso                 dense, caramel-like  espresso.jpg       10
                                   sweetness with a sof
                                   t acidity. Roasted s
                                   omewhat darker than
                                   traditional Italian.
```

```
IDPRODUCT   ACTIVE   FEATURED FEATUREST FEATUREEN T IDDEPARTMENT   STOCK   ORDERED   REORDER
--------- --------- --------- --------- --------- - ------------ --------- --------- ---------
        1         1                               E            2        23         0        12
        2         1                               E            2        15         0         9
        3         1                               E            2        30         0        15
        4         1                               E            2        26         0        25
        5         1                               C            1        41         0        45
        6         1                               C            1        42         0        35
        7         1                               C            1        61         0        35
        8         1                               C            1        53         0        35
        9         1                               C            1        54         0        35
       10         1                               C            1        50        50        50
```

Figure A-8   Data contained in BB_PRODUCT

BB_PRODUCTOPTION

```
Name                                        Null?     Type
------------------------------------------- --------- ---------------------------
IDPRODUCTOPTION                             NOT NULL  NUMBER(3)
IDOPTION                                              NUMBER(2)
IDPRODUCT                                             NUMBER(2)
```

Figure A-9   Structure of BB_PRODUCTOPTION

```
IDPRODUCTOPTION   IDOPTION   IDPRODUCT
---------------  ----------  ----------
             1           1           5
             2           2           5
             3           3           5
             4           4           5
             5           1           6
             6           2           6
             7           3           6
             8           4           6
             9           1           7
            10           2           7
            11           3           7
            12           4           7
            13           1           8
            14           2           8
            15           3           8
            16           4           8
            17           1           9
            18           2           9
            19           3           9
            20           4           9
            21           1          10
            22           2          10
            23           3          10
            24           4          10
```

Figure A-10　Data contained in BB_PRODUCTOPTION

BB_PRODUCTOPTIONDETAIL

```
Name                                     Null?     Type
---------------------------------------  --------  ---------------------------
IDOPTION                                 NOT NULL  NUMBER(2)
OPTIONNAME                                         VARCHAR2(25)
IDOPTIONCATEGORY                                   NUMBER(2)
```

Figure A-11　Structure of BB_PRODUCTOPTIONDETAIL

```
IDOPTION OPTIONNAME                    IDOPTIONCATEGORY
---------- ------------------------    ----------------
         1 1/2 LB.                                    1
         2 1 LB.                                      1
         3 Whole Bean                                 2
         4 Regular Grind                              2
```

Figure A-12　Data contained in BB_PRODUCTOPTIONDETAIL

BB_ PRODUCTOPTIONCATEGORY

```
Name                                      Null?    Type
---------------------------------------- -------- ----------------------------
IDOPTIONCATEGORY                          NOT NULL NUMBER(2)
CATEGORYNAME                                       VARCHAR2(25)
```

Figure A-13   Structure of BB_PRODUCTOPTIONCATEGORY

```
IDOPTIONCATEGORY CATEGORYNAME
---------------- ------------------------
               1 Size
               2 Form
```

Figure A-14   Data contained in BB_PRODUCTOPTIONCATEGORY

BB_DEPARTMENT

```
Name                                      Null?    Type
---------------------------------------- -------- ----------------------------
IDDEPARTMENT                              NOT NULL NUMBER(2)
DEPTNAME                                           VARCHAR2(25)
DEPTDESC                                           VARCHAR2(100)
DEPTIMAGE                                          VARCHAR2(25)
```

Figure A-15   Structure of BB_DEPARTMENT

```
IDDEPARTMENT DEPTNAME                DEPTDESC             DEPTIMAGE
------------ ----------------------- -------------------- ------------------------
           1 Coffee                  Many types of coffee coffee.gif
                                      beans

           2 Equipment and Supplies  Cofee Makers to coff machines.gif
                                      ee filters available

           3 Coffee Club             What the benefits of club.gif
                                      our club membership
                                      ?
```

Figure A-16   Data contained in BB_DEPARTMENT

BB_BASKETSTATUS

```
Name                                                  Null?      Type
------------------------------------------------      --------   --------------------
IDSTATUS                                              NOT NULL   NUMBER(5)
IDBASKET                                                         NUMBER(5)
IDSTAGE                                                          NUMBER(1)
DTSTAGE                                                          DATE
NOTES                                                            VARCHAR2(50)
SHIPPER                                                          VARCHAR2(5)
SHIPPINGNUM                                                      VARCHAR2(20)
```

Figure A-17   Structure of BB_BASKETSTATUS

```
IDSTATUS   IDBASKET    IDSTAGE DTSTAGE    NOTES              SHIPP SHIPPINGNUM
---------- ---------- ---------- --------- --------------- ----- --------------------
        1          3            1 24-JAN-03
        2          3            5 25-JAN-03 Customer called UPS   ZW845584GD89H569
                                           to confirm shi
                                           pment

        3          4            1 13-FEB-03
        4          4            5 14-FEB-03
       15         12            3
```

Figure A-18   Data contained in BB_BASKETSTATUS

BB_TAX

```
Name                                        Null?      Type
-----------------------------------------   --------   ------------------------------
IDSTATE                                     NOT NULL   NUMBER(2)
STATE                                                  CHAR(2)
TAXRATE                                                NUMBER(4,3)
```

Figure A-19   Structure of BB_TAX

```
IDSTATE ST    TAXRATE
---------- -- ----------
        1 VA       .045
        2 NC        .03
        3 SC        .06
```

Figure A-20   Data contained in BB_TAX

BB_SHIPPING

```
Name                                    Null?     Type
-------------------------------------   --------  -----------------------
IDRANGE                                 NOT NULL  NUMBER(2)
LOW                                               NUMBER(3)
HIGH                                              NUMBER(3)
FEE                                               NUMBER(6,2)
```

Figure A-21   Structure of BB_SHIPPING

```
IDRANGE        LOW        HIGH         FEE
----------  ----------  ----------  ----------
         1           1           5           5
         2           6          10           8
         3          11          99          11
```

Figure A-22   Data contained in BB_SHIPPING

B

PROCEDURE BUILDER

Procedure Builder is a tool included in the Oracle Developer 6*i* suite and is used to assist in creating and testing program units. As we have seen, program units can be developed using SQL*Plus; however, Procedure Builder eases the creation of program units and provides more robust debugging features.

One feature of Procedure Builder is automatic color coding of the programming statements to simplify reading of code and determining syntax problems. Another feature is a debugger that enables the pausing of program unit execution to allow the developer to check variable values during execution.

This appendix is an introduction to using Procedure Builder to create and debug a program unit.

REBUILDING YOUR DATABASE

In this section, you rebuild your database. This has to be done to ensure the needed tables and procedures we will use to test Procedure Builder are available.

To rebuild the Brewbean's database:

1. Open SQL*Plus.

2. Create a spool file so that SQL*Plus keeps a copy of all the messages received from running the file that creates the database. On the main menu in SQL*Plus, click **File**, point to **Spool**, and then click **Spool File**. A Select File dialog box appears.

3. Browse to a folder used to contain your working files and in the File name text box, type **DB_log_appb**, and then click **Save**. Now, whatever text is seen in our SQL*Plus session is saved to this file for future reference.

4. Now let's create the database. In SQL*Plus, type the following command to run all the statements contained in the appbDbcreate.sql file. Messages verifying the creation and data insertion steps scroll on the SQL*Plus screen. This may take a couple minutes to complete.

   `@<pathname to PL/SQL files>\appendix.b\appbDbcreate`

5. Now, we will turn the spooling off so that we can review the results in the spool file created. On the main menu, click **File**, point to **Spool**, and then click **Spool Off**.

6. Open a text editor.

7. Open the file named **DB_log_appb.lst** in your working folder.

8. Review the messages for any errors.

CREATING A PROCEDURE USING PROCEDURE BUILDER

B

The Brewbean's application presented a need for logic that calculates order shipping costs, and we created a procedure named SHIP_COST_SP in SQL*Plus to fulfill this need. Let's create the same SHIP_COST_SP procedure again using Procedure Builder.

Starting Procedure Builder

Upon starting Procedure Builder, you need to connect to the database as an explicit action because it does not automatically prompt you to log on.

To start Procedure Builder:

1. Click **Start**, point to **Programs**, and click **Oracle Forms & Reports 6*i***.

2. Click the **Procedure Builder** subentry to start the software.

3. When the screen shown in Figure B-1 appears, you need to connect to a database. On the main menu, click **File**, and then click **Connect**.

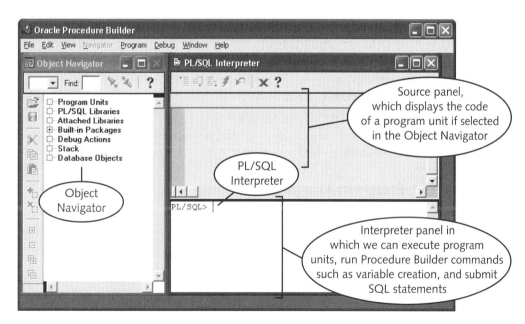

Figure B-1   Procedure Builder initial window

4. A logon screen much the same as with SQL*Plus appears. Enter your logon information, including the host string. A default host string provided for the Personal Oracle9*i* database is tcp–loopback.world.

5. Click the plus sign to expand the Database Objects node in the Object Navigator and verify Procedure Builder now appears, as shown in Figure B-2.

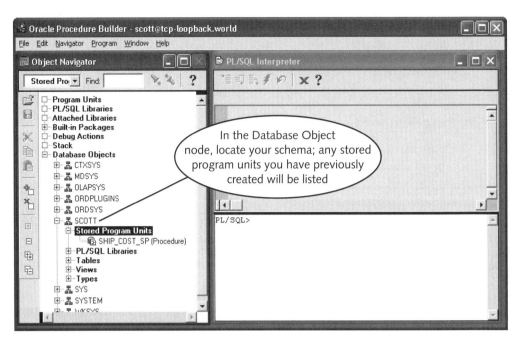

Figure B-2 Procedure Builder after a connection has been made to the database

6. Click the **SHIP_COST_SP** procedure name so that it is highlighted and the code is displayed.

Now that we have started Procedure Builder, in the next section we explore how to create a procedure within this tool.

Using the Program Unit Editor

Procedure Builder has a program unit editor in which we will enter and compile the code for our procedure. You will delete the existing procedure so that you can start from scratch.

To create the procedure:

1. Click the **SHIP_COST_SP** procedure name, if necessary, in the Object Navigator to highlight it and you will see that the Delete button, which has a red X as its icon, is available on the vertical tool bar on the left side of the window.

2. Click the **Delete** button with the SHIP_COST_SP procedure highlighted, and then click **Yes**. The procedure is deleted.

3. Click the **Stored Program Units** node in the Object Navigator so that it is highlighted.

4. Click the green **+** sign on the left toolbar to prompt the creation of a new program unit. A New Program Unit dialog box appears, as shown in Figure B-3.

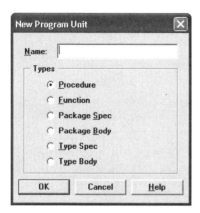

Figure B-3   New Program Unit dialog box

5. In the Name text box, type **SHIP_COST_SP**.

6. The Procedure option button should already be selected. Click **OK**.

7. The Stored Program Unit editor appears, as shown in Figure B-4. Note the editor automatically inserts the basic key word structure needed to create a procedure.

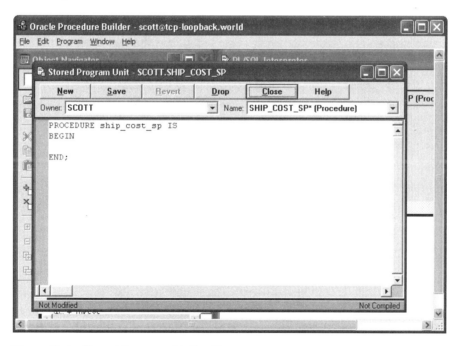

Figure B-4   Stored Program Unit editor

8. Enter the following procedure statement into the editor. Note that this code contains an error so that we can see what happens with an error in the editor.

```
PROCEDURE ship_cost_sp
  (p_qty IN NUMBER,
   p_ship OUT NUMBER)
  IS
BEGIN
  IF p_qty > 10 THEN
     p_ship := 11.00;
  ELSIF g_qty > 5 THEN
     p_ship := 8.00;
  ELSE
     p_ship := 5.00;
  END IF;
END;
```

Note that the CREATE OR REPLACE phrase is not used in Procedure Builder because this tool automatically attaches this phrase to the procedure code. In addition, no forward slash is needed on the last line, as is used in SQL*Plus.

9. Click the **Save** button at the top of the editor pane and then click **No** to the warning dialog box, if it appears. This saves and compiles the procedure. In this case, you misspelled the parameter name on line 8 as G_QTY rather than P_QTY, so an error message is displayed, as shown in Figure B-5.

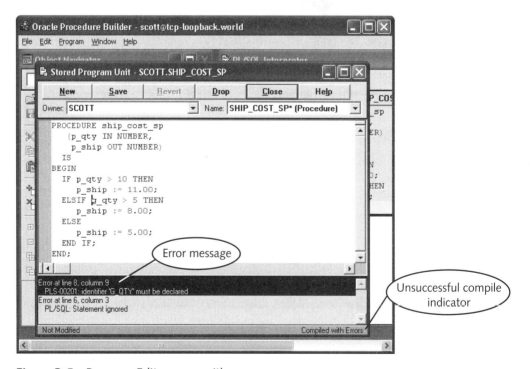

Figure B-5   Program Editor pane with errors

10. To correct the error, change the G_QTY reference on line 8 to **P_QTY**.

11. Click **Save** and the message on the status line at the bottom of the editor pane states "Successfully Compiled," as shown in Figure B-6.

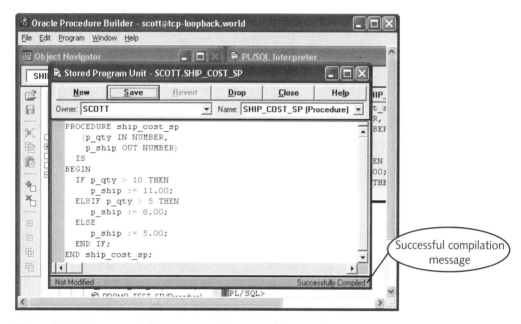

Figure B-6   Program Editor pane with successful compile

12. Close the Program Editor pane.

 Notice that when you have program unit code showing in the Program Editor pane, the statements are color coded. For example, key words appear in blue and operators appear in red. This is quite valuable in identifying syntax errors as you enter code.

Now that we have created a procedure in Procedure Builder, let's move on to testing it.

Testing the Procedure in the Interpreter Panel

The process of testing the procedure includes creating a host variable to hold the return value, executing the procedure, and then checking the results. The differences in commands used in SQL*Plus versus Procedure Builder to accomplish these tasks are listed in Table B-1.

Table B-1   Differences in commands used to test procedures

Task	SQL*Plus	Procedure Builder
Create a host variable	Use the VARIABLE command	Use the .CREATE command
Execute a procedure	Use the EXECUTE command and call the procedure by name passing appropriate arguments	Call the procedure by name passing appropriate arguments
Display contents of the host variable	Use the PRINT command	Use the TEXT_IO.PUT_LINE Oracle-supplied procedure

To test the procedure:

1. At the PL/SQL> prompt in the Interpreter panel, type **.CREATE NUMBER g_ship_cost**, and then press **Enter**. Be sure to begin the command with a period. This creates a variable in Procedure Builder, which is now our calling environment. The only information needed is the data type and name of the variable.

2. Now, we are ready to run the procedure. In the Interpreter panel, type **ship_cost_sp(7,:g_ship_cost);**, and then press **Enter** to execute the procedure.

3. Now check if the correct value was returned to the bind variable. In the Interpreter panel, type **TEXT_IO.PUT_LINE(:g_ship_cost);**, and then press **Enter** to display the value in the host variable, as shown in Figure B-7. The value of 8 should be displayed to the screen as the cost of shipping seven items.

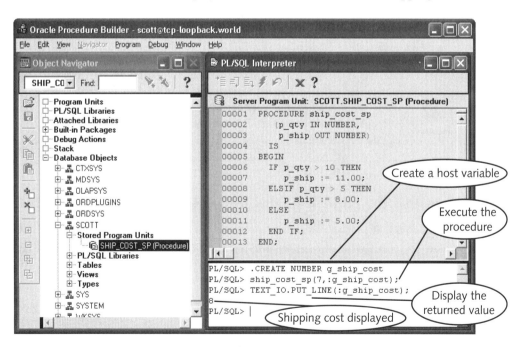

Figure B-7   Testing a procedure in Procedure Builder

> The TEXT_IO.PUT_LINE is a built-in facility within Procedure Builder that displays values to the screen, much like the DBMS_OUTPUT.PUT_LINE used in SQL*Plus.

B

DEBUGGING WITH PROCEDURE BUILDER

Two topics are very important to debugging successfully within Procedure Builder: breakpoints and displaying values to the screen with the TEXT_IO.PUT_LINE command. The Procedure Builder debugger features allow us to pause the execution of a program unit to view variable values at any point in the execution, check the flow of the execution, and display values of variables to the screen during execution. We can also embed TEXT_IO.PUT_LINE statements into program units to display values to the screen to assist in debugging.

In this section of the chapter, we first explore using breakpoints in debugging program units and then we explore the TEXT_IO.PUT_LINE statement.

Working with Breakpoints

Breakpoints allow the programmer to set flags on executable lines within a program unit to pause the processing on that line and check the flow and values of variables during the execution. After pausing the program execution, the programmer can then control stepping through the remaining executable lines to evaluate the processing. The breakpoint must be set on an executable line, which includes SQL statements.

Let's look at a procedure that contains a CURSOR FOR loop to demonstrate the breakpoints. Recall that a CURSOR FOR loop is a cursor that automatically sets up the necessary record and counter variable for the looping action. Review the procedure listed in Figure B-8 called PROMO_TEST_SP. It uses a CURSOR FOR loop to review the total purchases by each shopper for a given month and year.

Brewbean's is experimenting with setting up some purchase incentives to encourage repeat shoppers. If a shopper spends more than $25 in a month, a free shipping offer is extended for his or her next purchase over $25. If a shopper has spent more than $50 in a month, a free shipping offer for his or her next purchase is extended regardless of the total. The subtotal column is summed in the procedure to reflect the actual product purchase total. The procedure updates the BB_PROMOLIST table with the shopper information, which is used to e-mail shoppers the appropriate incentive. The PROMO_FLAG variable is used to assign the correct promotion code.

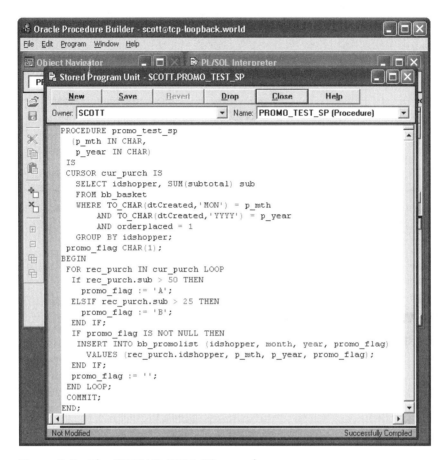

Figure B-8   The PROMO_TEST_SP procedure

Let's set up and run this procedure in order to see how we can use the breakpoints feature of Procedure Builder.

To create and test the PROMO_TEST_SP procedure:

1. In a text editor, open **promotest.txt** from the Appendix.B folder.

2. Copy all the code.

3. Open or return to Procedure Builder. Click the **Stored Program Units** node in the Object Navigator to highlight it.

4. Click the **Create** button to create a new procedure.

5. In the Name text box of the New Program Unit dialog box, type **PROMO_TEST_SP**. Click **OK**.

6. Highlight the text already in the editor pane and press the **Delete** key to erase it.

7. On the main menu, click **Edit**, and then click **Paste** to paste the procedure code from the text file into the pane.

8. Click the **Save** button and then click the **Close** button.

9. Now execute the procedure so that we can test the results. At the PL/SQL prompt in the Interpreter panel, type **promo_test_sp('FEB', '2003');**, and then press **Enter**.

10. You now need to run a query to check what the procedure inserted into the BB_PROMOLIST table. At the PL/SQL prompt, type **SELECT * FROM bb_promolist;**, and then press **Enter**. Notice that five rows were inserted into the BB_PROMOLIST table. However, notice that no PROMO_FLAG value was inserted for shopper 23, as shown in Figure B-9.

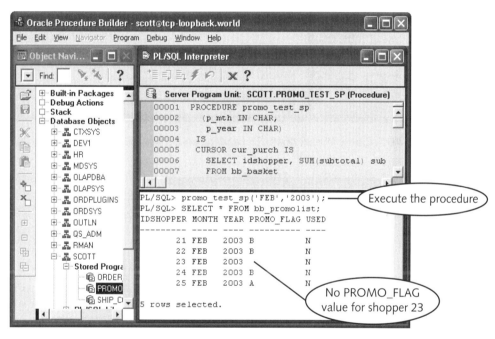

Figure B-9   Querying the BB_PROMOLIST table

11. To verify the results, let's query the BB_BASKET table to verify the order totals of each shopper and who is eligible for the promotion. Enter the SELECT statement shown in Figure B-10. Notice that shopper 23 only has a total of 21.60 and is not eligible for the promotion.

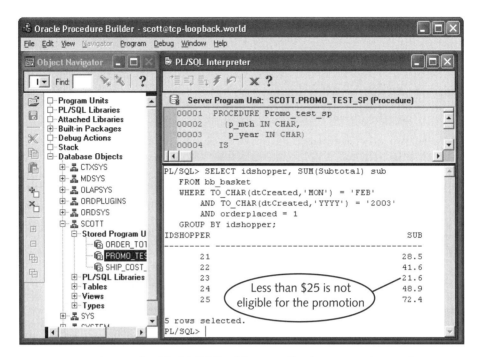

Figure B-10 Query the BB_BASKET table to verify shopper order totals

Now we have a problem. Our procedure has included shopper 23 in the BB_PROMOLIST table; however, this shopper is not eligible for the promotion and, therefore, should not be inserted into the table. How can we determine the problem with this procedure? Let's explore breakpoints to solve this problem.

To investigate the shopper 23 issue using breakpoints:

1. In Procedure Builder, expand the Stored Program Units node in the Object Navigator and click the **PROMO_TEST_SP** procedure entry to highlight it. Notice the source code for this procedure now appears in the source area at the top part of the Interpreter panel.

2. Notice the line numbers listed on the left side of the source area; we will use these to set breakpoints. Double-click the **00014** line number to set a breakpoint. Notice the line number changes to B(01) with a large red circle next to it, as shown in Figure B-11. A breakpoint must be placed on an executable statement line in the BEGIN section of a block.

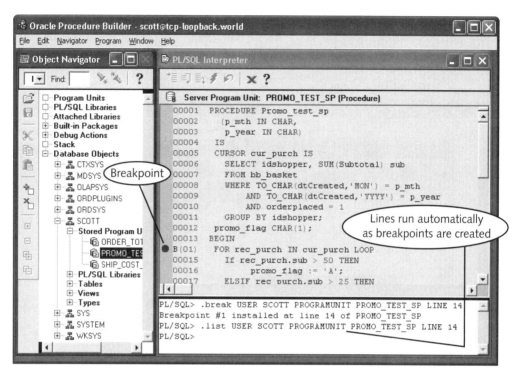

Figure B-11   Setting a breakpoint in Procedure Builder

3. Now, we need to invoke the procedure again to use the breakpoint to assist in debugging. First, delete the rows that were inserted into the BB_PROMOLIST table from our initial run by typing **DELETE FROM bb_promolist;** at the PL/SQL prompt of the Procedure Builder Interpreter panel.

4. At the PL/SQL> prompt in the Interpreter panel, invoke the procedure by typing **promo_test_sp('FEB','2003');**. Notice that the yellow arrow in the source panel indicates the processing is currently stopped on the breakpoint line, as shown in Figure B-12. Also notice a set of buttons immediately above the source panel is now available. These five buttons pertain to debugging actions using breakpoints.

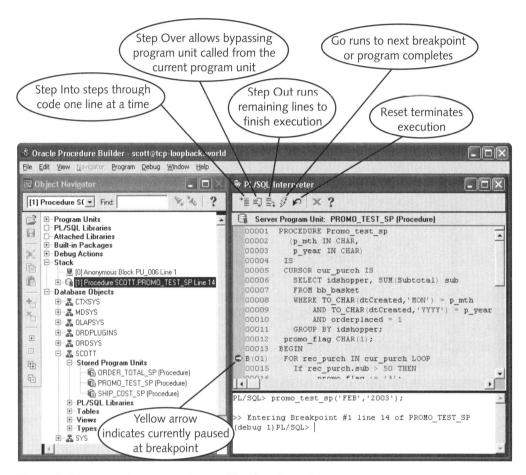

Figure B-12   Invoking a procedure with a breakpoint

5. In the Object Navigator, expand all the nodes within the stack node by clicking the + signs; your screen should look similar to Figure B-13. You may need to widen the Object Navigator window to see all the stack entries.

B

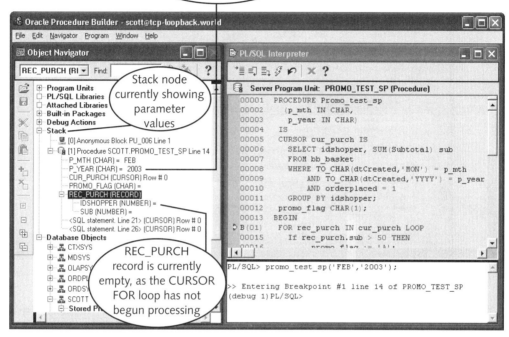

Figure B-13 The stack node

6. We can step through the procedure executable statements one at a time by clicking the Step Into button. Click the **Step Into** button twice and notice a yellow arrow in the source panel points to the executable line on which Procedure Builder is currently paused. Your two clicks will run the cursor setup and the FOR LOOP statements.

7. Review the stack in the Object Navigator. As shown in Figure B-14, the record variable now has values. Shopper number 21 has a total of $28.50 in purchases for the month and will be the first customer we test for the promotion. Note the purchase sum is greater than $25, so this customer should receive a promotion.

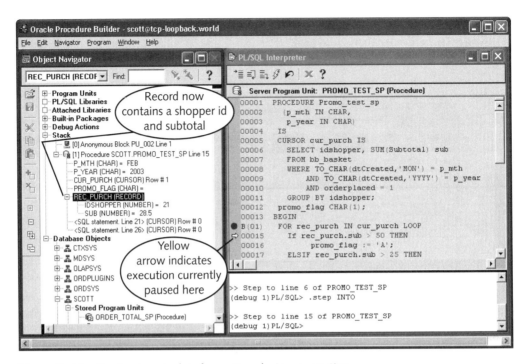

Figure B-14   Reviewing stack information during execution

8. Click the **Step Into** button again and, because the > 50 test is false, focus will jump to the > 25 test line.

9. Continue clicking the **Step Into** button and watch the values in the Object Navigator. For shopper number 21, you will see the PROMO_FLAG variable get set to "B" and then to blank after the insert. Stop when you reach shopper 23, the one with which we had the problem.

10. As you click the **Step Into** button, notice that while on the "IF promo_flag IS NOT NULL THEN" line with shopper number 23, the process moves to the INSERT statement that tells us that the PROMO_FLAG was not NULL and, therefore, the INSERT runs even though no promotion should be given to this shopper. Now, we know where the problem exists. Setting the PROMO_FLAG to empty using the command **PROMO_FLAG := ' ';** is not the same as setting the variable to NULL. Remember that a blank (empty spaces) and a zero are both treated as an actual value and not a NULL value, which indicates nothing has been entered into the field.

11. Click the **Reset** button to terminate the execution.

12. Let's make the needed correction and run the procedure again. First, type **DELETE FROM bb_promolist;** in the Interpreter panel to remove any rows in the BB_PROMOLIST table.

13. In the source panel, remove the breakpoint by double-clicking the **B(01)** line number.

14. Double-click the **PROMO_TEST_SP** icon to open it in the editor pane. Modify the PROMO_FLAG assignment statement at the end of the procedure to `promo_flag := NULL;`.

15. Click **Save** and **Close** to compile and close the editor.

16. Type `promo_test_sp('FEB','2003');` in the Interpreter panel to execute the procedure again.

17. Type `SELECT * FROM bb_promolist;` in the Interpreter panel to check which shoppers are assigned free shipping. The INSERT in the procedure inserts shoppers 21, 22, 24, and 25 to the BB_PROMOLIST table, because they should receive a free shipping offer.

Displaying Values to the Screen

Another important mechanism used to debug programs is displaying values to the screen to confirm values or whether statements have executed. This is popular to test exception handlers for which we have similarly used DBMS_OUTPUT.PUT_LINE in SQL*Plus.

Let's experiment with a new procedure named SHOPPER_QUERY_SP that contains a TEXT_IO.PUT_LINE statement in the EXCEPTION handler area.

To test TEXT_IO.PUT_LINE:

1. In a text editor, open **shop.txt** from the Appendix.B folder.

2. Copy all the code.

3. Open or return to Procedure Builder. Click the **Stored Program Units** node in the Object Navigator to highlight it.

4. Click the **Create** button to create a new procedure.

5. In the Name text box of the New Program Unit dialog box, type **SHOPPER_QUERY_SP**. Click **OK**.

6. Highlight the text already in the editor pane and press the **Delete** key to erase it.

7. On the main menu, click **Edit**, and then click **Paste** to paste the procedure code from our text file into the editor pane.

8. Click the **Save** button to compile and you will receive the compile error, as shown in Figure B-15.

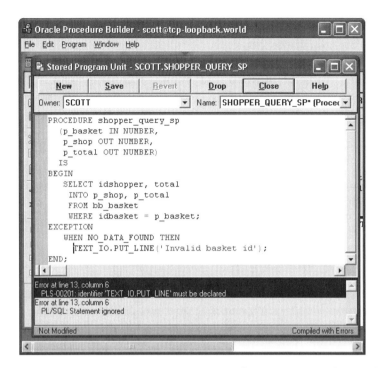

Figure B-15 TEXT_IO.PUT_LINE compile error in Procedure Builder

    9. Click the **Close** button and you will see an asterisk beside the procedure name in the Object Navigator indicating the unit has not compiled successfully.

So, how do we display lines of data to the screen from a PL/SQL block as we do using DBMS_OUTPUT.PUT_LINE in SQL*Plus for debugging assistance? We must use a TEXT_IO.PUT_LINE statement; however, this is only recognized for local (not stored) program units in Procedure Builder. Local program units are saved client-side, whereas stored program units are saved server-side. As such, including DBMS_OUTPUT.PUT_LINE statements in server-side program units will not display anything in Procedure Builder.

Let's move a copy of the SHOPPER_QUERY_SP procedure to the local program unit area to verify how the TEXT_IO.PUT_LINE statement works.

To move a stored program unit to the local program unit area:

    1. Drag the **SHOPPER_QUERY_SP** procedure in the Object Navigator from the Stored Program Units' area to the Program Units' node (the first node in the Object Navigator). A copy is made in the local Program Units' area. The Object Navigator area should now look like Figure B-16.

    2. In the local Program Units' node, double-click the **program unit** icon next to the SHOPPER_QUERY_SP procedure node in the Object Navigator to open it in the Program Unit editor.

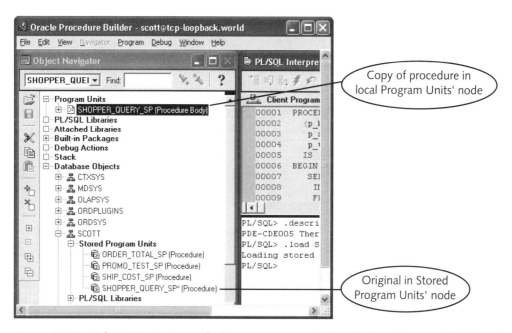

Figure B-16   Object Navigator with the procedure copied to the local Program Units' node

3. Click the **Compile** button and you will see that it now compiles successfully as the PL/SQL engine in Procedure Builder recognizes the TEXT_IO.PUT_LINE statement.

4. You will also notice a node expand + marker next to the procedure name. Use it to expand all the nodes under the SHOPPER_QUERY_SP procedure. There will be three areas of information, as shown in Figure B-17.

5. Let's execute the procedure to test the TEXT_IO.PUT_LINE statement. We will use a basket id of 99, which is not in the BB_BASKET table, to check if the exception area will display the error message we provided. First, we need to create two host variables for the two OUT parameters in the procedure. In the Interpreter panel of Procedure Builder, type **.CREATE NUMBER g_shop**, and then press **Enter**.

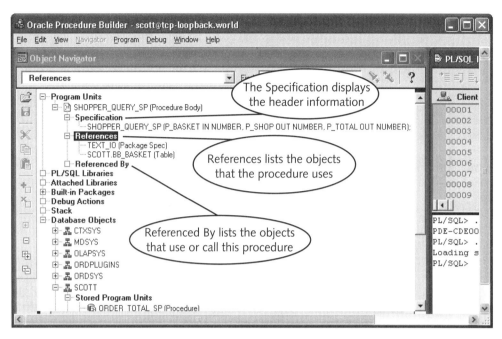

Figure B-17 Expanded local Program Units' node

6. Then, type **.CREATE NUMBER g_total** and press **Enter**.

7. In the Interpreter panel, invoke the procedure by typing **shopper_query_sp(99, :g_shop, :g_total);**, and then press **Enter**. Results will match Figure B-18.

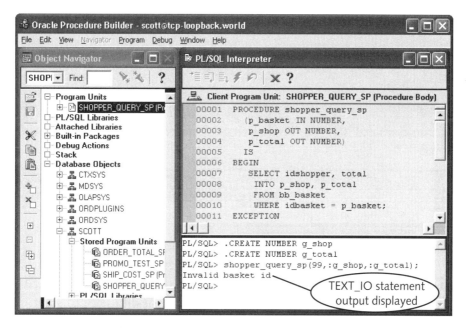

Figure B-18   Test local procedure containing a TEXT_IO.PUT_LINE statement

APPENDIX SUMMARY

So, we have confirmed that in Procedure Builder we can debug program units using breakpoints and display values to the screen using TEXT_IO.PUT_LINE. This appendix demonstrated moving stored program units to the local Program Units' area. If you prefer using TEXT_IO.PUT_LINE for testing, you can also create the procedure initially in the local Program Units' node, test it, delete or comment out the TEXT_IO.PUT_LINE statement, and then click and drag to the stored Program Units' node.

C

TOAD (TOOL FOR ORACLE APPLICATION DEVELOPERS)

TOAD is a tool offered by Quest Software (*www.quest.com*) and is used to assist in creating and testing SQL and PL/SQL code. TOAD is rich with features for both the developer and the DBA and is offered in a variety of versions. Reference the feature descriptions outlined on the Quest Web site to learn about all the capabilities of this software. This appendix uses the TOAD Professional version, which includes a PL/SQL debugger. Trial downloads of TOAD are available at the Quest Web site.

As we have seen, program units can be developed using SQL*Plus; however, TOAD eases the creation of program units and provides more robust debugging features. One feature is automatic color coding of the programming statements to simplify reading of code and determining syntax problems. Another feature is a debugger that enables the pausing of program unit execution to allow the developer to check variable values during execution. This appendix is an introduction to using TOAD to create and debug a program unit.

REBUILDING YOUR DATABASE

To rebuild the Brewbean's database:

1. Open SQL*Plus.

2. Create a spool file so that SQL*Plus keeps a copy of all the messages received from running the file that creates the database. On the main menu in SQL*Plus, click **File**, point to **Spool**, and then click **Spool File**. A Select File dialog box appears.

3. Browse to a folder used to contain your working files and in the File name text box, type **DB_log_appc**, and then click **Save**. Now, whatever text is seen in our SQL*Plus session is saved to this file for future reference.

4. Now let's create the database. In SQL*Plus, enter the following command, which runs all the statements contained in the appcDbcreate.sql file. Messages verifying the creation and data insertion steps scroll on the SQL*Plus screen. This may take a couple minutes to complete.

   `@<pathname to PL/SQL files>\appendix.c\appcDbcreate`

5. Now, we will turn the spooling off so that we can review the results in the spool file created. On the main menu, click **File**, point to **Spool**, and then click **Spool Off**.

6. Open a text editor.

7. Open the file named **DB_log_appc.lst** in your working folder.

8. Review the messages for any errors.

CREATE A PROCEDURE USING TOAD

In the main part of this book, the Brewbean's application presented a need for logic that calculates order shipping costs, and we created a procedure named SHIP_COST_SP in SQL*Plus to fulfill this need. Let's create the same SHIP_COST_SP procedure again using TOAD.

Starting TOAD

In this section, you will learn how to start TOAD.

To start TOAD:

1. Click the Windows **Start** menu, point to **All Programs**, and point to the **Quest Software** entry.

2. Point to **TOAD**, and click the **TOAD** subentry to start the software.

3. Once the initial screen, shown in Figure C-1, appears, we need to connect to a database. Select an appropriate connection string from the Database drop-down list box.

Figure C-1   TOAD initial window

4. Enter your user name and password. Click **Connect as**, click **Normal**, and then click **OK**.
5. After you are logged on, TOAD appears, as shown in Figure C-2. By default, the SQL Editor screen opens when TOAD is started.

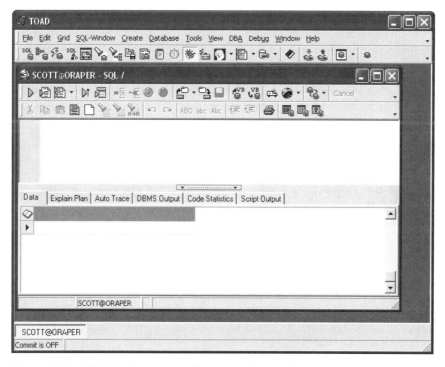

Figure C-2   TOAD after connection is made to a database

Now that we can start up TOAD, in the next section, we will explore how to create a procedure within this tool.

Using the Procedure Editor

TOAD has a procedure editor in which we can enter and compile the code for our procedure.

To create a procedure:

1. To open the procedure editor, click **Database**, and then click **Procedure Editor** from the main menu. The editor appears as shown in Figure C-3.

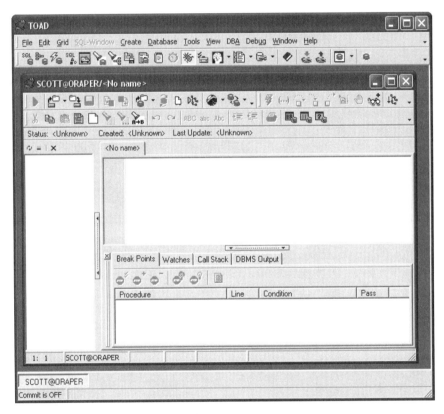

Figure C-3   TOAD procedure editor

2. Click the **Create new PLSQL object** button on the editor toolbar and enter the program unit type and name, as shown in Figure C-4. Click **OK**.

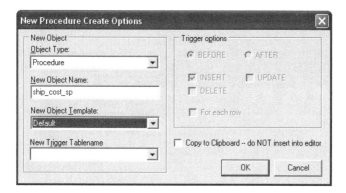

Figure C-4   Create a new program unit

3. The editor source pane now contains a starting template. Highlight all the template entries and delete them.

4. Type the following procedure in the source pane area shown in Figure C-5. This procedure contains an error so that we can see how compilation errors are presented, so enter it as is.

```
CREATE OR REPLACE PROCEDURE ship_cost_sp
   (p_qty IN NUMBER,
    p_ship OUT NUMBER)
  IS
BEGIN
  IF p_qty > 10 THEN
      p_ship := 11.00;
  ELSIF g_qty > 5 THEN
      p_ship := 8.00;
  ELSE
      p_ship := 5.00;
  END IF;
END;
```

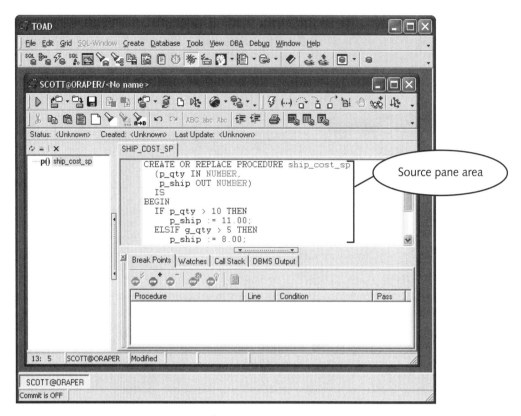

Figure C-5 Enter program unit code

5. Click the **Compile** button on the editor toolbar. This saves and compiles the procedure. Notice an error message displays in a box at the bottom of the editor pane. There is also the message "Compiled with errors, processing stopped!", as shown in Figure C-6. In this case, you misspelled the parameter name on line 8 as G_QTY rather than P_QTY.

 Notice that when you have program unit code showing in the editor pane, the statements are color coded. For example, key words appear in blue and operators appear in red. This is quite valuable in identifying syntax errors as you enter code.

6. To correct the error, change the G_QTY reference on line 8 to **P_QTY**.

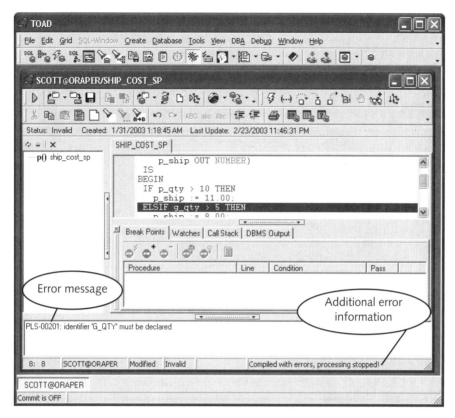

Figure C-6   Program editor pane with compile errors

7. Click the **Compile** button again and the message at the bottom of the editor pane disappears and a status of Valid is indicated, as shown in Figure C-7.

8. Click the **Close** button to close the editor. You will receive a message box asking if you want to save the procedure code. If you respond Yes, you are allowed to save the code to a text file. Click **No**.

 The preceding steps created and saved the SHIP_COST_SP procedure to the database as a stored program unit. The last option allowed you to write the procedure code to a text file if you wanted to maintain a copy of the source code.

Now that we have created a procedure in TOAD, let's move on to running and testing the procedure.

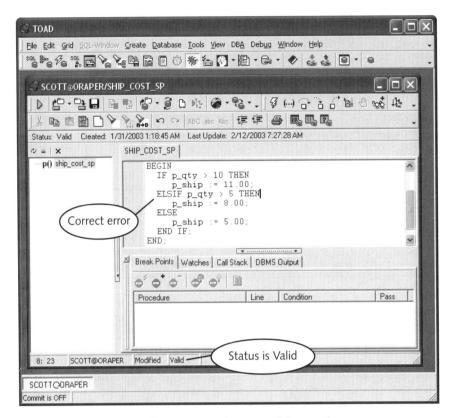

Figure C-7 Program editor pane with successful compile

Testing a Procedure with TOAD

The process of testing the procedure includes creating a host variable to hold the return value, executing the procedure, and then checking the results. The main differences in the execution process used in SQL*Plus versus TOAD to accomplish these tasks are that TOAD provides a mechanism that generates an autonomous block to test the procedure based on input for setting parameter values.

To test the procedure:

1. Click **Database** and then **Procedure Editor** from the main menu.

2. Retrieve the SHIP_COST_SP procedure by clicking the **Load source from existing object** button on the editor toolbar. A selection pane listing existing program units appears, as shown in Figure C-8.

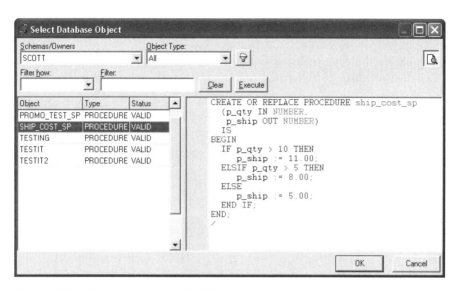

Figure C-8   Select program unit listing

3. Click the **SHIP_COST_SP** procedure in the list on the left, and then click **OK**.

4. Click the **Set Parameters** button on the editor toolbar and a dialog box, shown in Figure C-9, appears.

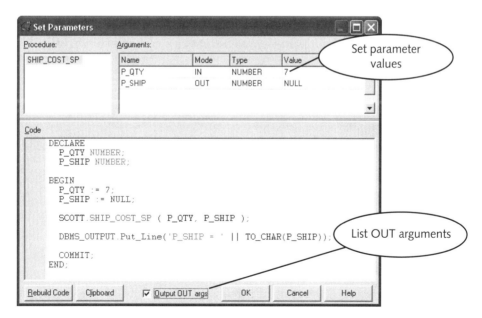

Figure C-9   Set Parameters dialog box

5. In the Arguments box at the top, click in the NULL value area for the P_QTY parameter and type **7**.

6. At the bottom, click the **Output OUT args** check box to turn this option on. Notice that TOAD creates the needed anonymous block to test the procedure.

7. Click **OK** to close the dialog box.

8. Click the **Run** button on the editor toolbar to execute the procedure. You are asked if you want to use DEBUG info. Click **Yes**.

9. A message box appears that states execution has terminated. Click **OK**.

10. At the bottom of the editor pane, click the **DBMS Output** tab. The P_SHIP OUT parameter correctly contains 8, as shown in Figure C-10.

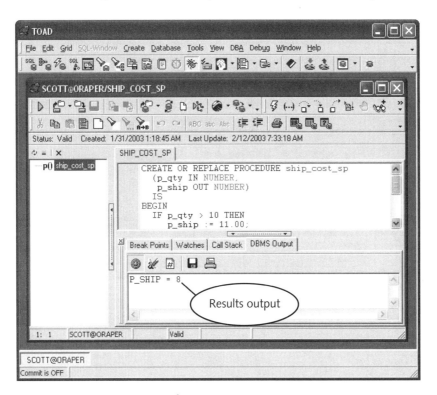

Figure C-10   DBMS output from execution

11. Close the editor.

Debugging with TOAD

Two features are very important to debugging successfully: setting breakpoints and displaying variable values to the screen during execution. The TOAD debugger is part of the Procedure Editor pane and allows us to pause the execution of a program unit to check variable values at any point in the execution, check the flow of the execution, and display values of variables to the screen during execution.

Creating Breakpoints

Breakpoints allow the programmer to set flags on executable lines within a program unit to pause the processing on that line and check the flow and values of variables during the execution. After pausing the program execution, you can then control stepping through the remaining executable lines to evaluate the processing. The breakpoint must be set on an executable line, which includes SQL statements.

Let's look at a procedure that contains a CURSOR FOR loop to demonstrate the breakpoints. Recall that a CURSOR FOR loop is a cursor that automatically sets up the necessary record and counter variable for the looping action. Review the procedure listed in Figure C-11 called PROMO_TEST_SP, which uses a CURSOR FOR loop to review the total purchases by each shopper for a given month and year.

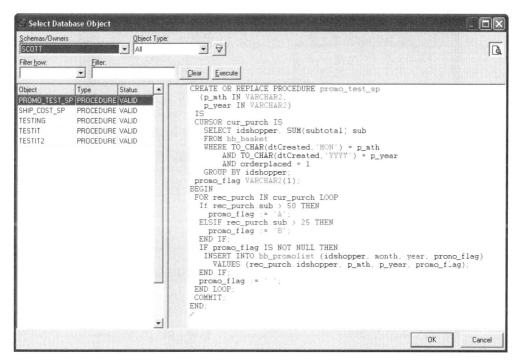

Figure C-11   The PROMO_TEST_SP procedure

Brewbean's is experimenting with setting up some purchase incentives to encourage repeat shoppers. If a shopper spends more than $25 in a month, a free shipping offer is extended for their next purchase over $25. If a shopper has spent more than $50 in a month, a free shipping offer for their next purchase is extended regardless of the total. The subtotal column is summed in the procedure to reflect the actual product purchase total. The procedure updates the promotions table with the shopper information, which is used to e-mail shoppers the appropriate incentive. The PROMO_FLAG variable is used to assign the correct promotion code.

Let's run this procedure in order to see how we can use the breakpoints feature of TOAD.

To test the PROMO_TEST_SP procedure:

1. Open the SQL Editor, if necessary, to execute the procedure. On the main menu, click the **Database** button, and then click **SQL Editor**.

2. Enter an anonymous block to execute the procedure, as shown in Figure C-12. Then click the **Execute Statement** button on the SQL Editor toolbar.

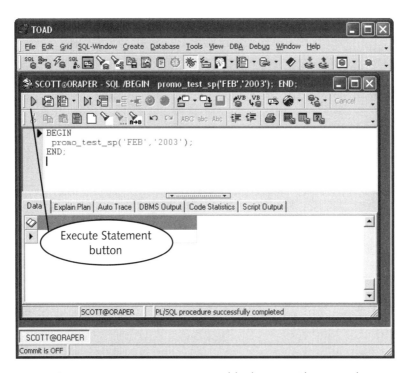

Figure C-12   Entering an anonymous block to test the procedure

3. Now, you need to check what the procedure inserted into the BB_PROMOLIST table. Open the Schema Browser by clicking **Database** and then **Schema Browser** from the main menu. On the left side of the browser pane is a list of tables. Scroll down and click the **BB_PROMOLIST** table.

4. On the right side, click the **Data** tab to view the contents of the table. Notice that no PROMO_FLAG value was inserted for shopper 23, as shown in Figure C-13.

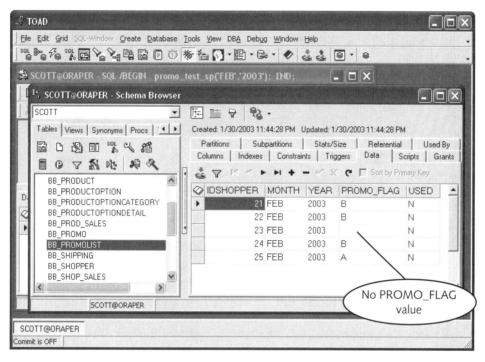

Figure C-13   View the BB_PROMOLIST table data

5. To verify the results, let's query the BB_BASKET table to verify the order totals of each shopper and who is eligible for the promotion. Return to the SQL Editor pane and click the **Clear All** button on the toolbar to erase the anonymous block entered earlier.

6. Type the SELECT command, as shown in Figure C-14. Click the **Execute Statement** button on the toolbar. Notice in the results displayed at the bottom that shopper 23 has a total of only 21.60 and is not eligible for the promotion.

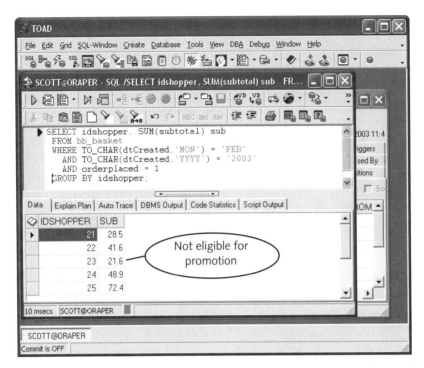

Figure C-14 Query the BB_BASKET table to verify shopper order totals

Now we have a problem. Our procedure has included shopper 23 in the BB_PROMOLIST table; however, this shopper is not eligible for the promotion and, therefore, should not be inserted into the table. How can we determine the problem with this procedure? Let's explore breakpoints to solve this problem.

To investigate the shopper 23 issue using breakpoints:

1. Open the procedure editor by clicking **Database** and then **Procedure Editor** from the main menu.

2. Retrieve the PROMO_TEST_SP procedure by clicking the **Load source from existing object** button on the editor toolbar. A selection pane listing existing program units appears.

3. Click the PROMO_TEST_SP procedure in the list on the left, and then click **OK**.

4. To set a breakpoint on the LOOP statement on line 14, click in the gray area to the left of the line and a breakpoint will be identified by a stop sign in the gray margin and highlighting on the line, as shown in Figure C-15.

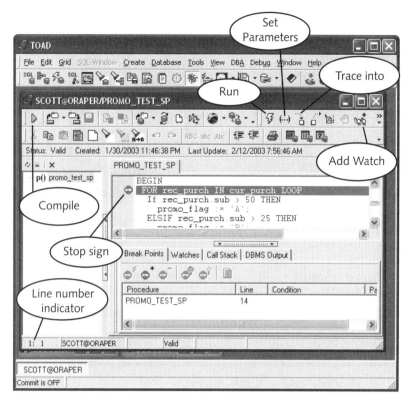

Figure C-15   Setting a breakpoint

5. Now, we need to invoke the procedure again to use the breakpoint to assist in debugging. First, delete the rows that were inserted into the BB_PROMOLIST table from our initial run by returning to the Schema Browser pane and clicking the **Delete record** button on the toolbar five times to delete all five rows previously inserted, as shown in Figure C-16. Then click the **Commit** button to save the modifications.

6. Return to the Procedure Editor pane and click the **Set Parameters** button. Be sure P_MTH is set to FEB and P_YEAR is set to 2003, as shown in Figure C-17. Click **OK**.

7. Now, we will set a watch on the PROMO_FLAG variable so that we can view the value of this variable during procedure execution. Click anywhere on the PROMO_FLAG variable name on line 20 and click the **Add Watch** button on the toolbar. If you click the Watches tab at the bottom, the PROMO_FLAG variable is now listed, as shown in Figure C-18.

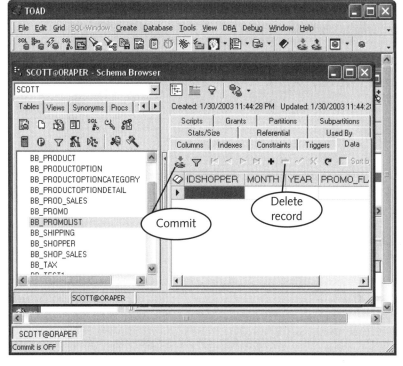

Figure C-16   Delete rows previously inserted

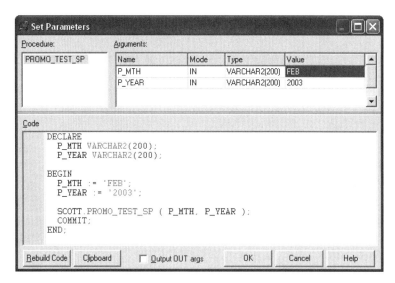

Figure C-17   Set parameter values

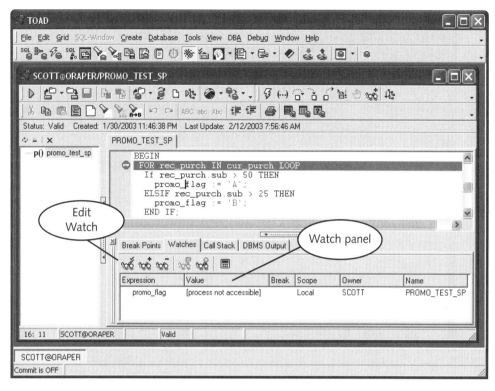

Figure C-18   The Watch list in the procedure editor

8. Click the **Edit Watch** button on the toolbar of the Watches tab at the bottom.

9. Check the **Break on value change** check box to turn this feature on, and then click **OK**.

10. Click the **Trace into** button on the procedure editor toolbar to begin executing the procedure.

11. You are asked if you want to compile referenced objects with debugging information. Click **Yes**. Note that execution starts and then stops at the breakpoint line, as shown in Figure C-19.

12. Continue to click the **Trace into** button to step through the execution of the procedure. Notice the PROMO_FLAG variable watch indicates the changes in value. Note that the PROMO_FLAG is a blank at one point and an INSERT is still processed. A dialog box appears when the execution is completed. Click **OK**.

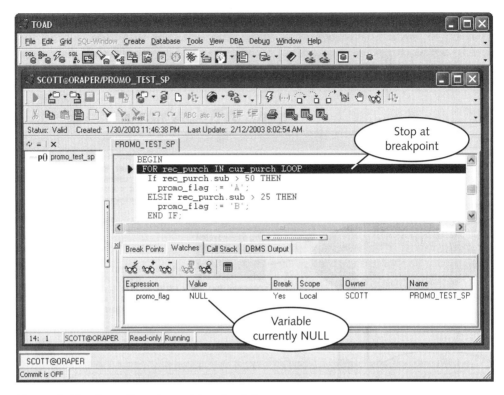

Figure C-19   Execution stops at breakpoint

Now, we know where the problem exists. Setting the PROMO_FLAG to empty using `promo_flag := ' ';` is not the same as setting the variable to NULL. Remember that a blank (empty spaces) and a zero are both treated as an actual value, not a NULL value, which indicates nothing has been entered into the field. Let's make the correction and test again.

To correct the procedure:

1. Delete the rows that were inserted into the BB_PROMOLIST table from our execution by returning to the Schema Browser pane and clicking the **Delete record** button on the toolbar five times to delete all five rows previously inserted. Then click the **Commit** button to save the modifications. Click the **Refresh data** button on the Schema Browser's toolbar to see the newly inserted rows.

2. Return to the procedure editor and change the PROMO_FLAG assignment statement at the end of the procedure to **promo_flag := NULL;**.

3. Click the **breakpoint** to remove it.

4. Click the **Delete Watch** button to eliminate the watch on the PROMO_FLAG variable.

5. Click the **Run** button on the editor toolbar and respond with **Yes** to the compile questions.

6. Click **OK** in the Confirm dialog box.

7. Return to the Schema Browser and click the **Refresh the detail panel** button (right side of browser) to see the newly inserted data. The data confirms that the procedure correctly inserts shoppers 21, 22, 24, and 25 to the BB_PROMOLIST table, because they should receive a free shipping offer.

The data type VARCHAR2 was used for the parameters and variables. TOAD version 7.4.0.3 is almost 100% compatible with Oracle9*i*, but using the CHAR in this example caused the debugger to not process the cursor. All Oracle9*i* compatibility issues are quickly being addressed by Quest Software.

Displaying Variables

Many developers like to use the DBMS_OUTPUT.PUT_LINE command to display variables from procedure execution. The TOAD procedure editor contains a tab at the bottom that holds output from this command for easy review. Let's add a DBMS_OUTPUT.PUT_LINE statement to the PROMO_TEST_SP procedure to see how this works.

To use DBMS_OUTPUT.PUT_LINE:

1. Return to the Schema Browser, if necessary.

2. Delete the rows that were inserted into the BB_PROMOLIST table from our execution by returning to the Data tab pane and clicking the **Delete record** button on the toolbar four times to delete all four rows previously inserted. Then click the **Commit** button to save the modifications.

3. Return to the procedure editor.

4. Insert a DBMS_OUTPUT.PUT_LINE statement before and after the IF clause at the top of the procedure, as shown in Figure C-20.

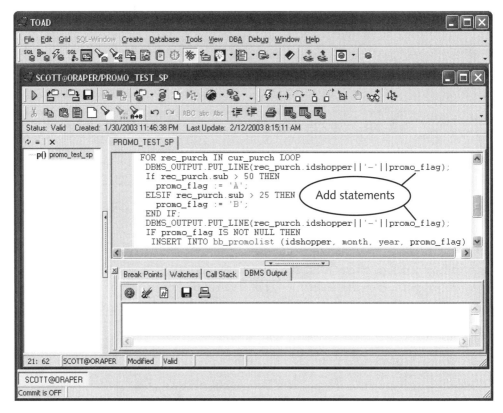

Figure C-20   Add DBMS_OUTPUT.PUT_LINE statements

5. Click the **DBMS Output** tab at the bottom to make this panel visible.

6. Click **Run** to execute the procedure. Respond with **Yes** to the compile questions.

7. Click **OK** in the Execution Terminated dialog box and review the output available in the DBMS_OUTPUT tab, as shown in Figure C-21.

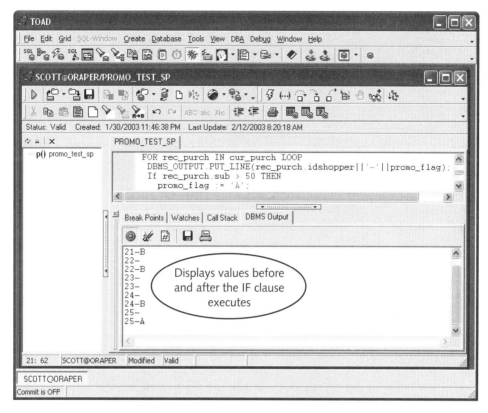

Figure C-21   DBMS_OUTPUT.PUT_LINE output listing

APPENDIX SUMMARY

This appendix introduced the procedure creation and debugging features of TOAD. I encourage you to explore all the available features by reviewing product information on the Quest Software Web site. In addition, notice that you have a user's manual available as one of the selections under the TOAD entry on your Windows Start menu.

Glossary

actual parameters — Values from the application that pass into parameters.

anchored data type — Using the %TYPE attribute to instruct the system to look up the data type of a database column and use this data type for the declared variable.

anonymous blocks — Block of code that is not stored to be reused. Every time an anonymous block of code is executed, it must be entered, compiled, and executed.

autonomous transactions — Transactions created within another transaction called a parent transaction which can be treated independently of the parent transaction.

basic loop — Uses the LOOP and END LOOP markers to begin and end the loop.

body — Text contained in an email message.

BROKEN parameter — A Boolean value where a TRUE indicates the job is broken and should not be executed by the job queue.

CASE expression — Evaluates conditions and returns a value in an assignment statement.

client-side — Program code that resides on the user or client machine code, which includes any statements that are to be repeated.

collection — A variable that can store and handle multiple values of the same data type as one unit.

compiler hint — A request a programmer includes within his or her code that asks Oracle to modify the default processing in some manner.

composite data type — A data type that can store and handle multiple values of different data types as one unit.

conditional predicates — Clauses available within a trigger to allow separate handling of different database triggering events within a single trigger.

constraining table — A table that is referenced via a foreign key constraint on the table that a trigger is modifying.

control structures — Statements used to control the flow of logic processing in our programs. The control structures provide the capability to perform conditional logic to determine which statements should be run, how many times the statements should be run, and the overall sequence of events.

correlation identifiers — Qualifiers, which are special PL/SQL bind variables that refer to values associated with the DML action that fires a trigger. The correlation identifiers are powerful in that they actually allow us to refer to and use the row data values of the DML action.

cursor — Represents a work area or section of memory in which an SQL statement is being processed in the Oracle server. This memory area is also referred to as the context.

cursor variables — References or pointers to a work area.

database trigger — A block of PL/SQL code that runs automatically when a particular database event occurs.

debugging — Process of identifying and removing errors from within program code.

dependency tree utility — A mechanism available in the Oracle system to provide a map to visualize both direct and indirect dependencies amongst the database objects.

Dynamic SQL — Enables greater flexibility in SQL code contained in a PL/SQL block such as providing table names at run time.

envelope — Sender and recipient information in an e-mail.

event — Can range from a user action such as clicking a button on the screen to a table UPDATE statement that automatically calls a database trigger.

exception handling — Area of a block that determines what happens if an error occurs.

explicit cursors — Work areas programmatically declared and manipulated in the PL/SQL block code for handling a set of rows returned by a SELECT statement.

FOR loop — Indicates a numeric range in the opening LOOP clause that dictates exactly how many times the loop should run.

formal parameters — Parameters that are listed in the program unit.

forward declaration — A declaration of a program unit in a package body by placing the header code at the top of the package body code.

function — Similar to a procedure in that it is a program unit that achieves a task, can receive input values, and returns values to the calling environment. The main difference between functions and procedures is that a function is part of an expression. It cannot serve as an entire statement.

global constructs — Values of the element persist throughout a user session and, therefore, can be referenced (used) in code within various parts of the application during a user session.

GOTO statement — Also called a jumping control in that it instructs the program to "jump to" some specific area of the code. It is used to branch logic in a manner where only certain portions of the code in the block are processed based on some condition.

header — Information about an email message such as the date and subject.

implicit cursors — Work areas declared automatically for all DML and SELECT statements issued within a PL/SQL block.

Index-by table — A variable that can handle many rows of data but only one field.

indirect dependencies — Involve references to database objects that in turn reference other database objects making the chain of dependencies less obvious to track.

infinite loop — A loop that is never instructed when to stop.

Instead-Of trigger — A PL/SQL block that executes in place of a DML action on a database view.

INTERVAL parameter — An Oracle date expression in a character string that will indicate how often the job should be executed.

JOB parameter — A unique integer that is assigned to each job in the queue.

key-preserved — A table is key-preserved if every key of the table can also be a key of the result of the join.

looping constructs — Repeats processing of a desired portion of code.

mode — Indicates which way the value provided for the parameter flows: into the program unit, out of the program unit, or both.

mutating table — A table that is being modified by a DML action when a trigger is fired.

named association — Associate a value to each parameter by name in the program unit invoke statement.

named program unit — A block of PL/SQL code that has been named so that is can be saved (stored) and reused.

NEXT_DATE parameter — A date instructing the job queue as to when the job should be executed next.

one time only procedure — A procedure in a package that runs only once—when the package is initially invoked.

overloading — Ability to use the same name on multiple program units within the same package.

package — A type of PL/SQL construct that is a container that can hold multiple program units, such as procedures and functions.

package body — The program unit that contains the code for any procedures and functions declared in the specification. The package body must be created using the same name of an existing specification.

package scope — Range of visibility for a particular element or construct contained in a package.

package specification — Declares all the contents of the package and is referred to as the package header. The specification is required and must be created before the body.

parameters — Mechanisms to send values in and out of a program unit.

passed by reference — A pointer to the value in the actual parameter is created instead of copying the value from the actual parameter to the formal parameter.

passed by value — The value is copied from the actual to the formal parameter.

positional method — Manner in which arguments used in the procedure execution call were passed. That is, when invoking the procedure, the first argument value is matched up with the first parameter in the procedure, the second argument value is matched up with the second parameter in the procedure, and so on.

PRAGMA — Instructs Oracle to use some additional information provided when compiling and executing the block.

predefined exceptions — Exception names associated with common Oracle errors.

private — Can be called only from other program units within the same package.

procedural language — Allows a programmer to code a logical sequence of steps to make decisions and to instruct the computer as to the tasks which need to be accomplished.

program unit — Denotes that blocks are typically created to perform a specific task that may be needed within a number of applications.

program unit dependencies — The inter-relationships of objects as they relate to procedures, functions and packages.

programming language — Allows the actions of an end user to be converted into instructions that a computer can understand.

public — Elements declared in a package specification are considered public, which means they can be referenced from outside of the package.

purity level — Defines what type of data structures the function reads or modifies.

purity level — Is identified with a set of acronyms that indicate the restrictions on using the function.

record — A data type similar to the structure of a row in a database table. A variable declared with a record data type can hold one row of data consisting of a number of column values.

remote database connections — Used to link to another database.

remote databases — Connections to other Oracle database servers.

row level — Indicates firing the trigger code for each row affected in the DML statements.

Searched CASE statement — Does not use a selector but individually evaluates conditions that are placed in WHEN clauses. The conditions checked in the WHEN clauses must evaluate to a Boolean value of TRUE or FALSE.

server-side — Program code that resides on the server.

signature model — Compares the mode, data type, and the order of parameters, to determine if invalidation occurs.

Simple Mail Transfer Protocol (SMTP) — The protocol used to send email across networks and the Internet.

statement level — Indicates firing the trigger only once for the event regardless of the number of rows affected by the DML statement.

stored — Indicates the program unit is saved in the database and, therefore, can be used or shared by different applications.

stored program units — Denotes that the program unit has been saved in the database.

subprogram — Program unit defined within another program unit.

table attributes — Functions that can be used in conjunction with table variables and allow greater ability to manipulate table values.

table of records — A type of composite data type which is the same as a record data type except that it can handle more than one record or row of data.

timestamp model — Compares the last modification date and time of objects to determine if invalidation occurs.

transaction scope — Refers to the logical group of DML actions that is affected by a transaction control statement.

user-defined exception — An exception that a developer explicitly raises in the block.

variable scope — Area of a program block that can identify a particular variable.

WHAT parameter — The PL/SQL code to be executed. Most commonly this parameter contains a call to a PL/SQL stored program unit.

WHILE loop — Includes a condition to check at the top of the loop in the LOOP clause itself. For each iteration of the loop, the condition is checked and if it is TRUE, the loop continues. If the condition is FALSE, the looping action stops.

Index

Special Characters

% ROWTYPE attribute, 46–47
% TYPE attribute, 43

A

actual parameters
 constraints, 194–196
 defined, 146, 194
Ada, 4
ALTER_COMPILE
 described, 366–367
 exceptions, 367
ALTER TRIGGER statement, 320–321
ANALYZE_OBJECT
 described, 367–368
 exceptions, 368
 parameters of, 368
anchored data type, 43
anonymous blocks, 29, 142
application functions, 144
application model
 components
  database, 5
  program logic, 5
  user interface (screens), 5
 three-tier model, 7–8
 two-tier model, 6–7

application procedures, 144
application programming, 2–4
application trigger, 144
attribute, *See specific attribute*
autonomous transactions, 166
AUTOTRACE tool
 explain plan and, 434–437
 options, 434
 statistic definition, 437

B

basic loop,
 process conditional logic with, 52
 repeat code, 98–100
BEGIN section
 in general, 30
 role of scalar variables, 34
bind variables, 42–43
blocks. *See also* PL/SQL blocks
 enclosing, 121, 122
 nested, 121, 122
body (e-mail), 342
breakpoints
 in
  Procedure Builder, 485–493
  TOAD, 509–517
BROKEN parameter, 360

C

CASE expression, 97
CASE statements, 93
 searched, 96
 structure of, 94–95
 use of CASE expression, 97
client/server
 application model, 6–7
 considerations in named program units, 143
client-side. *See also* client/server
 defined, 143
code
 comments in, 124–125
 identify problem areas in
  in general, 427
  set server parameters, 429–432
  SQL TRACE facility, 429
  V_$SQL AREA view, 427–428
collection
 defined, 53
 index-by table, 53–56
 in general, 44, 53
 nested tables, 56
 VARRAYS, 56
color coding, reading code and, 477, 499
column
 display width, 465
 selection, avoid unnecessary, 439

comments, in code, 124–125

communications

DBMS_ALERT, 340–341

DBMS_PIPE package, 338–340

in general, 337

UTL_HTTP, 344–345

UTL_SMTP, 341–344

UTL_TCP, 345–346

compiler hint, 196

composite data type

% ROWTYPE attribute, 46–47

defined, 44

record data type, 44–46

table of records, 47–51

concatenated index, 444–445

conditional predicates

defined, 303

described, 303–304

conditions, order by frequency, 454

CONSTANT option, 33

constraining table, 320

control structures

CASE statements, 93–97

defined, 85

IF statements, 85–93

correlation identifier

availability, 300

defined, 300

described, 300–301

cost-based Optimizer

described, 434

v. rule-based, 439–441

CREATE PROCEDURE

command, 145

cursor

attributes, 57

defined, 56

explicit, 56, 58–66

implicit, 56, 57–58

in general, 56

variables, 57, 66–67

CURSOR FOR loop, 63–65

cursor variables

defined, 57

working with, 66–67

D

database

rebuild for

database trigger, 296

dynamic SQL and object
technology, 382–383

functions, 181–182

Oracle-supplied packages, 337

packages, 214–215

performance tuning, 426

PL/SQL processing, 84–85

Program Builder, 478

procedures, 141–142

program unit dependencies, 256

TOAD, 500

remote, 260

connections, 272

database administrator. See DBA

database trigger. See also system
trigger

ALTER TRIGGER statement,
320–321

apply to address processing
needs, 313–316

create and test

DML trigger in SQL*Plus,
304–307

Instead-Of trigger, 307–311

data dictionary information,
321–322

defined, 293

delete, 321

described, 7

in general, 293

introduction, 296–297

named program unit, 144

rebuild database, 296

restrictions of, 316–320

row level, 299

statement level, 299

syntax and options

code example, 298

conditional predicates, 303–304

events, 301

in general, 297

timing and correlation identifiers, 298–301

trigger body, 301–303

uses of, 313–314

data dictionary

information for

packages, 246

program units, 201–202

triggers, 321–322

views on

dependencies, 262–263

privileges, 280–281

data types
anchored, 43
collection, 44
composite, 44–51
parameter requirement, 146–147
scalar variable, 31
variable comparisons with
   same —, 453–454
DBA
changing database settings, 32
familiarity with PL/SQL
   language, 5
DBMS_ALERT package, 340–341
DBMS_ DDL package
ALTER_COMPILE, 366–367
ANALYZE_OBJECT, 367–368
in general, 366
DBMS_ JAVA package, 369
DBMS_ JOB package
check settings, 361–366
in general, 359
parameters, 360
programs, 359–360
DBMS_LOB package
in general, 355
manipulate images, 355–357
programs, 355
DBMS_METADATA package, 369
DBMS_OUTPUT package
check values, 34–35
described, 346–351
procedures, 346
DBMS_PIPE package
described, 338–340
program units, 340

DBMS_RANDOM package, 369
DBMS_SESSION package, 369
DBMS_SQL package
DDL statements, 387–390
DML statements, 385–387
in general, 385
queries
   described, 390–394
   steps to perform, 391
v. native dynamic SQL, 398–399
DBMS_UTILITY package, 369
DBMS_XMLGEN, package, 369
DDL statements
steps to perform, 388
with DBMS_SQL, 387–390
debugging
defined, 157
in SQL*Plus, 157–161
with Procedure Builder
   display values to screen,
      493–497
   in general, 485
   work with breakpoints,
      485–493
DECLARE section, 29
error, 122
dependencies. See program unit
   dependencies
dependency tree utility
defined, 265
described, 263–268
DESCRIBE command, 156–157
DevPartner DB, 13
direct dependency
defined, 261
identify, 261–262

DML statements
steps to perform, 385–386
with DBMS_SQL package,
   385–387
DML trigger. See also database
   trigger
create and test in SQL*Plus,
   304–307
documentation
CD-ROM-based, 9
Web-based, 8–9
dynamic SQL
and PL/SQL, 357–359
DBMS_SQL package
   DDL statements, 387–390
   DML statements, 385–387
   in general, 385
   queries, 390–394
   v. native dynamic SQL,
      398–399
in general, 381, 383–385
native dynamic SQL
   EXECUTIVE IMMEDIATE,
      394–396
   in general, 394
   OPEN FOR, 396–398
rebuild database, 382–383

E
END; statement, 29
envelope (e-mail), 342
error handling, using RAISE_
   APPLICATION_ERROR,
   167–169
event, 8

exception handlers
  error in, 123
  in general, 105
  non-predefined Oracle errors,
    112–113
  predefined Oracle errors
    examples, 106–109
    exception handler coding,
      109–111
    in general, 105–106
  procedures and, 163–167
  propagation, 120–124
  RAISE_APPLICATION_
    ERROR, 119–120
  SQLCODE, 117–119
  SQLERRM, 117–119
  user-defined, 113–114
  WHEN OTHERS, 115–116
exception propagation, 120–124
  timing of error, 124
EXCEPTION section, 30
EXECUTE IMMEDIATE
  statement, 394
execution plan. *See* explain plan
EXIT clause, 103–104
explain plan
  AUTOTRACE and, 434–437
    statistic definitions, 437
    tool options, 434
  execution order, 436
Explicit cursor
  actions using, 60
  CURSOR FOR loop, 63–65
  defined, 56
  subqueries and parameters, 66
  working with, 58–62

F

foreign keys v. REF pointers,
  408–409
FOR loop
  process conditional logic, 52
  repeat code, 101–103
formal parameters
  constraints, 194–196
  defined, 146, 194
Forms 6*i* Procedure Builder, 12
Forms 9*i*, 12
forward declaration
  defined, 232
  in packages, 231–234
frequency, order conditions by, 454
function
  create stored function, 184–185
    build and test, 188–190
    invoke and test, 185–186
    use in SQL statement, 187–188
  described, 7
  introduction, 182–183
  manage package — SQL
    requirements
    default purity level, 243
    in general, 240–241
    PRAGMA RESTART_
      REFERENCES, 242–243
    purity level indications,
      241–242
  parameter constraints, actual
    and formal, 194–196
  passing parameter values
    controlling technique used,
      196–198
    techniques, 196

program units
  data dictionary information,
    201–202
  deleting, 202–203
  purity levels, 198–201
  rebuild database, 181–182
  RETURN statements
    multiple, 192–193
    use in procedure, 193–194
    use OUT parameter mode,
      190–192
    written in external
      languages, 243

G

global
  defined, 224
  package constructs
    in general, 224–225
    specifications with no body,
      227–228
    test persistence of package
      variables, 225–227
  variable, 42–43
GOTO statement, 104–105

H

header (e-mail), 342
host variables, 150
  working with, 42–43

I

IF statements
  IF/THEN/ELSE statements,
    87–88
  IF/THEN/ELSIF/ELSE, 89–91

IF/THEN v. IF, 88

logic, 85–87

operators, 92–93

simple, 51

IF/THEN/ELSE statements, 87–88

IF/THEN/ELSIF/ELSE, 89–91

IF/THEN statements, 88

images, manipulate with DBMS_LOB, 355–357

Implicit cursor

defined, 56

working with, 57–58

index

concatenated, 444–445

suppression, 441–444

index-by table, 53–56

attributes, 54

characteristics, 53

indirect dependency

defined, 261

identifying, 261–262

IN mode, 145, 146

IN OUT parameter, 152–153

mode, 146

type, 145

Instead-Of trigger

create and test, 307–311

defined, 307

Interpreter panel, 483–485

INTERVAL parameter, 360

J

JOB parameter, 360

join operation, 447–448

jumping control, 104

K

key-preserved table, 308

L

languages, external, 243

large objects (LOBs)

DBMS_LOB, 355

manipulate images, 355–357

described, 354–355

types, 354

LOBs. See large objects

looping constructs

basic, 52, 98

common errors

EXIT clause, 103–104

in general, 103

static statement, 104

CURSOR FOR, 63–65

FOR, 52, 101–103

in general, 98

WHILE, 100–101

loops. See looping constructs

M

messages, debug in SQL*Plus by displaying, 157–161

mode (parameter), 146

mutating table

defined, 316

restriction of trigger use, 316–320

N

named association, 151

named program units

client and server considerations, 143

defined, 7, 142

types of, 143–144

application procedures and functions, 144

application trigger, 144

database trigger, 144

package, 144

stored procedures and functions, 144

naming conventions, 146

native dynamic SQL

DBMS_SQL v., 398–399

EXECUTE IMMEDIATE statement, 394

OPEN FOR statement, 396–398

nested tables, 56

NEXT_DATE parameter, 360

NOT NULL option, 33

O

object-oriented programming. See object technology

Object technology

create object type, 399–400

in general, 399

methods, 403–406

REF pointers v. foreign keys, 408–409

relations, 406–408

sort and compare object type columns, 412–414

use object type, 400–403

views, 410–412

object type

create, 399–400

sort and compare — columns, 412–414

use, 400–403

one time only procedure

defined, 234

in PL/SQL packages, 234–237

OPEN FOR statement, 396–398

Optimizer

cost-based (CBO), 433, 434, 439–441

hints, 448–451

parameter values, 433

processing and, 432–434

rule-based, 433, 439–441

options, database trigger

code example, 298

conditional predicates, 303–304

events, 301

in general, 297

timing and correlation identifiers, 298–301

trigger body, 301–303

Oracle9i. *See also* Oracle-supplied functions; Oracle-supplied packages

naming standards, 146

object types, 406

tool suite, 162

Oracle Developer 6i suite, 10–11, 477

Oracle errors

non-predefined, 112–113

predefined

examples, 106–109

exception handler coding, 109–111

in general, 105–106

Oracle Forms, 4

deployment of, 7

Oracle object technologies. *See* object technologies

Oracle-supplied functions, 144

Oracle-supplied packages

communications

DBMS_ALERT, 340–341

DBMS_PIPE, 338–340

in general, 337

UTL_HTTP, 344–345

UTL_SMTP, 341–344

UTL_TCP, 345–346

dynamic SQL and PL/SQL, 357–359

generate output

DBMS_OUTPUT, 346–351

in general, 346

UTL_FILE, 351–354

in general, 335

large objects

DBMS_LOB, 355

in general, 354–355

manipulate images with DBMS_LOB, 355–357

miscellaneous packages

additional packages, 368–369

DBMS_DDL, 366–368

DBMS_JOB, 359–366

in general, 359

rebuild database, 337

OUT mode

in a function, 190–192

mode type, 145, 146

output, generate

DBMS_OUTPUT, 346–351

in general, 346

UTL_FILE, 351–354

overloading

defined, 237

program units in packages, 237–240

P

package body

defined, 217

described, 217–220

specifications with no —, 225–227

packages. *See also* Oracle-supplied packages

body, 217–220

data dictionary information, 244–246

defined, 213

delete, 246

dependencies, 268–272

described, 7

execute privileges, 244

forward declaration, 231–234

global constructs

improve processing efficiency, 228–231

in general, 224–225

specifications with no body, 225–227

test persistence of package variables, 225–227